高等学校计算机专业规划教材

计算机英语
（第2版）

邱仲潘　曾思亮　薛伟胜　等　编著

清华大学出版社
北京

内 容 简 介

本书主要介绍计算机硬件、软件、系统、网络、应用程序等相关知识，共分为20章，新版增加了智能手机、无线Wi-Fi、机器学习等领域最新的科技文章，增加了时代感和趣味性。

本书由一线教师编写，这些教师了解学生的知识水平、接受能力和需求点，而且翻译过大量计算机图书，有丰富的翻译经验。本书强调阅读理解，对一些难句进行了深入的解析。

本书可作为高等院校"计算机英语"课程的教材或教学参考书，也可供有一定英语基础的广大计算机用户学习计算机英语时使用。

本书封面贴有清华大学出版社防伪标签，无标签者不得销售。
版权所有，侵权必究。举报：010-62782989，beiqinquan@tup.tsinghua.edu.cn。

图书在版编目（CIP）数据

计算机英语/邱仲潘等编著. —2版. —北京：清华大学出版社，2017（2022.1重印）
（高等学校计算机专业规划教材）
ISBN 978-7-302-46255-2

Ⅰ.①计… Ⅱ.①邱… Ⅲ.①电子计算机–英语–高等学校–教材 Ⅳ.①TP3

中国版本图书馆CIP数据核字（2017）第021180号

责任编辑：龙启铭　李　晔
封面设计：何凤霞
责任校对：梁　毅
责任印制：杨　艳

出版发行：清华大学出版社
网　　址：http://www.tup.com.cn, http://www.wqbook.com
地　　址：北京清华大学学研大厦A座　　邮　编：100084
社　总　机：010-62770175　　邮　购：010-83470235
投稿与读者服务：010-62776969，c-service@tup.tsinghua.edu.cn
质　量　反　馈：010-62772015，zhiliang@tup.tsinghua.edu.cn
课　件　下　载：http://www.tup.com.cn,010-83470236
印　装　者：北京建宏印刷有限公司
经　　销：全国新华书店
开　　本：185mm×260mm　　印　张：24.75　　字　数：587千字
版　　次：2013年7月第1版　2017年6月第2版　印　次：2022年1月第6次印刷
定　　价：49.00元

产品编号：069641-01

前言

计算机技术的发展,最初是从英语国家开始的,目前美国具有绝对优势。从事计算机行业的人,难免会遇到大量英文资料,无论是外版教材、技术手册、联机说明,还是阅读或者发表高水平的专业论文,都必须使用英语。因此,学好专业英语对计算机专业学生来说非常重要。

本书是针对计算机专业学生编写的。本书的第一版得到了广大师生的许多好评,同时也收到了许多改进建议,新版本吸收了这些丰富而宝贵的教学实践经验。此外,新版还收录了智能手机、无线Wi-Fi、机器学习等领域最新的科技文章,增加了时代感和趣味性。

对计算机专业学生的基本要求是读懂英文的软件需求文档和在编程中根据要求插入简单的注释文本,因此在本书编写过程中,作者一直认为应该强调阅读理解、强调简单文本写作以及强调专业术语和基本科技英语语法。同时,为了提高效率和便于工作中的资料积累与交流,应该介绍一些翻译技巧,使学生能够把看懂的内容用比较准确和流畅的中文表达出来,能够把软件设计与实现中的思路翻译成简单英文。为此,特意挑选一些难句,在给出准确翻译的同时选择学生常见的翻译错误进行剖析,增加学生的理解深度。课文后面还用英语提供了关键术语的解释,以便有兴趣的学生可以了解到许多相关专业知识和有趣的词源知识。相关知识包括翻译技巧以及技术方面和语言方面的知识,非常实用。文章后面还有参考读物,难度略大于课文。建议老师在保证让学生掌握课文内容的前提下,根据学生的接受情况和兴趣决定教学内容的深浅。俗话说:"兴趣是成功之母。"本书努力通过各种背景知识和词源知识增加趣味性,老师还可以通过调动学生积极参与课堂教学活动激发学生的学习兴趣,可以鼓励学生自己从网络和其他地方寻找相关资料,扩大视野,并且把学到的专业英语知识应用到其他专业课程的学习中,学以致用,切实体会计算机英语的作用,变"要我学"为"我要学"。

本书第1~11章由邱仲潘负责,作者翻译了大量计算机科学图书,积累了许多素材,辅助材料大部分是由邱仲潘提供的。刘新钰、薛伟胜、王若涵同志负责第12章和第20章,曾思亮、王水德、洪镇宇同志负责第13~

19章。在本书的写作过程中，宋智军、王帅、刘文红、邹文、邓欣欣、王润涛、周丹丹、朱敏、张朋丽、刘文琼等同志也完成了大量工作，在此深表感谢。由于时间仓促，书中难免存在错误和缺漏之处，期待各位老师和同学不吝赐教，以便今后修订时改正和增补。

编 者
2016年11月

目 录

Chapter 1 PC Basic /1

1.1 Storage 1
1.2 Outer Hardware 3
1.3 Smartphone, Tablet and Laptop 5
1.4 Exercise 1 7
1.5 Further Reading: Flash Memory 8
    1.5.1 The Basics 8
    1.5.2 Removable Flash Memory Cards 9
    1.5.3 SmartMedia 10
    1.5.4 CompactFlash 10

Chapter 2 How Computer Monitors Work /12

2.1 The Basics 12
    2.1.1 Display Technology Background 13
    2.1.2 Display Technologies: VGA 13
    2.1.3 Display Technology: DVI 13
    2.1.4 Viewable Area 14
    2.1.5 Maximum Resolution and Dot Pitch 14
    2.1.6 Dot Pitch 14
    2.1.7 Refresh Rate 15
    2.1.8 Color Depth 16
    2.1.9 Power Consumption 17
    2.1.10 Monitor Trends: Flat Panels 17
2.2 Exercise 2 18
2.3 Further Reading: Liquid Crystal Display 19
    2.3.1 Brief History 21
    2.3.2 Transmissive and Reflective Displays 21
    2.3.3 Color Displays 21
    2.3.4 Passive-matrix and Active-matrix 22
    2.3.5 Quality Control 22

2.3.6　Zero-power Displays 23
2.3.7　Drawbacks 23

Chapter 3　How Cell Phones Work　/ 25

3.1　Cell-phone Frequencies 25
3.2　Cell-phone Channels 27
3.3　Analog Cell Phones 28
3.4　Along Comes Digital 29
3.5　Inside a Digital Cell Phone 29
3.6　Exercise 3 32
3.7　Further Reading: Cell Phone 33
　　3.7.1　History 33
　　3.7.2　Handsets 35

Chapter 4　Digital Camera Basics　/ 40

4.1　How does Digital Camera Work 40
4.2　CCD and CMOS: Filmless Cameras 40
4.3　Digital Camera Resolution 41
4.4　Capturing Color 43
4.5　Digital Photography Basics 43
4.6　Megapixel Ratings 44
4.7　Digital Camera Settings and Modes 44
4.8　Shutter Speed 45
4.9　Exercise 4 47
4.10　Further Reading: How to Take Good Photos 48
　　4.10.1　Digital Camera Problems 49
　　4.10.2　Image Editing Software 50

Chapter 5　How Bits and Bytes Work　/ 53

5.1　Decimal Numbers 53
5.2　Bits 54
5.3　Bytes 55
5.4　Bytes: ASCII 55
5.5　Standard ASCII Character Set 56
5.6　Lots of Bytes 57
5.7　Binary Math 57
5.8　Quick Recap 58
5.9　Exercise 5 58

5.10　Further Reading: How Boolean Logic Works 60
　　5.10.1　Simple Gates 60
　　5.10.2　Simple Adders 63
　　5.10.3　Flip Flops 66
　　5.10.4　Implementing Gates 68

Chapter 6　Microprocessors　　/71

6.1　Microprocessor History 71
6.2　Microprocessor Progression 72
6.3　Inside a Microprocessor 73
6.4　Microprocessor Instructions 75
6.5　Decoding Microprocessor Instructions 78
6.6　Microprocessor Performance 79
6.7　Microprocessor Trends 79
6.8　64-bit Processors 80
6.9　Exercise 6 81
6.10　Further Reading: E-commerce 82
　　6.10.1　Commerce 83
　　6.10.2　The Elements of Commerce 84
　　6.10.3　Why the Hype 85
　　6.10.4　The Dell Example 86
　　6.10.5　The Lure of E-commerce 87
　　6.10.6　Easy and Hard Aspects of E-commerce 89
　　6.10.7　Building an E-commerce Site 89
　　6.10.8　Affiliate Programs 90
　　6.10.9　Implementing an E-commerce Site 90

Chapter 7　Application Software　　/92

7.1　What is Software 92
7.2　Programming Languages 92
　　7.2.1　Assemblers 92
　　7.2.2　Compilers and Interpreters 93
　　7.2.3　Nonprocedural Languages 94
7.3　Libraries 94
7.4　The Program Development Process 95
　　7.4.1　Problem Definition 95
　　7.4.2　Planning 95
　　7.4.3　Writing the Program 96

       7.4.4   Debug and Documentation 96
       7.4.5   Maintenance 96
   7.5 Writing your Own Programs 97
   7.6 Exercise 7 98
   7.7 Further Reading: Computer Software 99
       7.7.1   Relationship to Hardware 99
       7.7.2   System and Application Software 100
       7.7.3   Users See Three Layers of Software 100
       7.7.4   Software Creation 101
       7.7.5   Software in Operation 101
       7.7.6   Software Reliability 101
       7.7.7   Software Patents 101
       7.7.8   System Software 101

Chapter 8   Compiler   / 103

   8.1 Introduction and History 103
   8.2 Types of Compilers 104
   8.3 Compiled vs. Interpreted Languages 105
   8.4 Compiler Design 105
   8.5 Compiler Front End 106
   8.6 Compiler Back End 106
   8.7 Exercise 8 108
   8.8 Further Reading: Assembly Language 109
       8.8.1   Assemblers 110
       8.8.2   Assembly Language 111
       8.8.3   Machine Instructions 112
       8.8.4   Assembly Language Directives 113
       8.8.5   Usage of Assembly Language 114
       8.8.6   Cross Compiler 115
       8.8.7   Compiling a Gcc Cross Compiler 115

Chapter 9   How Java Works   / 116

   9.1 A Little Terminology 116
   9.2 Downloading the Java Compiler 117
   9.3 Your First Program 119
   9.4 Understanding What Just Happened 121
   9.5 Exercise 9 124
   9.6 Further Reading: How Perl Works 125

9.6.1 Getting Started 125
9.6.2 Hello World 126
9.6.3 Variables 127
9.6.4 Loops and Ifs 128
9.6.5 Functions 129
9.6.6 Reading 130

Chapter 10 Database & C++ / 131

10.1 Text 131
10.2 Exercise 10 134
10.3 Further Reading: C++ 135
    10.3.1 Technical Overview 136
    10.3.2 Features Introduced in C++ 136
    10.3.3 C++ Library 137
    10.3.4 Object-oriented Features of C++ 137
    10.3.5 Design of C++ 140
    10.3.6 History of C++ 141
    10.3.7 C++ is not a Superset of C 143

Chapter 11 Artificial Intelligence / 145

11.1 Overview 145
11.2 Strong AI and Weak AI 145
    11.2.1 Strong Artificial Intelligence 146
    11.2.2 Weak Artificial Intelligence 146
    11.2.3 Philosophical Criticism and Support of Strong AI 146
11.3 History Development of AI Theory 148
11.4 Experimental AI Research 149
11.5 Exercise 11 151
11.6 Further Reading: Alan Turing 153
    11.6.1 Childhood and Youth 153
    11.6.2 College and his Work on Computability 154
    11.6.3 Cryptanalysis (Code Breaking) 155
    11.6.4 Work on Early Computers and the Turing Test 156
    11.6.5 Work on Pattern Formation and Mathematical Biology .. 157
    11.6.6 Prosecution for Homosexuality and Turing's Death 157

Chapter 12 Machine Learning / 158

12.1 Overview 159

    12.1.1 Types of problems and tasks 159

    12.1.2 History and relationships to other fields 160

    12.1.3 Theory 161

    12.1.4 Approaches 162

  12.2 Exercise 12 166

  12.3 Further Reading: Applications for machine learning 168

    12.3.1 Adaptive websites 168

    12.3.2 Affective computing 168

    12.3.3 Bioinformatics 168

    12.3.4 Brain-machine interfaces 169

    12.3.5 Cheminformatics 169

    12.3.6 Classifying DNA sequences 169

    12.3.7 Computational anatomy 170

    12.3.8 Computational finance 171

    12.3.9 Computer vision, including object recognition 171

    12.3.10 Detecting credit card fraud 171

    12.3.11 Software 172

Chapter 13 How DSL Works / 174

  13.1 Overview 174

  13.2 Telephone Lines 175

  13.3 Asymmetrical DSL 175

  13.4 Distance Limitations 176

  13.5 Splitting the Signal: CAP 177

  13.6 Splitting the Signal: DMT 177

  13.7 DSL Equipment 178

    13.7.1 DSL Equipment: Transceiver 179

    13.7.2 DSL Equipment: DSLAM 179

  13.8 Exercise 13 180

  13.9 Further Reading: How Telephones Work 182

    13.9.1 A Simple Telephone 182

    13.9.2 A Real Telephone 183

    13.9.3 The Telephone Network: Wires and Cables 184

    13.9.4 The Telephone Network: Digitizing and Delivering 184

    13.9.5 Creating Your Own Telephone Network 185

    13.9.6 Calling Someone 185

    13.9.7 Tones 186

Chapter 14　Internet Infrastructure　　/188

14.1　A Network Example 188
14.2　Bridging The Divide 189
14.3　Backbones 190
14.4　Internet Protocol: IP Addresses &　Domain Name System 190
14.5　Uniform Resource Locators 191
14.6　Clients, Servers and Ports 192
14.7　Exercise 14 194
14.8　Further Reading: Modem 195

Chapter 15　How Internet Search Engines Work　　/199

15.1　Looking at the Web 199
15.2　Building the Index 202
15.3　Building a Search 203
15.4　Future Search 203
15.5　Exercise 15 205
15.6　Further Reading: Web crawler 206
　　　15.6.1　Nomenclature 206
　　　15.6.2　Overview 207
　　　15.6.3　Crawling policy 207
　　　15.6.4　Architectures 209
　　　15.6.5　Security 210
　　　15.6.6　Crawler identification 210
　　　15.6.7　Crawling the deep web 210
　　　15.6.8　Visual vs programmatic crawlers 211

Chapter 16　Encryption　　/212

16.1　In the Key of... 212
16.2　Hash This 213
16.3　Are You Authentic 214
16.4　Exercise 16 216
16.5　Further Reading: Identity Theft 217
　　　16.5.1　Types of Identity Theft 217
　　　16.5.2　Stealing Your Identity 217
　　　16.5.3　Accessing Your Personal Information 218
　　　16.5.4　Public Information 219
　　　16.5.5　How To Protect Yourself 219

	16.5.6	Internet Transactions	219
	16.5.7	If It Happens To You	220
	16.5.8	What Congress Is Doing About It	222
	16.5.9	What the Future Holds	222

Chapter 17 Taking a Closer Look at the DCE   /223

17.1	Common Threads	223
17.2	Remote Calls	224
17.3	Directory Services	224
17.4	Distributed Security Service	225
17.5	Distributed File System	225
17.6	Distributed Time Service	226
17.7	Extending and Using the DCE	226
17.8	Exercise 17	227
17.9	Further Reading: How to Kerberize Your Site	228

	17.9.1	Introduction	228
	17.9.2	Pick a Kerberos Server Machine (KDC)	229
	17.9.3	DCE and Kerberos	229
	17.9.4	Install the Kerberos Server	229
	17.9.5	Obtain the Necessary Code	229
	17.9.6	Do You Need More Code	230
	17.9.7	Building the Gnu Tools	230
	17.9.8	Building the Gnu C Compiler	231
	17.9.9	Compiling Kerberos	232
	17.9.10	For All Platforms	233
	17.9.11	Configuring the Kerberos KDC	233
	17.9.12	Setting Up a Host Server	236
	17.9.13	Domain Names	237
	17.9.14	Kerberos Clients	237
	17.9.15	Getting a Ticket for Another Realm	237
	17.9.16	Kerberos Security Problems	238
	17.9.17	Kerberos Authentication Option in SSL	238
	17.9.18	Available Kerberized Goodies	239
	17.9.19	CygnusKerbnet for NT, Macs, and UNIX	239

Chapter 18 What is Wi-Fi and How does it work   /240

18.1	How does Wi-Fi work	240
18.2	Uses	241

18.3　Frequencies 242
18.4　Advantages and Challenges 243
18.5　Network Security 245
18.6　Exercise 18 248
18.7　Further Reading: Wireless Revolution:　The History of Wi-Fi 249

Chapter 19　Shockwave 3-D Technology　／251

19.1　Uses of Shockwave Technology 252
19.2　Making 3-D Content Accessible 254
19.3　Developing New 3-D Content 255
19.4　Exercise 19 257
19.5　Further Reading: Computer Viruses 258
　　　19.5.1　Types of Infection 258
　　　19.5.2　What's a "Virus" 259
　　　19.5.3　What's a "Worm" 259
　　　19.5.4　Code Red 259
　　　19.5.5　Early Cases: Executable Viruses 260
　　　19.5.6　Boot Sector Viruses 261
　　　19.5.7　E-mail Viruses 261
　　　19.5.8　Prevention of Virus 262
　　　19.5.9　Origins of Virus 264
　　　19.5.10　History of Virus 264

Chapter 20　Kinect　／266

20.1　Technology 266
20.2　History 268
20.3　Launch 270
20.4　Reception 271
20.5　Sales 272
20.6　Awards 272
20.7　Exercise 20 273
20.8　Further Reading: Software of Kinect 274
　　　20.8.1　Kinect for Windows 274
　　　20.8.2　Software 276

附录 A　部分参考译文　／279

第 1 章　电脑基本组件 279
第 2 章　计算机显示器是如何工作的 283

第3章 手机如何工作 288
第4章 数码相机基础知识 293
第5章 位和字节是怎样工作的 297
第6章 微处理器概述 302
第7章 应用软件 310
第8章 编译器 314
第9章 Java 是如何工作的 318
第10章 数据库与 VC++ 325
第11章 人工智能 328
第12章 机器学习 334
第13章 DSL 是如何工作的 341
第14章 Internet 基础结构 345
第15章 网络搜索引擎工作原理 350
第16章 加密 355
第17章 近看 DCE 358
第18章 什么是 Wi-Fi 以及它是如何运作的 362
第19章 Shockwave 三维技术 366
第20章 3D 体感摄影机：Kinect 370

习题答案 /377

Chapter 1
PC Basic

As a college student or person who engages in IT, you must know PC and its components first; include its storage equipments and I/O devices. Now let's see these components first:

1.1 Storage

The purpose of storage in a computer is to hold data or information and get that data to the CPU as quickly as possible when it is needed. Computers use disks for storage: hard disks that are located inside the computer, and floppy or compact disks that are used externally.

1. Hard Disks

Your computer uses two types of memory: primary memory which is stored on chips located on the motherboard, and secondary memory that is stored in the hard drive. Primary memory holds all of the essential memory that tells your computer how to be a computer. Secondary memory holds the information that you store in the computer.

Inside the hard disk drive case you will find circular disks that are made from polished steel.

On the disks, there are many tracks or cylinders. Within the hard drive, an electronic reading/writing device called the head passes back and forth over the cylinders, reading information from the disk or writing information to it. Hard drives spin at 3600 or more rpm (Revolutions Per Minute)—that means that in one minute, the hard drive spins around over 3600 times! Today's hard drives can hold a great deal of information-sometimes over 20GB!

2. Floppy Disks

When you look at a floppy disk, you'll see a plastic case that measures 3.5 by 5 inches. Inside that case is a very thin piece of plastic (see Figure 1.1) that is coated with microscopic iron particles.

Floppy Disk

inside view　　　　back view
Figure 1.1　Floppy disk

This disk is much like the tape inside a video or audio cassette. Take a look at the floppy disk pictured. At one end of it is a small metal cover with a rectangular hole in it. That cover can be moved aside to show the flexible disk inside.

But never touch the inner disk—you could damage the data that is stored on it. On one side of the floppy disk is a place for a label. On the other side is a silver circle with two holes in it. When the disk is inserted into the disk drive, the drive hooks into those holes to spin the circle. This causes the disk inside to spin at about 300 rpm! At the same time, the silver metal cover on the end is pushed aside so that the head in the disk drive can read and write to the disk.

Floppy disks are the smallest type of storage, holding only 1.44MB.

3. How Hard and Floppy Disks Work

The process of reading and writing to a hard or floppy disk is done with electricity and magnetism. The surfaces of both types of disks can be easily magnetized. The electromagnetic head of the disk drive records information to the disk by creating a pattern of magnetized and non-magnetized areas on the disk's surface.

Do you remember how the binary code uses on and off commands to represent information? On the disk, magnetized areas are on and non-magnetized areas are off, so that all information is stored in binary code. This is how the electronic head can both write to or read from the disk surface. It is very important to always keep magnets away from floppy disks and away from your computer! The magnets can erase information from the disks!

4. Compact Disks

Instead of electromagnetism, CDs(see Figure 1.2) use pits (microscopic indentations) and lands (flat surfaces) to store information much the same way floppies and hard disks use magnetic and non-magnetic storage. Inside the CD-ROM is a laser that reflects light off of the surface of the disk to an electric eye. The pattern of reflected light (pit) and no reflected light (land) creates a code that represents data.

Figure 1.2 CD

CDs usually store about 650MB. This is quite a bit more than the 1.44MB that a floppy disk stores. A DVD or Digital Video Disk holds even more information than a CD, because the DVD can store information on two levels, in smaller pits or sometimes on both sides.

5. Uses of Floppy Disks

You might wonder: If all the information is stored safely inside my computer, why would I need to store it outside? There are several reasons why portable storage is so important.

Floppies make it possible to backup important information in case it is lost by the computer. RAM loses its memory each time the computer is turned off, but ROM keeps information stored even when the computer is not turned on. Well, sometimes computers have problems that can cause them to crash. No, that doesn't mean they jump off the desk and smash on the floor.

A crash is something that happens inside the computer's circuits and can make it forget things. Some crashes can even make ROM forget everything! Having important information backed up on disks will allow you to put it back into your computer's memory. Backup disks can save you lots of time and headaches!

Disks also allow information to be transferred between different computers. Let's say that you are working on a project using a computer at the library, but you haven't finished it by closing time. There's your project sitting in the computer. How do you get it home to finish it on your computer? You write the information to a disk, take it home and upload the information into your computer from the disk. What an easy way to transfer information!

6. Uses of Compact Disks

The most common use for compact disks (aside from playing music) is storage of software programs. When you purchase a computer game, the program that tells your computer how to run the game stored on a CD. You move the program into your computer's memory by installing it. Some programs are transferred completely into your computer's hard drive. However, many programs are very large and would take up lots of memory space on your hard drive. To keep that from happening, these programs are designed to only upload part of the program onto your computer. The rest of the program stays on the software. The program cannot be run from your computer unless you have the CD in the disk drive so that RAM can read the rest of the program from it.

With the introduction of CD-RW (disk drives that can write to compact disks as well as read from them), CDs can now be used for storage much like floppies. Using a CD-RW, computer data can be backed up to a CD. All kinds of information that was too large to fit on floppy disks can now be saved on CD. Many people store music files or family photos on CD.

1.2 Outer Hardware

Take a look at the computer (see Figure 1.3) in front of you. No, not just the screen. Look at all of the other parts. Do you know what they are? Do you know what they do? If you already know—great! Give yourself a big pat on the back! But if you don't know about all the gadgets surrounding your computer, then read on and find out!

1. The Basics

Let's start with the center of any computer system. Do you see something shaped like a box nearby? It will have a power switch and a light or two. It should also have a place or places to insert disks. This is the case that houses all of the important computer components. If it stands up tall, it is a tower case (see Figure 1.4). If it sits flat, it is a desktop case.

Figure 1.3　Outer Hardware

Figure 1.4　Tower case

Look at the back of the computer, you will see lots of cords and cables coming out of the back of the case and going to other computer parts like the monitor.

Your computer case probably has a place to insert floppy disks or CDs. These are called the floppy disk drive and the CD-ROM. The floppy disk drive reads information from a very thin disk that is inside a flat, square plastic case.

You can also write information to these disks and "save" it. CD-ROM is short for Compact Disk-Read Only Memory. A compact disk is a shiny, circular disk that stores information. A CD-ROM can only read information from the disk.

Many new computers have a CD-RW (RW stands for ReWrite) instead of a CD-ROM. CD-RW allows you to write information to the disk as well as read from it. Also, some new computers have a DVD (Digital Video Disk) drive instead of a CD-ROM or CD-RW. A DVD looks just like a CD, but it holds much more information. You can watch movies, listen to music, or play computer games from DVDs. One important thing to know is that you can play CDs in a DVD player, but you cannot play DVDs in a CD player!

2. Input Devices (see Figure 1.5)

There are several ways to get new information or input into a computer. The two most common ways are the keyboard and the mouse. The keyboard has keys for characters (letters, numbers and punctuation marks) and special commands, Pressing the keys tells the computer what to do or what to write.

Figure 1.5　Input Devices

The mouse has a special ball that allows you to roll it around on a pad or desk and move the cursor around on screen. By clicking on the buttons on the mouse, you give the computer directions on what to do. There are other devices similar to a mouse that can be used in its place. A trackball has the ball on top and you move it with your finger. A touchpad allows you to move your finger across a pressure sensitive pad and press to click.

Other types of input devices allow you to put images into the computer. A scanner copies a picture or document into the computer. There are several types of scanners and some look very different, but most look like a flat tray with a glass pane and a lid to cover it. You can input photographs into a computer with a digital camera. Photos are taken with the camera away from the computer and stored on a memory chip. Then the camera is plugged into the computer, so that the images can be downloaded. Another input device is a graphics tablet. A pressure sensitive pad is plugged into the computer. When you draw on the tablet with the special pen (never use an ink pen or pencil), the drawing appears on the screen. The tablet and pen can also be used like a mouse to move the cursor and click.

3. Output Devices

Output devices display information in a way that you can understand. The most common output device is a monitor. It looks like a TV and houses the computer screen. The monitor allows you to see what you and the computer are doing together.

Speakers (see Figure 1.6) are output devices that allow you to hear sound from your computer. Computer speakers are just like stereo speakers. There are usually two of them and they come in various sizes.

A printer is another common part of a computer system. It takes what you see on the computer screen and prints it on paper. There are two types of printers. The inkjet printer uses inks to print. It is the most common printer used with home computers and it can print in either black and white or color. Laser printers run much faster because they use lasers to print.

Speakers are an Output Device

Figure 1.6 speaker

Laser printers are mostly used in businesses. Black and white laser printers are the most common, but some print in color, too.

1.3 Smartphone, Tablet and Laptop

Smartphone, Tablet and Laptop are the most popular mobile devices. Next, we will see the differences between the three.

Though a smartphone is basically a device made for calling and receiving calls, it can be thought of as a handheld mini computer as against simple phones, it makes use of an independent operating system(os) to install and run advanced and complex applications. Most smartphones have a virtual keyboard that the user can operate easily with the help of a highly capacitive touchscreen.

Smartphones have fast processors and large internal memory, big display screens (around 3.5") and OS are very user friendly. Two OS that have dominated the smartphone market are Apple's iOS and Google's Android. While iOS is used by smartphones made by Apple only,

Android is an open source OS used by nearly all the other manufacturers of smartphones.

Tablet PC, as it is called makes available capabilities of an enriched multimedia device allowing the user to experience audio and video files on a bigger screen which is normally around 10 inches, just a little bit smaller than a laptop. As tablets make use of virtual keyboard, they are good for little typing work such as sending e-mails.

All tablets are Wi-Fi, enabled meaning they can be used to surf the web and they can be used to play games also. Today, tablets are coming equipped with dual camera both for capturing HD (high definition) videos and to make video chatting and video calling possible. However, as there are compromises in hardware, the functions such as multimedia tasking and other complex operations are difficult to perform in tablets.

Of the three mobile devices, laptop is the most powerful when it comes to computing and also for browsing the net. It has the fastest processor and also largest capacity of internal memory. A laptop is basically a PC that can be carried along at all places and integrates all the capabilities of a computer. Instead of a mouse the user has a touchpad and the speakers are inbuilt to make it a complete package. In addition, a laptop can be operated on a battery, and without power, it can run for 3-5 hours. With a display of 14" or more, a laptop is capable of performing all the tasks your computer theoretically can.

Keywords

primary memory	主存
hard disk	硬盘
floppy disk	软盘
secondary memory	辅存
pits	光盘点
crash	死机
supercomputer	超级计算机
cursor	光标
RAM	随机存储器
ROM	只读存储器
keyboard	键盘
mouse	鼠标
touch pad	触摸板
monitor	显示器
speaker	音箱
printer	打印机
smartphone	智能手机
tablet PC	平板电脑
laptop	手提电脑

1.4 Exercise 1

Multiple or single choices.

1. The purpose of storage in a computer is_____.
   A. hold data
   B. hold information
   C. get data
   D. transform data

2. Primary memory stored on_____.
   A. chips
   B. drive
   C. motherboard
   D. Disks

3. A crash is something that happens inside the computer's circuits which may_____.
   A. make ROM forget everything
   B. save computer's memory
   C. reboot the computer
   D. damage the computer

4. The two most common ways to get new information into a computer are_____.
   A. the keyboard
   B. the printer
   C. the mouse
   D. the speaker

5. Output devices include_____.
   A. monitor
   B. chips
   C. hard device
   D. speaker

True or False.

1. The purpose of storage in a computer is to hold data or information.

2. Using a CD-RW, computer data can be backed up to a CD.

3. You can play DVDs in a CD player.

4. An inkjet printer print more faster than A Laser printer.

5. A smartphone can be look like a mini computer.

1.5 Further Reading: Flash Memory

Electronic memory comes in a variety of forms to serve a variety of purposes. Flash memory is used for easy and fast information storage in such devices as digital cameras and home video game consoles. It is used more as a hard drive than as RAM. In fact, Flash memory is considered a solid state storage device. Solid state means that there are no moving parts-everything is electronic instead of mechanical.

Here are a few examples of Flash memory:
- Your computer's BIOS chip.
- CompactFlash (most often found in digital cameras).
- SmartMedia (most often found in digital cameras).
- Memory Stick (most often found in digital cameras).
- PCMCIA Type Ⅰ and Type Ⅱ memory cards (used as solid-state disks in laptops).
- Memory cards for video game consoles.

In this article, we'll find out how Flash memory works and look at some of the forms it takes and types of devices that use it.

1.5.1 The Basics

Flash memory is a type of EEPROM chip. It has a grid of columns and rows with a cell that has two transistors at each intersection. The two transistors are separated from each other by a thin oxide layer. One of the transistors is known as a floating gate, and the other one is the control gate. The floating gate's only link to the row, or wordline, is through the control gate. As long as this link is in place, the cell has a value of 1. To change the value to 0 requires a curious process called Fowler-Nordheim tunneling.

Tunneling is used to alter the placement of electrons in the floating gate. An electrical charge, usually 10 to 13 volts, is applied to the floating gate. The charge comes from the column, or bitline, enters the floating gate and drains to a ground.

This charge causes the floating-gate transistor to act like an electron gun. The excited electrons are pushed through and trapped on other side of the thin oxide layer, giving it a negative charge. These negatively charged electrons act as a barrier between the control gate and the floating gate. A special device called a cell sensor monitors the level of the charge passing through the floating gate. If the flow through the gate is greater than 50 percent of the charge, it has a value of 1. When the charge passing through drops below the 50 percent threshold, the value changes to 0. A blank EEPROM has all of the gates fully open, giving each cell a value of 1.

The electrons in the cells of a Flash-memory chip can be returned to normal (1) by the application of an electric field, a higher-voltage charge. Flash memory uses in-circuit wiring to apply the electric field either to the entire chip or to predetermined sections known as blocks. This erases the targeted area of the chip, which can then be rewritten. Flash memory works much faster than traditional EEPROMs because instead of erasing one byte at a time, it erases a block or the entire chip, and then rewrites it.

You may think that your car radio has Flash memory, since you are able to program the presets and the radio remembers them. But it is actually using Flash RAM. The difference is that Flash RAM has to have some power to maintain its contents, while Flash memory will maintain its data without any external source of power. Even though you have turned the power off, the car radio is pulling a tiny amount of current to preserve the data in the Flash RAM. That is why the radio will lose its presets if your car battery dies or the wires are disconnected.

In the following sections, we will concentrate on removable Flash memory products.

1.5.2 Removable Flash Memory Cards

While your computer's BIOS chip is the most common form of Flash memory, removable solid-state storage devices are becoming increasingly popular. SmartMedia and CompactFlash cards are both well-known, especially as "electronic film" for digital cameras. Other removable Flash memory products include Sony's Memory Stick, PCMCIA memory cards, and memory cards for video game systems such as Nintendo's N64, Sega's Dreamcast and Sony's PlayStation. We will focus on SmartMedia and CompactFlash, but the essential idea is the same for all of these products. Every one of them is simply a form of Flash memory.

There are several reasons to use Flash memory instead of a hard disk:
- Flash memory is noiseless.
- It allows faster access.
- It is smaller in size.
- It is lighter.
- It has no moving parts.

So why don't we just use Flash memory for everything? Because the cost per megabyte for a hard disk is drastically cheaper, and the capacity is substantially more.

1.5.3 SmartMedia

The solid-state floppy-disk card (SSFDC), better known as SmartMedia, was originally developed by Toshiba.

SmartMedia cards are available in capacities ranging from 2 MB to 128 MB. The card itself is quite small, approximately 45 mm long, 37 mm wide and less than 1 mm thick. This is amazing when you consider what is packed into such a tiny package!

As shown in Figure 1.7, SmartMedia cards are elegant in their simplicity. A plane electrode is connected to the Flash-memory chip by bonding wires. The Flash-memory chip, plane electrode and bonding wires are embedded in a resin using a technique called over-molded thin package (OMTP). This allows everything to be integrated into a single package without the need for soldering.

The OMTP module is glued to a base card to create the actual card, as shown in Figure 1.8. Power and data is carried by the electrode to the Flash-memory chip when the card is inserted into a device. A notched corner indicates the power requirements of the SmartMedia card. Looking at the card with the electrode facing up, if the notch is on the left side, the card needs 5 volts. If the notch is on the right side, it requires 3.3 volts.

Figure 1.7　SmartMedia card

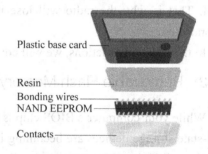

Figure 1.8　OMTP module

SmartMedia cards erase, write and read memory in small blocks (256- or 512-byte increments). This approach means that they are capable of fast, reliable performance while allowing you to specify which data you wish to keep. They are small, lightweight and easy to use. They are less rugged than other forms of removable solid-state storage, so you should be very careful when handling and storing them.

1.5.4 CompactFlash

CompactFlash cards were developed by Sandisk in 1994, and they are different from SmartMedia cards in two important ways:

They are thicker.

They utilize a controller chip.

CompactFlash consists of a small circuit board with Flash-memory chips and a dedicated controller chip, all encased in a rugged shell that is several times thicker than a SmartMedia card.

As shown in Figure 1.9, CompactFlash cards are 43 mm wide and 36 mm long, and come in two thicknesses: Type I cards are 3.3 mm thick, and Type II cards are 5.5 mm thick.

Figure 1.9 CompactFlash card

CompactFlash cards support dual voltage and will operate at either 3.3 volts or 5 volts.

The increased thickness of the card allows for greater storage capacity than SmartMedia cards. CompactFlash sizes range from 8 MB to 192 MB. The onboard controller can increase performance, particularly on devices that have slow processors. The case and controller chip add size, weight and complexity to the CompactFlash card when compared to the SmartMedia card.

Both of these types of removable storage, as well as PCMCIA Type I and Type II memory cards, adhere to standards developed by the Personal Computer Memory Card International Association (PCMCIA). Because of these standards, it is easy to use CompactFlash and SmartMedia products in a variety of devices. You can also buy adapters that allow you to access these cards through a standard floppy drive, USB port or PCMCIA card slot (like the one you find on a laptop computer). Sony's Memory Stick is available in a large array of products offered by Sony, and is now showing up in products from other manufacturers as well.

Although standards are flourishing, there are many Flash-memory products that are completely proprietary in nature, such as the memory cards in video game systems. But it is good to know that as electronic components become increasingly interchangeable and learn to communicate with each other (by way of technologies such as Bluetooth), standardized removable memory will allow you to keep your world close at hand.

Chapter 2
How Computer Monitors Work

What does "aspect ratio" mean? What is dot pitch? How much power does a display use? What is the difference between CRT and LCD? What does "refresh rate" mean?

In this article, we will answer all of these questions and many more. By the end of the article, you will be able to understand your current display and also make better decisions when purchasing your next one.

2.1 The Basics

Often referred to as a monitor when packaged in a separate case, the display is the most-used output device on a computer. The display provides instant feedback by showing you text and graphic images as you work or play. Most desktop displays use a cathode ray tube (CRT), while portable computing devices such as laptops incorporate liquid crystal display (LCD), light-emitting diode (LED), gas plasma or other image projection technology. Because of their slimmer design and smaller energy consumption, monitors using LCD technologies are beginning to replace the venerable CRT on many desktops.

When purchasing a display, you have a number of decisions to make. These decisions affect how well your display will perform for you, how much it will cost and how much information you will be able to view with it. Your decisions include:

- Display technology—Currently, the choices are mainly between CRT and LCD technologies.
- Cable technology—VGA and DVI are the two most common.
- Viewable area (usually measured diagonally).
- Aspect ratio and orientation (landscape or portrait).
- Maximum resolution.
- Dot pitch.
- Refresh rate.
- Color depth.
- Amount of power consumption.

In the following sections we will talk about each of these areas so that you can completely understand how your monitor works!

2.1.1 Display Technology Background

Displays have come a long way since the blinking green monitors in text-based computer systems of the 1970s. Just look at the advances made by IBM over the course of a decade:

In 1981, IBM introduced the Color Graphics Adapter (CGA), which was capable of rendering four colors, and had a maximum resolution of 320 pixels horizontally by 200 pixels vertically.

IBM introduced the Enhanced Graphics Adapter (EGA) display in 1984. EGA allowed up to 16 different colors and increased the resolution to 640×350 pixels, improving the appearance of the display and making it easier to read text.

In 1987, IBM introduced the Video Graphics Array (VGA) display system. Most computers today support the VGA standard and many VGA monitors are still in use.

IBM introduced the Extended Graphics Array (XGA) display in 1990, offering 800×600 pixel resolution in true color (16.8 million colors) and 1024×768 resolution in 65,536 colors.

2.1.2 Display Technologies: VGA

Once the display information is in analog form, it is sent to the monitor through a VGA cable. See Figure 2.1.

Figure 2.1 VGA cable

1: Red out; 2: Green out; 3: Blue out; 4: Unused; 5: Ground; 6: Red return (ground);
7: Green return (ground); 8: Blue return (ground); 9: Unused; 10: Sync return (ground);
11: Monitor ID 0 in; 12: Monitor ID 1 in or data from display; 13: Horizontal Sync out;
14: Vertical Sync; 15: Monitor ID 3 in or data clock

You can see that a VGA connector like this has three separate lines for the red, green and blue color signals, and two lines for horizontal and vertical sync signals. In a normal television, all of these signals are combined into a single composite video signal. The separation of the signals is one reason why a computer monitor can have so many more pixels than a TV set.

Since today's VGA adapters do not fully support the use of digital monitors, a new standard, Digital Video Interface (DVI) has been designed for this purpose.

2.1.3 Display Technology: DVI

Because VGA technology requires that the signal be converted from digital to analog for transmission to the monitor, a certain amount of degradation occurs. DVI keeps data in digital form from the computer to the monitor, virtually eliminating signal loss.

The DVI specification is based on Silicon Image's Transition Minimized Differential Signaling (TMDS) and provides a high-speed digital interface. TMDS takes the signal from the graphics adapter, determines the resolution and refresh rate that the monitor is using and spreads the signal out over the available bandwidth to optimize the data transfer from computer to monitor. DVI is technology-independent. Essentially, this means that DVI is going to perform properly with any display and graphics card that is DVI compliant. If you buy a DVI monitor, make sure that you have a video adapter card that can connect to it.

2.1.4 Viewable Area

Two measures describe the size of your display: the aspect ratio and the screen size. Most computer displays, like most televisions, have an aspect ratio of 4∶3 right now. This means that the ratio of the width of the display screen to the height is 4 to 3. The other aspect ratio in common use is 16∶9. Used in cinematic film, 16∶9 was not adopted when the television was first developed, but has always been common in the manufacture of alternative display technologies such as LCD. With widescreen DVD movies steadily increasing in popularity, most TV manufacturers now offer 16∶9 displays.

The display includes a projection surface, commonly referred to as the screen. Screen sizes are normally measured in inches from one corner to the corner diagonally across from it. This diagonal measuring system actually came about because the early television manufacturers wanted to make the screen size of their TVs sound more impressive.

Popular screen sizes are 15, 17, 19 and 21 inches. Notebook screen sizes are usually somewhat smaller, typically ranging from 12 to 15 inches. Obviously, the size of the display will directly affect resolution. The same pixel resolution will be sharper on a smaller monitor and fuzzier on a larger monitor because the same number of pixels is being spread out over a larger number of inches. An image on a 21-inch monitor with a 640×480 resolution will not appear nearly as sharp as it would on a 15-inch display at 640×480.

2.1.5 Maximum Resolution and Dot Pitch

Resolution refers to the number of individual dots of color, known as pixels, contained on a display. Resolution is typically expressed by identifying the number of pixels on the horizontal axis (rows) and the number on the vertical axis (columns), such as 640×480. The monitor's viewable area (discussed in the previous section), refresh rate and dot pitch all directly affect the maximum resolution a monitor can display.

2.1.6 Dot Pitch

As shown in Figure 2.2.

Briefly, the dot pitch is the measure of how much space there is between a display's pixels. When considering dot pitch, remember that smaller is better. Packing the pixels closer together is fundamental to achieving higher resolutions.

A display normally can support resolutions that match the physical dot (pixel) size as well as several lesser resolutions. For example, a display with a physical grid of 1280 rows by 1024 columns can obviously support a maximum resolution of 1280×1024 pixels. It usually also supports lower resolutions such as 1024×768, 800×600, and 640×480.

2.1.7 Refresh Rate

In monitors based on CRT technology, the refresh rate is the number of times that the image on the display is drawn each second. If your CRT monitor has a refresh rate of 72 Hertz (Hz), then it cycles through all the pixels from top to bottom 72 times a second. Refresh rates are very important because they control flicker, and you want the refresh rate as high as possible. Too few cycles per second and you will notice a flickering, which can lead to headaches and eye strain.

Televisions have a lower refresh rate than most computer monitors. To help adjust for the lower rate, they use a method called interlacing. This means that the electron gun in the television's CRT will scan through all the odd rows from top to bottom, then start again with the even rows. The phosphors hold the light long enough that your eyes are tricked into thinking that all the lines are being drawn together.

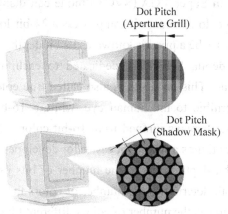

Figure 2.2 Dot pitch

Because your monitor's refresh rate depends on the number of rows it has to scan, it limits the maximum possible resolution. A lot of monitors support multiple refresh rates, usually dependent on the resolution you have chosen. Keep in mind that there is a tradeoff between flicker and resolution, and then pick what works best for you as shown in Figure 2.3.

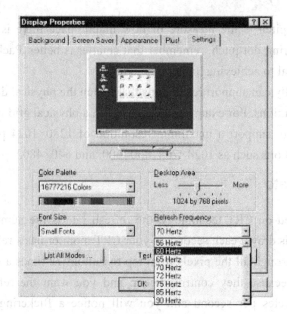

Figure 2.3 Display Properties

2.1.8 Color Depth

The combination of the display modes supported by your graphics adapter and the color capability of your monitor determine how many colors can be displayed. For example, a display that can operate in Super VGA (SVGA) mode can display up to 16,777,216 (usually rounded to 16.8 million) colors because it can process a 24-bit-long description of a pixel. The number of bits used to describe a pixel is known as its bit depth.

With a 24-bit bit depth, 8 bits are dedicated to each of the three additive primary colors-red, green and blue. This bit depth is also called true color because it can produce the 10,000,000 colors discernible to the human eye, while a 16-bit display is only capable of producing 65,536 colors. Displays jumped from 16-bit color to 24-bit color because working in 8-bit increments makes things a whole lot easier for developers and programmers.

Simply put, color bit depth refers to the number of bits used to describe the color of a single pixel. The bit depth determines the number of colors that can be displayed at one time. Take a look at Table 2.1 to see the number of colors different bit depths can produce:

Table 2.1 Bit-Depth & Number of Colors

Bit-Depth	Number of Colors
1	2(monochrome)
2	4(CGA)
4	16(EGA)
8	256(VGA)
16	65,536(High Color, XGA)

Continued

Bit-Depth	Number of Colors
24	16,777,216 (True Color, SVGA)
32	16,777,216 (True Color + Alpha Channel)

You will notice that the last entry in the chart is for 32 bits. This is a special graphics mode used by digital video, animation and video games to achieve certain effects. Nearly every monitor sold today can handle 24-bit color using a standard VGA connector, as discussed previously.

2.1.9 Power Consumption

Power consumption varies greatly with different technologies. CRTs are somewhat power-hungry, at about 110 watts for a typical display, especially when compared to LCDs, which average between 30 and 40 watts.

In a typical home computer setup with a CRT-based display, the monitor accounts for over 80 percent of the electricity used! the U.S. government initiated the Energy Star program in 1992. According to the EPA, if you use a computer system that is Energy Star compliant, it could save you approximately $400 a year on your electric bill! Similarly, because of the difference in power usage, an LCD monitor might cost more upfront but end up saving you money in the long run.

2.1.10 Monitor Trends: Flat Panels

CRT technology is still the most prevalent system in desktop displays. Because standard CRT technology requires a certain distance between the beam projection device and the screen, monitors employing this type of display technology tend to be very bulky. Other technologies make it possible to have much thinner displays, commonly known as flat-panel displays.

As shown in Figure 2.4, liquid crystal display (LCD) technology works by blocking light rather than creating it, while light-emitting diode (LED) and gas plasma work by lighting up display screen positions based on the voltages at different grid intersections. LCDs require far less energy than LED and gas plasma technologies and are currently the primary technology for notebook and other mobile computers. As flat-panel displays continue to grow in screen size and improve in resolution and affordability, they will gradually replace CRT-based displays.

Figure 2.4 Liquid crystal display (LCD)

Keywords

Cathode Ray Tube (CRT)	阴极射线管
Liquid Crystal Display (LCD)	液晶显示器
Light-emitting Diode (LED)	发光二极管
Gas Plasma	气体光栅
maximum resolution	最大分辨率
dot pitch	点距
refresh rate	刷新率
color depth	色度
Color Graphics Adapter (CGA)	彩色图形适配器
Enhanced Graphics Adapter (EGA)	增强型图形适配器
Video Graphics Array (VGA)	视频图像阵列
Extended Graphics Array	扩展型图形阵列

2.2 Exercise 2

Multiple or single choices.

1. When purchasing a display, you have a number of decisions about parameter to make include: ____.
   A. refresh rate
   B. material
   C. color depth
   D. maximum resolution

2. We usually describe the size of your display by ____.
   A. the computer size.
   B. the aspect ratio
   C. the screen size.
   D. the weight of the display

3. A display with a physical grid of 1280 rows by 1024 columns can obviously support ____.
   A. 1440×900
   B. 800×600
   C. 640×480
   D. 1600×900

4. Too few cycles per second and you will notice a flickering, which can lead to ____.
   A. myopia
   B. headaches
   C. dizzy
   D. eye strain

5. How many colors can be displayed is determined by the combination of ____.
   A. the display modes
   B. the display size
   C. the color capability
   D. the refresh rate

True or False.

1. All desktop displays use a cathode ray tube (CRT).

2. In 1981, IBM introduced the Color Graphics Adapter (CGA) which had a maximum resolution of 640 pixels horizontally by 200 pixels vertically.

3. VGA technology may bring a certain amount of degradation.

4. A display with a physical grid of 1280 rows by 1024 columns can support 1024×768, 800×600.

5. Liquid crystal display (LCD) technology works by blocking light rather than creating it.

2.3 Further Reading: Liquid Crystal Display

As shown in Figure 2.5, a liquid crystal display (LCD) is a thin, flat display device made up of any number of color or monochrome pixels arrayed in front of a light source or reflector. It is prized by engineers because it uses very small amounts of electric power, and is therefore suitable for use in battery-powered electronic devices.

Each pixel (picture element) consists of a column of liquid crystal molecules suspended between two transparent electrodes, and two polarizing filters, the axes of polarity of which are perpendicular to each other. Without the liquid crystals between them, light passing through one would be blocked by the other. The liquid crystal twists the polarization of light entering one filter to allow it to pass through the other.

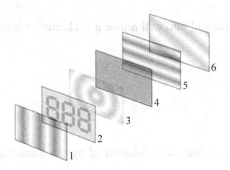

Figure 2.5　LCD

The molecules of the liquid crystal have electric charges on them. By applying small electrical charges to transparent electrodes over each pixel or subpixel, the molecules are twisted by electrostatic forces. This changes the twist of the light passing through the molecules, and allows varying degrees of light to pass (or not pass) through the polarizing filters.

Before applying an electrical charge, the liquid crystal molecules are in a relaxed state. Charges on the molecules cause these molecules to align themselves in a helical structure, or twist (the "crystal"). In some LCDs, the electrode may have a chemical surface that seeds the crystal, so it crystallizes at the needed angle. Light passing through one filter is rotated as it passes through the liquid crystal, allowing it to pass through the second polarized filter. A small amount of light is absorbed by the polarizing filters, but otherwise the entire assembly is transparent.

When an electrical charge is applied to the electrodes, the molecules of the liquid crystal align themselves parallel to the electric field, thus limiting the rotation of entering light. If the liquid crystals are completely untwisted, light passing through them will be polarized perpendicular to the second filter, and thus be completely blocked. The pixel will appear unlit. By controlling the twist of the liquid crystals in each pixel, light can be allowed to pass though in varying amounts, correspondingly illuminating the pixel.

Many LCDs are driven to darkness by an alternating current, which disrupts the twisting effect, and become light or transparent when no current is applied.

To save cost in the electronics, LCD displays are often multiplexed. In a multiplexed display, electrodes on one side of the display are grouped and wired together, and each group gets its own voltage source. On the other side, the electrodes are also grouped, with each group getting a voltage sink. The groups are designed so each pixel has a unique, unshared combination of source and sink. The electronics, or the software driving the electronics then turns on sinks in sequence, and drives sources for the pixels of each sink.

Important factors to consider when evaluating an LCD monitor include viewable size, response time (sync rate), matrix type (passive or active), viewing angle, color support, brightness and contrast ratio, resolution and aspect ratio, and input ports (e.g. DVI or VGA).

2.3.1 Brief History

The first operational LCD was based on the Dynamic Scattering Mode (DSM) and was introduced in 1968 by a group at RCA headed by George Heilmeier. Heilmeier founded Optel, which introduced a number of LCDs based on this technology. In 1969, the twisted nematic field effect in liquid crystals was discovered by James Fergason at Kent State University, and in 1971 his company (ILIXCO) produced the first LCDs based on it, which soon superseded the poor-quality DSM types.

However for a contrary view one should consider the pioneering work undertaken in the late 1960s by the UK's Radar Research Establishment at Malvern (part of DERA). The team at RRE supported ongoing work by George Gray and his team at the University of Hull-who ultimately discovered the cyanobiphenyl liquid crystals (which had all of the correct stability and temperature properties for application in LCDs).

2.3.2 Transmissive and Reflective Displays

LCDs can be either transmissive or reflective, depending on the location of the light source. A transmissive LCD is illuminated from the back by a backlight and viewed from the opposite side (front). This type of LCD is used in applications requiring high luminance levels such as computer displays, personal digital assistants, and mobile phones. The illumination device used to illuminate the LCD in such a product usually consumes much more power than the LCD itself.

Reflective LCDs, often found in digital watches and calculators, are illuminated by external light reflected by a (sometimes) diffusing reflector behind the display. This type of LCD has higher contrast than the transmissive type since light must pass through the liquid crystal layer twice and thus is attenuated twice. The absence of a lamp significantly reduces power consumption, allowing for longer battery life in battery-powered devices; small reflective LCDs consume so little power that they can rely on a photovoltaic cell, as often found in pocket calculators.

Transflective LCDs can work as either transmissive or reflective LCDs. They generally work reflectively when external light levels are high, and transmissively in darker environments via a low-power backlight.

2.3.3 Color Displays

In color LCDs each individual pixel is divided into three cells, or subpixels, which are colored red, green, and blue, respectively, by additional filters. Each subpixel can be controlled independently to yield thousands or millions of possible colors for each pixel. Older CRT monitors employ a similar method for displaying color. Color components may be arrayed in various pixel geometries, depending on the monitor's usage.

2.3.4 Passive-matrix and Active-matrix

LCDs with a small number of segments, such as those used in digital watches and pocket calculators, have a single electrical contact for each segment. An external dedicated circuit supplies an electric charge to control each segment. This display structure is unwieldy for more than a few display elements.

Small monochrome displays such as those found in personal organizers, or older laptop screens have a passive-matrix structure employing supertwist nematic (STN) or double-layer STN (DSTN) technology (DSTN corrects a color-shifting problem with STN). Each row or column of the display has a single electrical circuit. The pixels are addressed one at a time by row and column addresses. This type of display is called a passive matrix because the pixel must retain its state between refreshes without the benefit of a steady electrical charge. As the number of pixels (and, correspondingly, columns and rows) increases, this type of display becomes increasingly less feasible. Very slow response times and poor contrast are typical of passive-matrix LCDs.

For high-resolution color displays such as modern LCD computer monitors and televisions, an active-matrix structure is used. A matrix of thin-film transistors (TFTs) is added to the polarizing and color filters. Each pixel has its own dedicated transistor, which allows each column line to access one pixel. When a row line is activated, all of the column lines are connected to a row of pixels and the correct voltage is driven onto all of the column lines. The row line is then deactivated and the next row line is activated. All of the row lines are activated in sequence during a refresh operation. Active-matrix displays are much brighter and sharper than passive-matrix displays of the same size, and generally have quicker response times.

2.3.5 Quality Control

Some LCD panels have defective transistors, causing permanently lit or unlit pixels. Unlike integrated circuits, LCD panels with a few defective pixels are usually still usable. It is also economically prohibitive to discard a panel with just a few bad pixels because LCD panels are much larger than ICs. Manufacturers have different standards for determining a maximum acceptable number of defective pixels. Table 2.2 presents the maximum acceptable number of defective pixels for IBM's ThinkPad laptop line.

Table 2.2 Defective pixels for IBM's ThinkPad laptop line

Resolution	Bright Dots	Dark Dots	Total
QXGA	15	16	16
UXGA	11	16	16
SXGA+	11	13	16

| | | | Continued |
Resolution	Bright Dots	Dark Dots	Total
XGA	8	8	9
SVGA	5	5	9

LCD panels are more likely to have defects than most ICs due to their larger size. In this example, a 12 "SVGA LCD has 8 defects and a 6" wafer has only 3 defects, as shown in Figure 2.6. However, 134 of the 137 dies on the wafer will be acceptable, whereas rejection of the LCD panel would be a 0% yield (The standard is much higher now due to fierce competition between manufacturers and improved quality control. An LCD panel with 4 defective pixels is usually considered defective and customers can request an exchange for a new one). The location of a defective pixel is also important. Often manufacturers relax their requirements when defective pixels are in the center of the viewing area.

Some manufacturers offer a zero dead pixel policy.

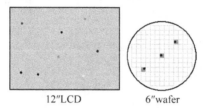

12"LCD   6"wafer

Figure 2.6 Defectives

2.3.6 Zero-power Displays

The zenithal bistable device, developed by Qinetiq (formerly DERA), can retain an image without power. The crystals may exist in one of two stable orientations-Black and "White". Power is only required to change the image. ZBD Displays is a spin-off company from Qinetiq who manufacture both grayscale and colour ZBD devices.

A French company, Nemoptic, has developed another zero-power, paper-like LCD technology which has been mass-produced in Taiwan since July 2003. This technology is intended for use in low-power mobile applications such as e-books and wearable computers. Zero-power LCDs are in competition with electronic paper.

2.3.7 Drawbacks

Despite their promising future LCD technology has a few drawbacks in comparison to CRTs. While CRTs are capable of displaying multiple video resolutions, each with the same quality, LCD monitors usually produce the crispest images in a "native" resolution. Secondly external light affects the brightness of the image in a LCD display while this is not a major concern in the CRT. Finally the viewing angle of a LCD is far less than the traditional alternative thus reducing the number of people who can conveniently view the same image. However this negative has been capitalised upon by an electronics company, allowing multiple TV outputs from the same LCD screen just by changing the angle from where the TV is seen.

Chapter 3
How Cell Phones Work

Billions of people in China and around the world use cellular phones. They are such great gadgets—with a cell phone, you can talk to anyone on the planet from just about anywhere!

These days, cell phones provide an incredible array of functions, and new ones are being added at a breakneck pace. Depending on the cell-phone model, you can:
- Store contact information.
- Make task or to-do lists.
- Keep track of appointments and set reminders.
- Use the built-in calculator for simple math.
- Send or receive e-mail.
- Get information (news, entertainment, stock quotes) from the Internet.
- Play games.
- Watch TV.
- Send text messages.
- Integrate other devices such as PDAs, MP3 players and GPS receivers.

But have you ever wondered how a cell phone works? What makes it different from a regular phone? What do all those terms like PCS, GSM, CDMA and TDMA mean? In this article, we will discuss the technology behind cell phones so that you can see how amazing they really are. If you are thinking about buying a cell phone, be sure to check out How Buying a Cell Phone Works to learn what you should know before making a purchase.

To start with, one of the most interesting things about a cell phone is that it is actually a radio—an extremely sophisticated radio, but a radio nonetheless. The telephone was invented by Alexander Graham Bell in 1876, and wireless communication can trace its roots to the invention of the radio by Nikolai Tesla in the 1880s (formally presented in 1894 by a young Italian named Guglielmo Marconi). It was only natural that these two great technologies would eventually be combined.

3.1 Cell-phone Frequencies

In the dark ages before cell phones, people who really needed mobile-communications ability installed radio telephones in their cars. In the radio-telephone system, there was one central antenna tower per city, and perhaps 25 channels available on that tower. This central antenna meant that the phone in your car needed a powerful transmitter—big enough to transmit 40 or 50 miles (about 70 km). It also meant that not many people could use radio telephones—there just were not enough channels.

The genius of the cellular system is the division of a city into small cells. This allows extensive frequency reuse across a city, so that millions of people can use cell phones simultaneously.

A good way to understand the sophistication of a cell phone is to compare it to a CB radio or a walkie-talkie.

- Full-duplex vs. half-duplex as shown in Figure 3.1 and Figure 3.2: Both walkie-talkies and CB radios are half-duplex devices. That is, two people communicating on a CB radio use the same frequency, so only one person can talk at a time. A cell phone is a full-duplex device. That means that you use one frequency for talking and a second, separate frequency for listening. Both people on the call can talk at once.
- Channels: A walkie-talkie typically has one channel, and a CB radio has 40 channels. A typical cell phone can communicate on 1664 channels or more!
- Range: A walkie-talkie can transmit about 1 mile (1.6 km) using a 0.25-watt transmitter. A CB radio, because it has much higher power, can transmit about 5 miles (8 km) using a 5-watt transmitter. Cell phones operate within cells, and they can switch cells as they move around. Cells give cell phones incredible range. Someone using a cell phone can drive hundreds of miles and maintain a conversation the entire time because of the cellular approach.

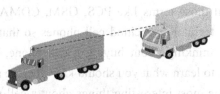

Figure 3.1 In half-duplex radio, both transmitters use the same frequency. Only one party can talk at a time

Figure 3.2 In full-duplex radio, the two transmitters use different frequencies, so both parties can talk at the same time. Cell phones are full-duplex

In a typical analog cell-phone system in the United States, the cell-phone carrier receives about 800 frequencies to use across the city. The carrier chops up the city into cells. Each cell is typically sized at about 10 square miles (26 square kilometers). Cells are normally thought of as hexagons on a big hexagonal grid, as shown in Figure 3.3.

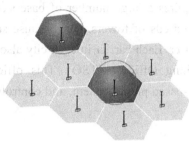

Figure 3.3 Because cell phones and base stations use low-power transmitters, the same frequencies can be reused in non-adjacent cells. The two purple cells can reuse the same frequencies

Each cell has a base station that consists of a tower and a small building containing the radio equipment. We'll get into base stations later. First, let's examine the "cells" that make up a cellular system.

3.2 Cell-phone Channels

A single cell in an analog cell-phone system uses one-seventh of the available duplex voice channels. That is, each cell (of the seven on a hexagonal grid) is using one-seventh of the available channels so it has a unique set of frequencies and there are no collisions:

- A cell-phone carrier typically gets 832 radio frequencies to use in a city.
- Each cell phone uses two frequencies per call—a duplex channel—so there are typically 395 voice channels per carrier (The other 42 frequencies are used for control channels—more on this later).

Therefore, each cell has about 56 voice channels available. In other words, in any cell, 56 people can be talking on their cell phone at one time. Analog cellular systems are considered first-generation mobile technology, or 1G. With digital transmission methods (2G), the number of available channels increases. For example, a TDMA-based digital system (more on TDMA later) can carry three times as many calls as an analog system, so each cell has about 168 channels available.

Cell phones have low-power transmitters in them. Many cell phones have two signal strengths: 0.6 watts and 3 watts (for comparison, most CB radios transmit at 4 watts). The base station is also transmitting at low power. Low-power transmitters have two advantages:

- The transmissions of a base station and the phones within its cell do not make it very far outside that cell. Therefore, in Figure 3.3, both of the purple cells can reuse the

same 56 frequencies. The same frequencies can be reused extensively across the city.
- The power consumption of the cell phone, which is normally battery-operated, is relatively low. Low power means small batteries, and this is what has made handheld cellular phones possible.

The cellular approach requires a large number of base stations in a city of any size. A typical large city can have hundreds of towers. But because so many people are using cell phones, costs remain low per user. Each carrier in each city also runs one central office called the Mobile Telephone Switching Office (MTSO). This office handles all of the phone connections to the normal land-based phone system, and controls all of the base stations in the region.

3.3 Analog Cell Phones

In 1983, the analog cell-phone standard called AMPS (Advanced Mobile Phone System) was approved by the FCC and first used in Chicago. As shown in Figure 3.4, AMPS uses a range of frequencies between 824 megahertz (MHz) and 894 MHz for analog cell phones. In order to encourage competition and keep prices low, the U. S. government required the presence of two carriers in every market, known as A and B carriers. One of the carriers was normally the local-exchange carrier (LEC), a fancy way of saying the local phone company.

Carriers A and B are each assigned 832 frequencies: 790 for voice and 42 for data. A pair of frequencies (one for transmit and one for receive) is used to create one channel. The frequencies used in analog voice channels are typically 30 kHz wide—30 kHz was chosen as the standard size because it gives you voice quality comparable to a wired telephone.

The transmit and receive frequencies of each voice channel are separated by 45 MHz to keep them from interfering with each other. Each carrier has 395 voice channels, as well as 21 data channels to use for housekeeping activities like registration and paging.

Figure 3.4 Photo courtesy Motorola, Inc. Old school: DynaTAC cell phone, 1983

A version of AMPS known as Narrowband Advanced Mobile Phone Service (NAMPS) incorporates some digital technology to allow the system to carry about three times as many calls as the original version. Even though it uses digital technology, it is still considered analog.

AMPS and NAMPS only operate in the 800-MHz band and do not offer many of the features common in digital cellular service, such as E-mail and Web browsing.

3.4 Along Comes Digital

Digital cell phones are the second generation (2G) of cellular technology. They use the same radio technology as analog phones, but they use it in a different way. Analog systems do not fully utilize the signal between the phone and the cellular network-analog signals cannot be compressed and manipulated as easily as a true digital signal. This is the reason why many cable companies are switching to digital—so they can fit more channels within a given bandwidth. It is amazing how much more efficient digital systems can be.

Digital phones convert your voice into binary information (1s and 0s) and then compress it (see How Analog-Digital Recording Works for details on the conversion process). This compression allows between three and 10 digital cell-phone calls to occupy the space of a single analog call.

Many digital cellular systems rely on frequency-shift keying (FSK) to send data back and forth over AMPS. FSK uses two frequencies, one for 1s and the other for 0s, alternating rapidly between the two to send digital information between the cell tower and the phone. Clever modulation and encoding schemes are required to convert the analog information to digital, compress it and convert it back again while maintaining an acceptable level of voice quality. All of this means that digital cell phones have to contain a lot of processing power.

Let's take a good look inside a digital cell phone.

3.5 Inside a Digital Cell Phone

On a "complexity per cubic inch" scale, cell phones are some of the most intricate devices people use on a daily basis. Modern digital cell phones can process millions of calculations per second in order to compress and decompress the voice stream.

If you take a basic digital cell phone apart, you find that it contains just a few individual parts as shown in Figure 3.5:

Figure 3.5   The parts of a cell phone

- An amazing circuit board containing the brains of the phone.
- An antenna.
- A liquid crystal display (LCD).
- A keyboard (not unlike the one you find in a TV remote control).
- A microphone.
- A speaker.
- A battery.

The circuit board is the heart of the system. Here is one from a typical Nokia digital phone, as shown in Figure 3.6 and Figure 3.7.

Figure 3.6　The front of the circuit board　　　　Figure 3.7　The back of the circuit board

In the photos above, you see several computer chips. Let's talk about what some of the individual chips do. The analog-to-digital and digital-to-analog conversion chips translate the outgoing audio signal from analog to digital and the incoming signal from digital back to analog. You can learn more about A-to-D and D-to-A conversion and its importance to digital audio in How Compact Discs Work. The digital signal processor (DSP) is a highly customized processor designed to perform signal-manipulation calculations at high speed.

The microprocessor as shown in Figure 3.8, handles all of the housekeeping chores for the keyboard and display, deals with command and control signaling with the base station and also coordinates the rest of the functions on the board.

Figure 3.8　The microprocessor

The ROM and Flash memory chips provide storage for the phone's operating system and customizable features, such as the phone directory. The radio frequency (RF) and power section handles power management and recharging, and also deals with the hundreds of FM channels. Finally, the RF amplifiers handle signals traveling to and from the antenna.

The display, as shown in Figure 3.9, has grown considerably in size as the number of features in cell phones have increased. Most current phones offer built-in phone directories, calculators and games. And many of the phones incorporate some type of PDA or Web browser.

Figure 3.9 The display and keypad contacts

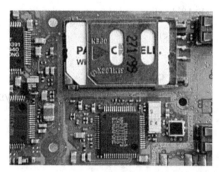

Figure 3.10 The Flash memory card on the circuit board

Some phones store certain information, such as the SID and MIN codes, in internal Flash memory, as shown in Figure 3.11, while others use external cards that are similar to Smart Media cards.

Figure 3.11 The Flash memory card removed

Figure 3.12 The cell-phone speaker, microphone and battery backup

Cell phones have such tiny speakers and microphones that it is incredible how well most of them reproduce sound. As you can see in the picture above, the speaker is about the size of a dime and the microphone is no larger than the watch battery beside it. Speaking of the watch battery, this is used by the cell phone's internal clock chip.

What is amazing is that all of that functionality—which only 30 years ago would have filled an entire floor of an office building—now fits into a package that sits comfortably in the palm of your hand!

Keywords

to-do lists	待办事项
sophisticated radio	无线电收发器
antenna tower	天线塔
walkie-talkie	步话机
Full-duplex	全双工传输
half-duplex	半双工传输
Channels	信道
watt	瓦
analog cell phones	模拟手机
AMPS (Advanced Mobile Phone System)	高级移动电话系统
bandwidth	频宽
microphone	麦克风
battery	电池
circuit board	电路板

3.6　Exercise 3

Multiple or single choices.

1. With a cell phone, you can _____.
    A. play games
    B. get information from internet
    C. contact with someone
    D. send or receive e-mail

2. A basic digital cell phone contains just a few individual parts, include_____.
    A. battery
    B. speaker
    C. LCD
    D. remote controller

True or False.

1. A cell phone is that it is actually a radio-an extremely sophisticated radio.

2. Both walkie-talkies and CB radios are Full-duplex devices.

3. Digital phones convert your voice into binary information.

4. A basic digital cell phone contains a few individual parts and the circuit board is the core.

5. Digital cellular systems rely on frequency-shift keying (FSK) to send data back and forth over AMPS.

3.7 Further Reading: Cell Phone

A mobile phone or mobile (also called cellphone and handphone) is an electronic device used for mobile telecommunications (mobile telephone, text messaging or data transmission) over a cellular network of specialized base stations known as cell sites. Mobile phones differ from cordless telephones, which only offer telephone service within limited range, e.g. within a home or an office, through a fixed line and a base station owned by the subscriber and also from satellite phones and radio telephones. As opposed to a radio telephone, a cell phone offers full duplex communication, automates calling to and paging from a public land mobile network (PLMN), and handoff (handover) during a phone call when the user moves from one cell (base station coverage area) to another. Most current cell phones connect to a cellular network consisting of switching points and base stations (cell sites) owned by a mobile network operator. In addition to the standard voice function, current mobile phones may support many additional services, and accessories, such as SMS for text messaging, e-mail, packet switching for access to the Internet, gaming, Bluetooth, infrared, camera with video recorder and MMS for sending and receiving photos and video, MP3 player, radio and GPS. The concept of a handheld phone was Martin Cooper's brainchild, and with the help of his Motorola team, the first handset was born in 1973 weighing in at two kilos.

The International Telecommunication Union estimated that mobile cellular subscriptions worldwide would reach approximately 4.6 billion by the end of 2009. Mobile phones have gained increased importance in the sector of Information and communication technologies for development in the 2000s and have effectively started to reach the bottom of the economic pyramid.

3.7.1 History

Radiophones have a long and varied history going back to Reginald Fessenden's invention and shore-to-ship demonstration of radio telephony, through the Second World War with military use of radio telephony links and civil services in the 1950s, while hand-held mobile radio devices have been available since 1973. A patent for the first wireless phone as we know today was issued in US Patent Number 3,449,750 to George Sweigert of Euclid, Ohio on June 10, 1969.

In 1960, the world's first partly automatic car phone system Mobile System A (MTA)|MTA was launched in Sweden. MTA phones were consisted of vacuum tubes and relays, and had a weight of 40 kg. In 1962, a more modern version called Mobile System B (MTB) was launched, which was a push-button telephone, and which used transistors in order to enhance the telephone's calling capacity and improve its operational reliability. In 1971 the MTD version was launched, opening for several different brands of equipment and gaining commercial success.

Martin Cooper, a Motorola researcher and executive is considered to be the inventor of the first practical mobile phone for hand-held use in a non-vehicle setting, after a long race against Bell Labs for the first portable mobile phone. Cooper is the first inventor named on "Radio telephone system" filed on October 17, 1973 with the US Patent Office and later issued as US Patent 3,906,166; other named contributors on the patent included Cooper's boss, John F. Mitchell, Motorola's chief of portable communication products, who successfully pushed Motorola to develop wireless communication products that would be small enough to use outside the home, office or automobile and participated in the design of the cellular phone. Using a modern, if somewhat heavy portable handset, Cooper made the first call on a hand-held mobile phone on April 3, 1973 to his rival, Dr. Joel S. Engel of Bell Labs.

1. Analog Cellular Telephony (1G)

The first commercially automated cellular network (the 1G generation) was launched in Japan by NTT in 1979, initially in the metropolitan area of Tokyo. Within five years, the NTT network had been expanded to cover the whole population of Japan and became the first nation-wide 1G networks.

The second launch of 1G networks was the simultaneous launch of the Nordic Mobile Telephone (NMT) system in Denmark, Finland, Norway and Sweden in 1981. NMT was the first mobile phone network featuring international roaming.

Several countries then followed in the early 1980s including the UK, Mexico and Canada. The first 1G network launched in the USA was Chicago based Ameritech in 1983 using the famous first hand-held mobile phone Motorola DynaTAC.

2. Digital Mobile Communication (2G)

The first "modern" network technology on digital 2G (second generation) cellular technology was launched by Radiolinja (now part of Elisa Group) in 1991 in Finland on the GSM standard which also marked the introduction of competition in mobile telecoms when Radiolinja challenged incumbent Telecom Finland (now part of TeliaSonera) who ran a 1G NMT network.

3. Wideband Mobile Communication (3G)

In 2001 the first commercial launch of 3G (Third Generation) was again in Japan by NTT

DoCoMo on the WCDMA standard.

One of the newest 3G technologies to be implemented is High-Speed Downlink Packet Access (HSDPA). It is an enhanced 3G (third generation) mobile telephony communications protocol in the High-Speed Packet Access (HSPA) family, also coined 3.5G, 3G+ or turbo 3G, which allows networks based on Universal Mobile Telecommunications System (UMTS) to have higher data transfer speeds and capacity.

3.7.2 Handsets

There are several categories of mobile phones, from basic phones to feature phones such as musicphones and cameraphones. There are also smartphones, the first smartphone was the Nokia 9000 Communicator in 1996 which incorporated PDA functionality to the basic mobile phone at the time. As miniaturisation and increased processing power of microchips has enabled ever more features to be added to phones, the concept of the smartphone has evolved, and what was a high-end smartphone five years ago, is a standard phone today. Several phone series have been introduced to address a given market segment, such as the RIM BlackBerry focusing on enterprise/corporate customer e-mail needs; the SonyEricsson Walkman series of musicphones and Cybershot series of cameraphones; the Nokia Nseries of multimedia phones, the Palm Pre the HTC Dream and the Apple iPhone.

1. Features

Mobile phones often have features extending beyond sending text messages and making voice calls, including call registers, GPS navigation, music (MP3) and video (MP4) playback, RDS radio receiver, alarms, memo and document recording, personal organiser and personal digital assistant functions, ability to watch streaming video or download video for later viewing, video calling, built-in cameras (1.0+ Mpx) and camcorders (video recording), with autofocus and flash, ringtones, games, PTT, memory card reader (SD), USB (2.0), infrared, Bluetooth (2.0) and Wi-Fi connectivity, instant messaging, Internet e-mail and browsing and serving as a wireless modem for a PC, and soon will also serve as a console of sorts to online games and other high quality games. Some phones also include a touchscreen.

Nokia and the University of Cambridge are demonstrating a bendable cell phone called the Morph.

2. Software and Applications

The most commonly used data application on mobile phones is SMS text messaging, with 74% of all mobile phone users as active users (over 2.4 billion out of 3.3 billion total subscribers at the end of 2007). SMS text messaging was worth over 100 billion dollars in annual revenues in 2007 and the worldwide average of messaging use is 2.6 SMS sent per day per person across the whole mobile phone subscriber base (source Informa 2007). The first SMS text message was sent from a computer to a mobile phone in 1992 in the UK, while the first person-to-person SMS from phone to phone was sent in Finland in 1993.

The other non-SMS data services used by mobile phones were worth 31 billion dollars in 2007, and were led by mobile music, downloadable logos and pictures, gaming, gambling, adult entertainment and advertising (source: Informa 2007). The first downloadable mobile content was sold to a mobile phone in Finland in 1998, when Radiolinja (now Elisa) introduced the downloadable ringing tone service. In 1999 Japanese mobile operator NTT DoCoMo introduced its mobile internet service, i-Mode, which today is the world's largest mobile internet service and roughly the same size as Google in annual revenues.

The first mobile news service, delivered via SMS, was launched in Finland in 2000. Mobile news services are expanding with many organisations providing "on-demand" news services by SMS. Some also provide "instant" news pushed out by SMS. Mobile telephony also facilitates activism and public journalism being explored by Reuters and Yahoo! and small independent news companies such as Jasmine News in Sri Lanka.

Companies are starting to offer mobile services such as job search and career advice. Consumer applications are on the rise and include everything from information guides on local activities and events to mobile coupons and discount offers one can use to save money on purchases. Even tools for creating websites for mobile phones are increasingly becoming available.

Mobile payments were first trialled in Finland in 1998 when two Coca-Cola vending machines in Espoo were enabled to work with SMS payments. Eventually the idea spread and in 1999 the Philippines launched the first commercial mobile payments systems, on the mobile operators Globe and Smart. Today mobile payments ranging from mobile banking to mobile credit cards to mobile commerce are very widely used in Asia and Africa, and in selected European markets. For example in the Philippines it is not unusual to have one's entire paycheck paid to the mobile account. In Kenya the limit of money transfers from one mobile banking account to another is one million US dollars. In India paying utility bills with mobile gains a 5% discount. In Estonia mobile phones are the most popular method of paying for public parking.

Mobile phones generally obtain power from rechargeable batteries. There are a variety of ways used to charge cell phones, including USB, portable batteries, mains power (using an AC adapter), cigarette lighters (using an adapter), or a dynamo. In 2009, wireless charging became a reality, and the first wireless charger was released for consumer use.

3. Standardization of Micro-USB Connector for Charging

Starting from 2010, many mobile phone manufacturers have agreed to use the Micro-USB connector for charging their phones. The mobile phone manufacturers who have agreed to this standard include:

- Apple
- LG
- Motorola

- Nokia
- Research In Motion
- Samsung
- Sony Ericsson

On 17 February 2009, the GSM Association announced that they had agreed on a standard charger for mobile phones. The standard connector to be adopted by 17 manufacturers in the Open Mobile Terminal Platform including Nokia, Motorola and Samsung is to be the micro-USB connector (several media reports erroneously reported this as the mini-USB). The new chargers will be much more efficient than existing chargers. Having a standard charger for all phones, means that manufacturers will no longer have to supply a charger with every new phone.

In addition, on 22 October 2009 the International Telecommunication Union (ITU) announced that it had embraced micro-USB as the Universal Charger Solution its "energy-efficient one-charger-fits-all new mobile phone solution", and added: "Based on the Micro-USB interface, UCS chargers will also include a 4-star or higher efficiency rating-up to three times more energy-efficient than an unrated charger."

4. Charger Efficiency

The world's five largest handset makers introduced a new rating system in November 2008 to help consumers more easily identify the most energy-efficient chargers.

The majority of energy lost in a mobile phone charger is in its no load condition, when the mobile phone is not connected but the charger has been left plugged in and using power. To combat this in November 2008 the top five mobile phone manufacturers Nokia, Samsung, LG Electronics, Sony Ericsson and Motorola set up a star rating system to rate the efficiency of their chargers in the no-load condition. Starting at zero stars for >0.5 W and going up to the top five star rating for <0.03 W (30 mW) no load power.

A number of semiconductor companies offering flyback controllers, such as Power Integrations and CamSemi, now claim that the five star standard can be achieved with use of their product.

5. Battery

Formerly, the most common form of mobile phone batteries were nickel metal-hydride, as they have a low size and weight. lithium ion batteries are sometimes used, as they are lighter and do not have the voltage depression that nickel metal-hydride batteries do. Many mobile phone manufacturers have now switched to using lithium-polymer batteries as opposed to the older Lithium-Ion, the main advantages of this being even lower weight and the possibility to make the battery a shape other than strict cuboid. Mobile phone manufacturers have been experimenting with alternative power sources, including solar cells and Coca Cola.

6. SIM Card

In addition to the battery, GSM mobile phones require a small microchip, called a

Subscriber Identity Module or SIM Card, to function. Approximately the size of a small postage stamp, the SIM Card is usually placed underneath the battery in the rear of the unit, and (when properly activated) stores the phone's configuration data, and information about the phone itself, such as which calling plan the subscriber is using. When the subscriber removes the SIM Card, it can be re-inserted into another phone that is configured to accept the SIM card and used as normal.

Each SIM Card is activated by use of a unique numerical identifier; once activated, the identifier is locked down and the card is permanently locked in to the activating network. For this reason, most retailers refuse to accept the return of an activated SIM Card.

Those cell phones that do not use a SIM Card have the data programmed in to their memory. This data is accessed by using a special digit sequence to access the "NAM" as in "Name" or number programming menu. From here, one can add information such as a new number for the phone, new Service Provider numbers, new emergency numbers, change their Authentication Key or A-Key code, and update their Preferred Roaming List or PRL. However, to prevent someone from accidentally disabling their phone or removing it from the network, the Service Provider puts a lock on this data called a Master Subsidiary Lock or MSL.

The MSL also ensures that the Service Provider gets payment for the phone that was purchased or "leased". For example, the Motorola RAZR V9C costs upwards of CAD $500. Depending on the carrier, such a phone may be available for as little as $200. The difference is paid by the customer in the form of a monthly bill. If the carrier did not use an MSL, then they may lose the $300-$400 difference that is paid in the monthly bill, since some customers would cancel their service and take the phone to another carrier.

The MSL applies to the SIM only so once the contract has been completed the MSL still applies to the SIM. The phone however, is also initially locked by the manufacturer into the Service Providers MSL. This lock may be disabled so that the phone can use other Service Providers SIM cards. Most phones purchased outside the US are unlocked phones because there are numerous Service Providers in close proximity to one another or have overlapping coverage. The cost to unlock a phone varies but is usually very cheap and is sometimes provided by independent phone vendors.

Having an unlocked phone is extremely useful for travelers due to the high cost of using the MSL Service Providers access when outside the normal coverage areas. It can cost sometimes up to 10 times as much to use a locked phone overseas as in the normal service area, even with discounted rates. T-Mobile will provide a SIM unlock code to account holders in good standing after 90 days according to their FAQ.

For example, in Jamaica, an AT&T subscriber might pay in excess of US$1.65 per minute for discounted international service while a B-Mobile (Jamaican) customer would pay US$0.20 per minute for the same international service. Some Service Providers focus sales on international sales while others focus on regional sales. For example, the same B-Mobile customer might pay more for local calls but less for international calls than a subscriber to the Jamaican national phone C&W (Cable & Wireless) company. These rate differences are mainly due to currency variations because SIM purchases are made in the local currency. In the US, this type of service competition does not exist because some of the major Service Providers do not offer Pay-As-You-Go services.

Chapter 4
Digital Camera Basics

4.1 How does Digital Camera Work

Let's say you want to take a picture and e-mail it to a friend. To do this, you need the image to be represented in the language that computers recognize—bits and bytes. Essentially, a digital image is just a long string of 1s and 0s that represent all the tiny colored dots—or pixels—that collectively make up the image.

If you want to get a picture into this form, you have two options:
- You can take a photograph using a conventional film camera, process the film chemically, print it onto photographic paper and then use a digital scanner to sample the print (record the pattern of light as a series of pixel values).
- You can directly sample the original light that bounces off your subject, immediately breaking that light pattern down into a series of pixel values, in other words, you can use a digital camera.

Digital cameras (as shown in Figure 4.1) focus light onto a semiconductor to create a digital image.

Just like a conventional camera, it has a series of lenses that focus light to create an image of a scene. But instead of focusing this light onto a piece of film, it focuses it onto a semiconductor device that records light electronically. A computer then breaks this electronic information down into digital data. All the fun and interesting features of digital cameras come as a direct result of this process.

Figure 4.1 Digital camera

In the next few sections, we'll find out exactly how the camera does all this.

4.2 CCD and CMOS: Filmless Cameras

Instead of film, a digital camera has a sensor that converts light into electrical charges.

The image sensor employed by most digital cameras is a charge coupled device (CCD). Some cameras use complementary metal oxide semiconductor (CMOS) technology instead. Both CCD and CMOS image sensors convert light into electrons. If you've read How Solar

Cells Work, you already understand one of the pieces of technology used to perform the conversion. A simplified way to think about these sensors is to think of a 2-D array of thousands or millions of tiny solar cells.

Once the sensor converts the light into electrons, it reads the value (accumulated charge) of each cell in the image. This is where the differences between the two main sensor types kick in:

- A CCD transports the charge across the chip and reads it at one corner of the array. An analog-to-digital converter (ADC) then turns each pixel's value into a digital value by measuring the amount of charge at each photo site and converting that measurement to binary form.
- CMOS devices use several transistors at each pixel to amplify and move the charge using more traditional wires.

Differences between the two types of sensors lead to a number of pros and cons:

- CCD sensors (as shown in Figure 4.2) create high-quality, low-noise images.
- CMOS sensors are generally more susceptible to noise.

Because each pixel on a CMOS sensor has several transistors located next to it, the light sensitivity of a CMOS chip is lower. Many of the photons hit the transistors instead of the photodiode.

CMOS sensors (as shown in Figure 4.3) traditionally consume little power. CCDs, on the other hand, use a process that consumes lots of power. CCDs consume as much as 100 times more power than an equivalent CMOS sensor.

Figure 4.2   A CCD sensor

Figure 4.3   A CMOS image sensor

CCD sensors have been mass produced for a longer period of time, so they are more mature. They tend to have higher quality pixels, and more of them.

Although numerous differences exist between the two sensors, they both play the same role in the camera—they turn light into electricity. For the purpose of understanding how a digital camera works, you can think of them as nearly identical devices.

4.3   Digital Camera Resolution

The amount of detail that the camera can capture is called the resolution, and it is measured in pixels. The more pixels a camera has, the more detail it can capture and the larger pictures can be without becoming blurry or "grainy."

Some typical resolutions (as shown in Figure 4.4) include:

Figure 4.4　Typical resolutions

The size of an image taken at different resolutions:

256×256—Found on very cheap cameras, this resolution is so low that the picture quality is almost always unacceptable. This is 65,000 total pixels.

640×480—This is the low end on most "real" cameras. This resolution is ideal for e-mailing pictures or posting pictures on a Web site.

1216×912—This is a "megapixel" image size—1,109,000 total pixels—good for printing pictures.

1600×1200—With almost 2 million total pixels, this is "high resolution." You can print a 4×5 inch print taken at this resolution with the same quality that you would get from a photo lab.

2240×1680—Found on 4 megapixel cameras-the current standard-this allows even larger printed photos, with good quality for prints up to 16×20 inches.

4064×2704—A top-of-the-line digital camera with 11.1 megapixels takes pictures at this resolution. At this setting, you can create 13.5×9 inch prints with no loss of picture quality.

Next, we'll look at how the camera adds color to these images.

You may have noticed that the number of pixels and the maximum resolution don't quite compute. For example, a 2.1-megapixel camera can produce images with a resolution of 1600×1200, or 1,920,000 pixels. But "2.1 megapixel" means there should be at least 2,100,000 pixels.

This isn't an error from rounding off or binary mathematical trickery. There is a real discrepancy between these numbers because the CCD has to include circuitry for the ADC to measure the charge. This circuitry is dyed black so that it doesn't absorb light and distort the image.

4.4 Capturing Color

Unfortunately, each photosite is colorblind. It only keeps track of the total intensity of the light that strikes its surface. In order to get a full color image, most sensors use filtering to look at the light in its three primary colors. Once the camera records all three colors, it combines them to create the full spectrum.

There are several ways of recording the three colors in a digital camera. The highest quality cameras use three separate sensors, each with a different filter. A beam splitter directs light to the different sensors. Think of the light entering the camera as water flowing through a pipe. Using a beam splitter would be like dividing an identical amount of water into three different pipes. Each sensor gets an identical look at the image; but because of the filters, each sensor only responds to one of the primary colors.

As shown in Figure 4.5, we can see the process of split a original image.

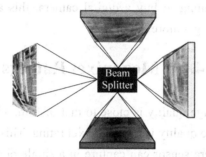

Figure 4.5 How the original (left) image is split in a beam splitter

The advantage of this method is that the camera records each of the three colors at each pixel location. Unfortunately, cameras that use this method tend to be bulky and expensive.

Another method is to rotate a series of red, blue and green filters in front of a single sensor. The sensor records three separate images in rapid succession. This method also provides information on all three colors at each pixel location; but since the three images aren't taken at precisely the same moment, both the camera and the target of the photo must remain stationary for all three readings. This isn't practical for candid photography or handheld cameras.

Both of these methods work well for professional studio cameras, but they're not necessarily practical for casual snapshots.

4.5 Digital Photography Basics

Digital photography has many advantages over traditional film photography. Digital photos are convenient, allow you to see the results instantly, don't require the costs of film and developing, and are suitable for software editing and uploading to the Internet.

While shooting on film will always have a place in the world of photography, digital models have taken over the consumer camera market almost completely. Just five years ago, buying a digital camera that could take photos of the same visual quality as a film camera could cost more than $1,000. But prices have dropped tremendously, and camera quality has increased. Today cameras in the $500 range are near-professional quality, and all but the cheapest digital cameras produce decent looking images.

There are also many additional features available on digital cameras, including image stabilization, on-board image editing, color correction functions, auto-bracketing and burst modes. A lot of these can be handled by image editing software, and so they can be unnecessary (and often inferior) when built into a camera. Burst mode, macro mode and image stabilization are probably the most useful extra features, but the best way to find out which camera is best for you is to explore any of the numerous digital photography magazines and Web sites that offer comparisons and user reviews of hundreds of different cameras.

If you own or are planning to buy a digital camera, this article will take you beyond point-and-shoot and help you get more out of it.

4.6 Megapixel Ratings

At five megapixels, image quality is close to that of film. The basic attribute of a digital camera that determines image quality is its megapixel rating. This number refers to the amount of information that the camera sensor can capture in a single photograph. Cameras with high megapixel ratings take larger pictures with more detail. Those photos will also look better when printed, especially in bigger sizes.

An example for digital camera as shown in Figure 4.6.

Figure 4.6　A Canon Camera

Next, let's learn about digital camera settings.

4.7 Digital Camera Settings and Modes

With a decent digital camera and a bit of practice, anyone can take acceptable quality photos with the camera set on full automatic. You can even take a bunch of so-so pictures and make them look acceptable later with image editing. But to really wring every ounce of ability out of your camera and produce truly beautiful photographs, you'll need to learn a few things about the manual settings. Keep in mind that lower-end cameras might not have manually adjustable settings.

When you're changing the settings on a camera, you're trying to find the proper exposure for the subject and lighting conditions. Exposure is the amount of light hitting the camera's sensor when you take a photo. Generally, you will want the exposure set so that the image captured by the camera's sensor closely matches what you see with your eyes. The camera tries to accomplish this when it's on full automatic mode, but the camera is easily fooled and a little slow, which is why manual settings usually produce better pictures.

As you get more familiar with your camera, you can play with different exposures for different effects. There are times when auto is better—something happens suddenly and you only have a few seconds to get your photo. Just flip to auto and take a picture.

To adjust exposure, you can tweak two different settings: aperture and shutter speed. Aperture is the diameter of the lens opening—a wider aperture means more light gets through. Aperture is measured in f-stops, higher f-stop numbers mean a smaller aperture. The aperture setting also affects depth of field, the amount of the photograph that is in focus. Smaller apertures (higher f-stops) give longer depth of field. A person in the foreground and the cars 20 feet behind her could all be in focus with a small enough aperture. A larger aperture results in a shallow depth of field, which you normally use for close-up shots and portraits. We'll take a closer look at shutter speed.

An example for digital camera setting as shown in Figure 4.7.

Figure 4.7   Our digital camera set at its slowest shutter speed-30 seconds

4.8   Shutter Speed

Shutter speed is the amount of time the shutter remains open to allow light through it. An extremely fast shutter speed is 1/2000 of a second, while camera settings usually allow up to about one second, which is very slow. One-sixtieth of a second is about as slow a shutter speed as you can use when taking a hand-held shot, and not get any blur. Some photographers force their camera shutters to stay open for much longer to create various special effects. Leaving a camera pointed at the night sky with the shutter open for several hours results in a photo of the paths the stars seem to take across the sky as the Earth rotates.

Practice and experience are the best ways to figure out which combinations of aperture and shutter speed are best for different kinds of photos. While a slow shutter speed lets in more light, it also makes it very difficult to get a crisp picture. Any movement at all (of either the subject or the camera) will result in blurring. Sometimes you might want this effect, but for a clear photo of a moving object, you need a fast shutter speed.

Many cameras have a semi-automatic mode that can be set to either aperture priority or shutter priority. You can either set the aperture or shutter speed (depending on which priority mode is enabled) to the desired setting, and the camera calculates the right settings to accommodate the lighting conditions. The camera might also have a variety of modes to choose from, such as sports mode or outdoor mode. These are aperture/shutter speed presets. Again, experience will let you know what conditions are right for each mode.

Keywords

digital camera	数码相机
film	胶片
scanner	扫描仪
original light	原始光
lenses	棱镜
semiconductor	半导体
electrical charges	电荷
sensor	传感器
pixel	像素
transistors	晶体管
wires	电线
sensitivity	灵敏度
consume	消耗
numerous	很大的
discrepancy	差异
colorblind	色盲
track	轨迹
surface	表面
spectrum	光谱
filter	过滤器
beam splitter	电子束分裂器
image stabilization	图像稳定
auto-bracketing	自动包围式曝光
manually adjustable	手动调节
exposure	曝光量

aperture	光圈
shutter speed	快门速度
diameter	直径

4.9 Exercise 4

Multiple or single choices.

1. The amount of detail that the camera can capture is called the resolution which is measured in_____.

    A. pixels

    B. color

    C. clarity

    D. shadow

2. Typical resolutions include:_____.

    A. 256×256

    B. 512×512

    C. 1216×912

    D. 1600×1200

3. Most sensors use filtering to look at the light in its_____ primary colors.

    A. one

    B. two

    C. three

    D. four

4. Digital photography has many advantages over traditional film photography include:____.

    A. convenient to carry

    B. good results

    C. see the results instantly

    D. do not require the costs of film

5. Digital cameras have many additional features such as:_____.

    A. more colorful

    B. image stabilization

    C. color correction functions

    D. auto-bracketing

True or False.

1. A digital camera has a sensor that converts light into electrical charges without using a film.

2. The image sensor employed by all digital cameras is a charge coupled device.

3. The two sensors, CCD and CMOS, though numerous differences exist between them, they both play the same role in the camera—they turn light into electricity.

4. Digital photos are convenient, allow you to see the results instantly.

5. To adjust exposure, you can tweak two different settings: aperture and shutter speed.

4.10 Further Reading: How to Take Good Photos

Taking photos with a digital camera follows many of the same techniques that make for successful film photographs. However, digital cameras differ in a few important ways.

There's usually a lag between the moment you press the shutter release button and when the camera takes the picture (except for the most expensive models). A longer lag time means that it's more difficult to capture a moment. Here are a few ways to minimize this problem:

Set your focus ahead of time. When using auto focus, pressing the shutter release halfway tells the camera to focus in on your target. You might have to wait for a few seconds with that button halfway down, but when you finally take the picture, the camera won't have to waste time focusing.

Use manual exposure settings. It takes time for the camera to calculate exposure settings in full automatic mode, so set them manually whenever you can.

Don't use flash unless it's absolutely necessary. The time it takes to charge the flash can create additional lag. If you need a flash, consider using an external flash unit, as shown in Figure 4.8.

Figure 4.8 An external flash unit

Use the viewfinder instead (as shown in Figure 4.9) of the LCD screen. This will save your batteries and reduce the amount of work the camera has to do.

Figure 4.9 The viewfinder on our camera is through-the-lens, so we shouldn't have a problem with parallax

Reduce image quality. Digital cameras allow you to adjust the size and resolution of the photos you are taking. Huge, uncompressed tiff files will look great, but they might create lag. If you are trying to capture action shots, try a lower quality setting with smaller images. Obviously you're sacrificing large, high-resolution images, but it will increase your chance of getting the shot you wanted. Experiment with your camera's settings to find the right balance between image quality and shutter lag.

Use burst mode. If your camera offers it, burst mode is a great way to get the precise moment you're shooting for by taking a series of quick photos over the course of a few seconds. Depending on the camera, burst mode (or continuous mode) may require a compromise in image quality.

4.10.1 Digital Camera Problems

Another problem with digital cameras is that they tend to need more light than a film camera to create a comparable exposure. As a result, slower shutter speeds are often used to get enough light. This can make it very difficult to take a photo without some blurring. The solution is simple: use a tripod. They're not very expensive, and you can mount almost all cameras to one. This step will result in a huge improvement in photo quality. If you have to take a photo with a slow shutter speed and you don't have your tripod handy, try using burst mode. It will take practice to hold the camera steady (hold it against your face and exhale as you press the button), but the burst will give you a better chance at getting one unblurred picture out of the bunch. When all else fails, a sturdy shelf or stack of books can help.

When in doubt, go for a greater depth of field. With digital images, it is always possible to use software to take certain areas of a photo out of focus, but you can never "fix" anything that's out of focus to begin with.

If your camera has a viewfinder in the upper left corner, you may have problems with parallax, especially for close-up shots. This means that the viewfinder is looking at a slightly different area than what the lens is seeing. Cameras with a through-the-lens viewfinder don't have this problem. Using the LCD screen to line up your shot can help, but the screens still don't usually show all of what the lens is seeing, and they can make it difficult to focus properly.

4.10.2 Image Editing Software

There are several image editing software packages available, and some cameras come with them. You can use these programs to manipulate, crop, combine and print your digital photos. There are no limits to what you can accomplish with this software, from a subtle contrast adjustment to elaborate works of art incorporating multiple images, textures and treatments. A few basic procedures will enhance digital photos and help fix problem areas:

- Once you start editing a digital photo, save the result as a separate file. You always want to have your original image available in an unedited form.
- As shown in Figure 4.10 and Figure 4.11. Adjust contrast and color levels. Giving the

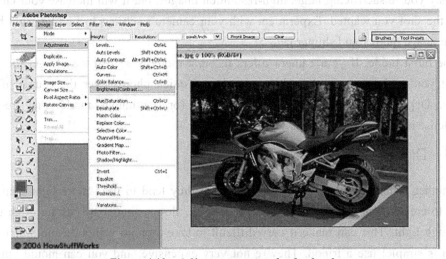

Figure 4.10 Adjust contrast and color levels

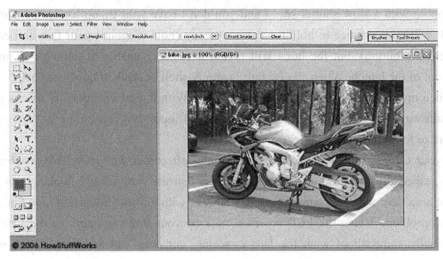

Figure 4.11 This image was too dark after we took it. But we used Adobe Photoshop, a popular brand of photo editing software, to adjust the brightness and contrast

contrast setting a nudge can really enhance a photograph and create a more dramatic look. If a photo looks washed out, increasing saturation levels can make the colors more vibrant. In cases where an incorrect white balance on the camera has given the image a colored tint, adjusting the color levels can bring the photo back to a realistic, untinted state.

- Rotate, crop and matte photos. Photo editing software makes it simple to rotate a photo (90 degrees if you took the picture with the camera turned to the side, or smaller amounts if the horizon is just slightly out of line). You can also easily crop off unwanted portions of a photo, making it possible to recompose a shot long after you took it. Most programs allow you to add a frame around the picture, making a built-in matte.

- Get rid of red-eye. Some programs have a "red-eye removal" function built in. You can also do it manually by selecting the subject's eyes and altering the color balance to reduce redness. More advanced users can change the entire color palette of an image, create a sepia or monochrome image, or make an image black and white.

- Remove unwanted objects as shown in Figure 4.12. You can use certain tools in a photo-editing program to remove parts of a photo, leaving what appears to be the plain background in its place. Say you've taken a photo of a centuries-old castle, but someone had parked a mini-van in front of it, ruining the mood of the scene. Though it will take some practice, you don't have to be a photo-editing expert to completely remove the van and leave just the castle behind it.

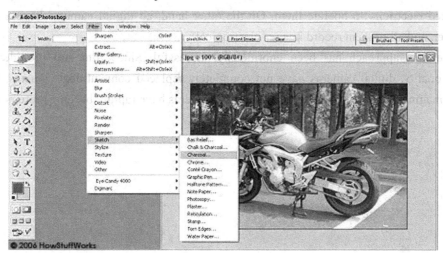

Figure 4.12 Remove unwanted objects

- Create works of art. Filters and plug-ins included with many programs allow you to change photos into artistic pieces with a few mouse clicks. Change can be drastic or subtle, such as making a photo of your back yard look like a watercolor painted by an

Impressionist master, or changing a photo of your motorcycle into a charcoal sketch. As shown in Figure 4.13, you can change your images with photo editing software.

By now you should have a good idea of what to look for in a digital camera and how to use it to get the images that you want. Above all, digital photography should be a fun experience. Feel free to experiment with settings and editing software to create your own works of art.

Figure 4.13 With photo editing software you can manipulate images in many different ways, like turning our motorcycle picture into a charcoal sketch

Conventional cameras depend entirely on chemical and mechanical processes—you don't even need electricity to operate them. On the other hand, all digital cameras have a built-in computer, and all of them record images electronically.

The new approach has been enormously successful. Since film still provides better picture quality, digital cameras have not completely replaced conventional cameras. But, as digital imaging technology has improved, digital cameras have rapidly become more popular.

Chapter 5
How Bits and Bytes Work

If you have used a computer for more than five minutes, then you have heard the words bits and bytes. Both RAM and hard disk capacities are measured in bytes, as are file sizes when you examine them in a file viewer.

You might hear an advertisement that says, "This computer has a 32-bit Pentium processor with 64 megabytes of RAM and 2.1 gigabytes of hard disk space." In this article, we will discuss bits and bytes so that you have a complete understanding.

5.1 Decimal Numbers

The easiest way to understand bits is to compare them to something you know: digits. A digit is a single place that can hold numerical values between 0 and 9. Digits are normally combined together in groups to create larger numbers. For example, 6357 has four digits. It is understood that in the number 6357, the 7 is filling the "1s place," while the 5 is filling the 10s place, the 3 is filling the 100s place and the 6 is filling the 1,000s place. So you could express things this way if you wanted to be explicit:

(6*1000)+(3*100)+(5*10)+(7*1)=6000+300+50+7=6357

Another way to express it would be to use powers of 10. Assuming that we are going to represent the concept of "raised to the power of " with the "^" symbol (so "10 squared" is written as "10^2"), another way to express it is like this:

(6*10^3)+(3*10^2)+(5*10^1)+(7*10^0)=6000+300+50+7=6357

What you can see from this expression is that each digit is a placeholder for the next higher power of 10, starting in the first digit with 10 raised to the power of zero.

That should all feel pretty comfortable, we work with decimal digits every day. The neat thing about number systems is that there is nothing that forces you to have 10 different values in a digit. Our base-10 number system likely grew up because we have 10 fingers, but if we happened to evolve to have eight fingers instead, we would probably have a base-8 number system. You can have base-anything number systems. In fact, there are lots of good reasons to use different bases in different situations.

5.2 Bits

Computers happen to operate using the base-2 number system, also known as the binary number system (just like the base-10 number system is known as the decimal number system). The reason computers use the base-2 system is because it makes it a lot easier to implement them with current electronic technology. You could wire up and build computers that operate in base-10, but they would be fiendishly expensive right now. On the other hand, base-2 computers are relatively cheap.

So computers use binary numbers, and therefore use binary digits in place of decimal digits. The word bit is a shortening of the words "Binary digIT." Whereas decimal digits have 10 possible values ranging from 0 to 9, bits have only two possible values: 0 and 1. Therefore, a binary number is composed of only 0s and 1s, like this: 1011. How do you figure out what the value of the binary number 1011 is? You do it in the same way we did it above for 6357, but you use a base of 2 instead of a base of 10. So:

$(1*2^3)+(0*2^2)+(1*2^1)+(1*2^0)=8+0+2+1=11$

You can see that in binary numbers, each bit holds the value of increasing powers of 2. That makes counting in binary pretty easy. Starting at zero and going through 20, counting in decimal and binary looks like this:

```
 0=    0
 1=    1
 2=    10
 3=    11
 4=    100
 5=    101
 6=    110
 7=    111
 8=    1000
 9=    1001
10=    1010
11=    1011
12=    1100
13=    1101
14=    1110
15=    1111
16=    10000
17=    10001
18=    10010
```

19= 10011
20= 10100

When you look at this sequence, 0 and 1 are the same for decimal and binary number systems. At the number 2, you see carrying first take place in the binary system. If a bit is 1, and you add 1 to it, the bit becomes 0 and the next bit becomes 1. In the transition from 15 to 16 this effect rolls over through 4 bits, turning 1111 into 10000.

5.3 Bytes

Bits are rarely seen alone in computers. They are almost always bundled together into 8-bit collections, and these collections are called bytes. Why are there 8 bits in a byte? A similar question is, "Why are there 12 eggs in a dozen?" The 8-bit byte is something that people settled on through trial and error over the past 50 years.

With 8 bits in a byte, you can represent 256 values ranging from 0 to 255, as shown here:

$$0=00000000$$
$$1=00000001$$
$$2=00000010$$
$$...$$
$$254=11111110$$
$$255=11111111$$

You known that a CD uses 2 bytes, or 16 bits, per sample. That gives each sample a range from 0 to 65,535, like this:

$$0=0000000000000000$$
$$1=0000000000000001$$
$$2=0000000000000010$$
$$...$$
$$65534=1111111111111110$$
$$65535=1111111111111111$$

5.4 Bytes: ASCII

Bytes are frequently used to hold individual characters in a text document. In the ASCII character set, each binary value between 0 and 127 is given a specific character. Most computers extend the ASCII character set to use the full range of 256 characters available in a byte. The upper 128 characters handle special things like accented characters from common foreign languages.

You can see the 127 standard ASCII codes below. Computers store text documents, both on disk and in memory, using these codes. For example, if you use Notepad in Windows 95/98 to create a text file containing the words, "Four score and seven years ago," Notepad would use 1 byte of memory per character (including 1 byte for each space character between the words—ASCII character 32). When Notepad stores the sentence in a file on disk, the file will also contain 1 byte per character and per space.

Try this experiment: Open up a new file in Notepad and insert the sentence, "Four score and seven years ago" in it. Save the file to disk under the name getty.txt. Then use the explorer and look at the size of the file. You will find that the file has a size of 30 bytes on disk: 1 byte for each character. If you add another word to the end of the sentence and resave it, the file size will jump to the appropriate number of bytes. Each character consumes a byte.

If you were to look at the file as a computer looks at it, you would find that each byte contains not a letter but a number—the number is the ASCII code corresponding to the character (see below). So on disk, the numbers for the file look like this:

F o u r   a n d   s e v e n
70 111 117 114 32 97 110 100 32 115 101 118 101 110

By looking in the ASCII table, you can see a one-to-one correspondence between each character and the ASCII code used. Note the use of 32 for a space—32 is the ASCII code for a space. We could expand these decimal numbers out to binary numbers (so 32=00100000) if we wanted to be technically correct—that is how the computer really deals with things.

5.5 Standard ASCII Character Set

The first 32 values (0 through 31) are codes for things like carriage return and line feed. The space character is the 33rd value, followed by punctuation, digits, uppercase characters and lowercase characters. See Table 5.1.

Table 5.1 ASCII Character Set

0	NUL	12	FF	24	CAN	36	$	48	0	60	<	72	H	84	T
1	SOH	13	CR	25	EM	37	%	49	1	61	=	73	I	85	U
2	STX	14	SO	26	SUB	38	&	50	2	62	>	74	J	86	V
3	ETX	15	SI	27	ESC	39	'	51	3	63	?	75	K	87	W
4	EOT	16	DLE	28	FS	40	(	52	4	64	@	76	L	88	X
5	ENQ	17	DC1	29	GS	41	)	53	5	65	A	77	M	89	Y
6	ACK	18	DC2	30	RS	42	*	54	6	66	B	78	N	90	Z
7	BEL	19	DC3	31	US	43	+	55	7	67	C	79	O	91	[
8	BS	20	DC4	32		44	,	56	8	68	D	80	P	92	\
9	TAB	21	NAK	33	!	45	-	57	9	69	E	81	Q	93	]
10	LF	22	SYN	34	"	46	.	58	:	70	F	82	R	94	^
11	VT	23	ETB	35	#	47	/	59	;	71	G	83	S	95	_

96	`	100	d	104	h	108	l	112	p	116	t	120	x	124	|
97	a	101	e	105	i	109	m	113	q	117	u	121	y	125	}
98	b	102	f	106	j	110	n	114	r	118	v	122	z	126	~
99	c	103	g	107	k	111	o	115	s	119	w	123	{	127	DEL

5.6 Lots of Bytes

When you start talking about lots of bytes, you get into prefixes like kilo, mega and giga, as in kilobyte, megabyte and gigabyte (also shortened to K, M and G, as in Kbytes, Mbytes and Gbytes or KB, MB and GB). Table 5.2 shows the multipliers:

Table 5.2 The multipliers

Name	Abbr.	Size
Kilo	K	2^{10}=1,024
Mega	M	2^{20}=1,048,576
Giga	G	2^{30}=1,073,741,824
Tera	T	2^{40}=1,099,511,627,776
Peta	P	2^{50}=1,125,899,906,842,624
Exa	E	2^{60}=1,152,921,504,606,846,976
Zetta	Z	2^{70}=1,180,591,620,717,411,303,424
Yotta	Y	2^{80}=1,208,925,819,614,629,174,706,176

You can see in this chart that kilo is about a thousand, mega is about a million, giga is about a billion, and so on. So when someone says, "This computer has a 2 gig hard drive," what he or she means is that the hard drive stores 2 gigabytes, or approximately 2 billion bytes, or exactly 2,147,483,648 bytes. How could you possibly need 2 gigabytes of space? When you consider that one CD holds 650 megabytes, you can see that just three CDs worth of data will fill the whole thing! Terabyte databases are fairly common these days, and there are probably a few petabyte databases floating around the Pentagon by now.

5.7 Binary Math

Binary math works just like decimal math, except that the value of each bit can be only 0 or 1. To get a feel for binary math, let's start with decimal addition and see how it works. Assume that we want to add 452 and 751:

$$\begin{array}{r} 452 \\ +751 \\ \hline 1203 \end{array}$$

To add these two numbers together, you start at the right: 2+1=3. No problem. Next, 5+5=10, so you save the zero and carry the 1 over to the next place. Next, 4+7+1 (because of the carry)=12, so you save the 2 and carry the 1. Finally, 0+0+1=1. So the answer is 1203.

Binary addition works exactly the same way:

$$\begin{array}{r} 010 \\ +111 \\ \hline 1001 \end{array}$$

Starting at the right, 0+1=1 for the first digit. No carrying there. You've got 1+1=10 for the second digit, so save the 0 and carry the 1. For the third digit, 0+1+1=10, so save the zero and carry the 1. For the last digit, 0+0+1=1. So the answer is 1001. If you translate everything over to decimal you can see it is correct: 2+7=9.

5.8   Quick Recap

To sum up this entire article, here's what we've learned about bits and bytes:

- Bits are binary digits. A bit can hold the value 0 or 1.
- Bytes are made up of 8 bits each.
- Binary math works just like decimal math, but each bit can have a value of only 0 or 1.

There really is nothing more to it—bits and bytes are that simple!

Keywords

bits	位
bytes	字节
decimal numbers	十进制数
powers	幂
binary numbers	二进制数
symbol	符号
placeholder	占位符
bundled	捆绑
character	字符
correspondence	对应

5.9   Exercise 5

Multiple or single choices.

1. Bits are binary digits. A bit can hold the value:_____.

A. 0
B. 1
C. 2
D. 3

2. Bytes are made up of ____ bits each.
   A. 2
   B. 4
   C. 8
   D. 16

3. Transform decimal number(55) into binary number____.
   A. 110001
   B. 101101
   C. 110111
   D. 101101

4. Transform binary number(11110011) into decimal number ____.
   A. 241
   B. 243
   C. 245
   D. 247

5. Inquire the standard ASCII Character Set, calculate M minus C, the result is ____.
   A. 7
   B. 8
   C. 9
   D. 10

True or False.

1. The space character is the 33rd value in a standard ASCII character set.

2. Bytes are made up of 16 bits each.

3. Each bit can have a value of only 0 or 1.

4. By looking in the ASCII table, you can see a one-to-one correspondence between each character and the ASCII code used.

5. kilo is about a thousand, mega is about a billion, giga is about a million.

5.10 Further Reading: How Boolean Logic Works

Have you ever wondered how a computer can do something like balance a check book, or play chess, or spell-check a document? These are things that, just a few decades ago, only humans could do. Now computers do them with apparent ease. How can a "chip" made up of silicon and wires do something that seems like it requires human thought?

If you want to understand the answer to this question down at the very core, the first thing you need to understand is something called Boolean logic. Boolean logic, originally developed by George Boole in the mid 1800s, allows quite a few unexpected things to be mapped into bits and bytes. The great thing about Boolean logic is that, once you get the hang of things, Boolean logic (or at least the parts you need in order to understand the operations of computers) is outrageously simple. In this article we will first discuss simple logic "gates," and then see how to combine them into something useful.

5.10.1 Simple Gates

There are three, five or seven simple gates that you need to learn about, depending on how you want to count them (you will see why in a moment). With these simple gates you can build combinations that will implement any digital component you can imagine. These gates are going to seem a little dry here, and incredibly simple, but we will see some interesting combinations in the following sections that will make them a lot more inspiring.

The simplest possible gate is called an "inverter," or a NOT gate. It takes one bit as input and produces as output its opposite. The table below shows a logic table for the NOT gate and the normal symbol for it in circuit diagrams:

NOT Gate

A	Q
0	1
1	0

A —▷o— Q

You can see in this figure that the NOT gate has one input called A and one output called Q ("Q" is used for the output because if you used "O," you would easily confuse it with zero). The table shows how the gate behaves. When you apply a 0 to A, Q produces a 1. When you apply a 1 to A, Q produces a 0. Simple.

The AND gate performs a logical "and" operation on two inputs, A and B:

AND Gate

A	B	Q
0	0	0
0	1	0
1	0	0
1	1	1

The idea behind an AND gate is, "If A AND B are both 1, then Q should be 1." You can see that behavior in the logic table for the gate. You read this table row by row, like this:

AND Gate

A	B	Q	
0	0	0	If A is 0 AND B is 0, Q is 0.
0	1	0	If A is 0 AND B is 1, Q is 0.
1	0	0	If A is 1 AND B is 0, Q is 0.
1	1	1	If A is 1 AND B is 1, Q is 1.

The next gate is an OR gate. Its basic idea is, "If A is 1 OR B is 1 (or both are 1), then Q is 1."

OR Gate

A	B	Q
0	0	0
0	1	1
1	0	1
1	1	1

Those are the three basic gates (that's one way to count them). It is quite common to recognize two others as well: the NAND and the NOR gate. These two gates are simply combinations of an AND or an OR gate with a NOT gate. If you include these two gates, then the count rises to five. Here's the basic operation of NAND and NOR gates—you can see they are simply inversions of AND and OR gates:

NOR Gate

A	B	Q
0	0	1
0	1	0
1	0	0
1	1	0

NAND Gate

A	B	Q
0	0	1
0	1	1
1	0	1
1	1	0

The final two gates that are sometimes added to the list are the XOR and XNOR gates, also known as "exclusive or" and "exclusive nor" gates, respectively. Here are their tables:

XOR Gate

A	B	Q
0	0	0
0	1	1
1	0	1
1	1	0

XNOR Gate

A	B	Q
0	0	1
0	1	0
1	0	0
1	1	1

The idea behind an XOR gate is, "If either A OR B is 1, but NOT both, Q is 1." The reason why XOR might not be included in a list of gates is because you can implement it easily using the original three gates listed. Here is one implementation:

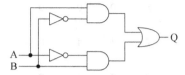

If you try all four different patterns for A and B and trace them through the circuit, you will find that Q behaves like an XOR gate. Since there is a well-understood symbol for XOR gates, it is generally easier to think of XOR as a "standard gate" and use it in the same way as AND and OR in circuit diagrams.

5.10.2 Simple Adders

In this section, you will learn how you can create a circuit capable of binary addition using the gates described in the previous section.

Let's start with a single-bit adder. Let's say that you have a project where you need to add single bits together and get the answer. The way you would start designing a circuit for that is to first look at all of the logical combinations. You might do that by looking at the following four sums:

```
    0      0      1      1
   +0     +1     +0     +1
   ---    ---    ---    ---
    0      1      1     10
```

That looks fine until you get to 1+1. In that case, you have that pesky carry bit to worry about. If you don't care about carrying (because this is, after all, a 1-bit addition problem), then you can see that you can solve this problem with an XOR gate. But if you do care, then you might rewrite your equations to always include 2 bits of output, like this:

```
    0      0      1      1
   +0     +1     +0     +1
   ---    ---    ---    ---
   00     01     01     10
```

From these equations you can form the logic table:

1-bit Adder with Carry-Out

A	B	Q	CO
0	0	0	0
0	1	1	0
1	0	1	0
1	1	0	1

By looking at this table you can see that you can implement Q with an XOR gate and CO (carry-out) with an AND gate. Simple.

What if you want to add two 8-bit bytes together? This becomes slightly harder. The easiest solution is to modularize the problem into reusable components and then replicate components. In this case, we need to create only one component: a full binary adder.

The difference between a full adder and the previous adder we looked at is that a full adder accepts an A and a B input plus a carry-in (CI) input. Once we have a full adder, then we can string eight of them together to create a byte-wide adder and cascade the carry bit from one adder to the next.

The logic table for a full adder is slightly more complicated than the tables we have used before, because now we have 3 input bits. It looks like this:

One-bit Full Adder with Carry-In and Carry-Out

CI	A	B	Q	CO
0	0	0	0	0
0	0	1	1	0
0	1	0	1	0
0	1	1	0	1
1	0	0	1	0
1	0	1	0	1
1	1	0	0	1
1	1	1	1	1

There are many different ways that you might implement this table. I am going to present one method here that has the benefit of being easy to understand. If you look at the Q bit, you can see that the top 4 bits are behaving like an XOR gate with respect to A and B, while the bottom 4 bits are behaving like an XNOR gate with respect to A and B. Similarly, the top 4 bits of CO are behaving like an AND gate with respect to A and B, and the bottom 4 bits behave like an OR gate. Taking those facts, the following circuit implements a full adder:

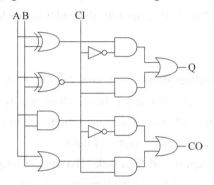

This definitely is not the most efficient way to implement a full adder, but it is extremely easy to understand and trace through the logic using this method. If you are so inclined, see what you can do to implement this logic with fewer gates.

Now we have a piece of functionality called a "full adder." What a computer engineer then does is "black-box" it so that he or she can stop worrying about the details of the component. A black box for a full adder would look like this:

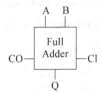

With that black box, it is now easy to draw a 4-bit full adder:

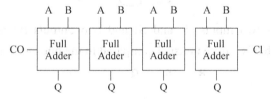

In this diagram the carry-out from each bit feeds directly into the carry-in of the next bit over. A 0 is hard-wired into the initial carry-in bit. If you input two 4-bit numbers on the A and B lines, you will get the 4-bit sum out on the Q lines, plus 1 additional bit for the final carry-out. You can see that this chain can extend as far as you like, through 8, 16 or 32 bits if desired.

The 4-bit adder we just created is called a ripple-carry adder. It gets that name because the carry bits "ripple" from one adder to the next. This implementation has the advantage of simplicity but the disadvantage of speed problems. In a real circuit, gates take time to switch states (the time is on the order of nanoseconds, but in high-speed computers nanoseconds matter). So 32-bit or 64-bit ripple-carry adders might take 100 to 200 nanoseconds to settle into their final sum because of carry ripple. For this reason, engineers have created more advanced adders called carry-lookahead adders. The number of gates required to implement carry-lookahead is large, but the settling time for the adder is much better.

5.10.3　Flip Flops

One of the more interesting things that you can do with Boolean gates is to create memory with them. If you arrange the gates correctly, they will remember an input value. This simple concept is the basis of RAM (random access memory) in computers, and also makes it possible to create a wide variety of other useful circuits.

Memory relies on a concept called feedback. That is, the output of a gate is fed back into the input. The simplest possible feedback circuit using two inverters is shown below:

If you follow the feedback path, you can see that if Q happens to be 1, it will always be 1. If it happens to be 0, it will always be 0. Since it's nice to be able to control the circuits we create, this one doesn't have much use-but it does let you see how feedback works.

It turns out that in "real" circuits, you can actually use this sort of simple inverter feedback approach. A more useful feedback circuit using two NAND gates is shown below:

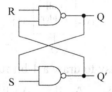

This circuit has two inputs (R and S) and two outputs (Q and Q'). Because of the feedback, its logic table is a little unusual compared to the ones we have seen previously:

R	S	Q	Q'
0	0		Illegal
0	1	1	0
1	0	0	1
1	1		Remembers

What the logic table shows is that:

If R and S are opposites of one another, then Q follows S and Q' is the inverse of Q.

If both R and S are switched to 1 simultaneously, then the circuit remembers what was previously presented on R and S.

There is also the funny illegal state. In this state, R and S both go to 0, which has no value in the memory sense. Because of the illegal state, you normally add a little conditioning logic on the input side to prevent it, as shown here:

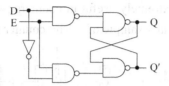

In this circuit, there are two inputs (D and E). You can think of D as "Data" and E as "Enable." If E is 1, then Q will follow D. If E changes to 0, however, Q will remember whatever was last seen on D. A circuit that behaves in this way is generally referred to as a flip-flop.

A very common form of flip-flop is the J-K flip-flop. It is unclear, historically, where the name "J-K" came from, but it is generally represented in a black box like this:

In this diagram, P stands for "Preset," C stands for "Clear" and Clk stands for "Clock." The logic table looks like this:

P	C	Clk	J	K	Q	Q'
1	1	1-to-0	1	0	1	0
1	1	1-to-0	0	1	0	1
1	1	1-to-0	1	1	Toggles	
1	0	X	X	X	0	1
0	1	X	X	X	1	0

Here is what the table is saying: First, Preset and Clear override J, K and Clk completely. So if Preset goes to 0, then Q goes to 1; and if Clear goes to 0, then Q goes to 0 no matter what J, K and Clk are doing. However, if both Preset and Clear are 1, then J, K and Clk can operate. The 1-to-0 notation means that when the clock changes from a 1 to a 0, the value of J and K are remembered if they are opposites. At the low-going edge of the clock (the transition from 1 to 0), J and K are stored. However, if both J and K happen to be 1 at the low-going edge, then Q simply toggles. That is, Q changes from its current state to the opposite state.

You might be asking yourself right now, "What in the world is that good for?" It turns out that the concept of "edge triggering" is very useful. The fact that J-K flip-flop only "latches" the J-K inputs on a transition from 1 to 0 makes it much more useful as a memory device. J-K flip-flops are also extremely useful in counters (which are used extensively when creating a digital clock). Here is an example of a 4-bit counter using J-K flip-flops:

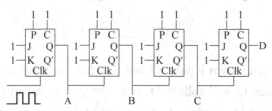

The outputs for this circuit are A, B, C and D, and they represent a 4-bit binary number. Into the clock input of the left-most flip-flop comes a signal changing from 1 to 0 and back to 1 repeatedly (an oscillating signal). The counter will count the low-going edges it sees in this signal. That is, every time the incoming signal changes from 1 to 0, the 4-bit number represented by A, B, C and D will increment by 1. So the count will go from 0 to 15 and then cycle back to 0. You can add as many bits as you like to this counter and count anything you like. For example, if you put a magnetic switch on a door, the counter will count the number of times the door is opened and closed. If you put an optical sensor on a road, the counter could count the number of cars that drive by.

Another use of a J-K flip-flop is to create an edge-triggered latch, as shown here:

In this arrangement, the value on D is "latched" when the clock edge goes from low to high. Latches are extremely important in the design of things like central processing units (CPUs) and peripherals in computers.

5.10.4 Implementing Gates

In the previous sections we saw that, by using very simple Boolean gates, we can implement adders, counters, latches and so on. That is a big achievement, because not so long ago human beings were the only ones who could do things like add two numbers together.

With a little work, it is not hard to design Boolean circuits that implement subtraction, multiplication, division... You can see that we are not that far away from a pocket calculator. From there, it is not too far a jump to the full-blown CPUs used in computers.

So how might we implement these gates in real life? Mr. Boole came up with them on paper, and on paper they look great. To use them, however, we need to implement them in physical reality so that the gates can perform their logic actively. Once we make that leap, then we have started down the road toward creating real computation devices.

The easiest way to understand the physical implementation of Boolean logic is to use relays. This is, in fact, how the very first computers were implemented. No one implements computers with relays anymore—today, people use sub-microscopic transistors etched onto silicon chips. These transistors are incredibly small and fast, and they consume very little power compared to a relay. However, relays are incredibly easy to understand, and they can implement Boolean logic very simply. Because of that simplicity, you will be able to see that mapping from "gates on paper" to "active gates implemented in physical reality" is possible and straightforward. Performing the same mapping with transistors is just as easy.

Let's start with an inverter. Implementing a NOT gate with a relay is easy: What we are going to do is use voltages to represent bit states. We will define a binary 1 to be 6 volts and a binary 0 to be zero volts (ground). Then we will use a 6-volt battery to power our circuits. Our NOT gate will therefore look like this:

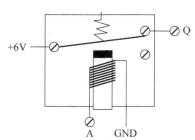

You can see in this circuit that if you apply zero volts to A, then you get 6 volts out on Q; and if you apply 6 volts to A, you get zero volts out on Q. It is very easy to implement an inverter with a relay!

It is similarly easy to implement an AND gate with two relays:

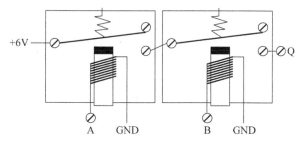

Here you can see that if you apply 6 volts to A and B, Q will have 6 volts. Otherwise, Q will have zero volts. That is exactly the behavior we want from an AND gate. An OR gate is even simpler—just hook two wires for A and B together to create an OR. You can get fancier than that if you like and use two relays in parallel.

You can see from this discussion that you can create the three basic gates-NOT, AND and OR-from relays. You can then hook those physical gates together using the logic diagrams shown above to create a physical 8-bit ripple-carry adder. If you use simple switches to apply A and B inputs to the adder and hook all eight Q lines to light bulbs, you will be able to add any two numbers together and read the results on the lights ("light on" =1, "light off" =0).

Boolean logic in the form of simple gates is very straightforward. From simple gates you can create more complicated functions, like addition. Physically implementing the gates is possible and easy. From those three facts you have the heart of the digital revolution, and you understand, at the core, how computers work.

Chapter 6
Microprocessors

The computer you are using uses a microprocessor to do its work. The microprocessor is the heart of any normal computer, whether it is a desktop machine, a server or a laptop. The microprocessor you are using might be a Pentium, a K6, a PowerPC, a Sparc or any of the many other brands and types of microprocessors, but they all do approximately the same thing in approximately the same way.

If you have ever wondered what the microprocessor in your computer is doing, or if you have ever wondered about the differences between types of microprocessors, then read on.

6.1 Microprocessor History

A microprocessor—also known as a CPU or central processing unit—is a complete computation engine that is fabricated on a single chip. The first microprocessor was the Intel 4004, introduced in 1971. The 4004 (as shown in Figure 6.1) was not very powerful—all it could do was add and subtract, and it could only do that 4 bits at a time. But it was amazing that everything was on one chip. Prior to the 4004, engineers built computers either from collections of chips or from discrete components (transistors wired one at a time). The 4004 powered one of the first portable electronic calculators.

The first microprocessor to make it into a home computer was the Intel 8080 (as shown in Figure 6.2), a complete 8-bit computer on one chip, introduced in 1974. The first microprocessor to make a real splash in the market was the Intel 8088, introduced in 1979 and

Figure 6.1   Intel 4004 chip

Figure 6.2   Intel 8080

incorporated into the IBM PC (which first appeared around 1982). If you are familiar with the PC market and its history, you know that the PC market moved from the 8088 to the 80286 to the 80386 to the 80486 to the Pentium to the Pentium II to the Pentium III to the Pentium 4 (as shown in Figure 6.3). All of these microprocessors are made by Intel and all of them are improvements on the basic design of the 8088. The Pentium 4 can execute any piece of code that ran on the original 8088, but it does it about 5000 times faster!

Figure 6.3   Photo courtesy Intel Corporation Intel Pentium 4 processor

6.2 Microprocessor Progression

Table 6.1 helps you to understand the differences between the different processors that Intel has introduced over the years.

Table 6.1 Differences between the different processors

Name	Date	Transistors	Microns	Clock speed	Data width	MIPS
8080	1974	6,000	6	2MHz	8bits	0.64
8088	1979	29,000	3	5MHz	16bits 8-bitbus	0.33
80286	1982	134,000	1.5	6MHz	16bits	1
80386	1985	275,000	1.5	16MHz	32bits	5
80486	1989	1,200,000	1	25MHz	32bits	20
Pentium	1993	3,100,000	0.8	60MHz	32bits 64-bitbus	100
Pentium II	1997	7,500,000	0.35	233MHz	32bits 64-bitbus	~300
Pentium III	1999	9,500,000	0.25	450MHz	32bits 64-bitbus	~510
Pentium 4	2000	42,000,000	0.18	1.5GHz	32bits 64-bitbus	~1,700

Information about Table 6.1:
- The date is the year that the processor was first introduced. Many processors are re-introduced at higher clock speeds for many years after the original release date.

- Transistors is the number of transistors on the chip. You can see that the number of transistors on a single chip has risen steadily over the years.
- Microns is the width, in microns, of the smallest wire on the chip. For comparison, a human hair is 100 microns thick. As the feature size on the chip goes down, the number of transistors rises.
- Clock speed is the maximum rate that the chip can be clocked at. Clock speed will make more sense in the next section.
- Data Width is the width of the ALU. An 8-bit ALU can add/subtract/multiply/etc. two 8-bit numbers, while a 32-bit ALU can manipulate 32-bit numbers. An 8-bit ALU would have to execute four instructions to add two 32-bit numbers, while a 32-bit ALU can do it in one instruction. In many cases, the external data bus is the same width as the ALU, but not always. The 8088 had a 16-bit ALU and an 8-bit bus, while the modern Pentiums fetch data 64 bits at a time for their 32-bit ALUs.
- MIPS stands for "millions of instructions per second" and is a rough measure of the performance of a CPU. Modern CPUs can do so many different things that MIPS ratings lose a lot of their meaning, but you can get a general sense of the relative power of the CPUs from this column.

From Table 6.1 you can see that, in general, there is a relationship between clock speed and MIPS. The maximum clock speed is a function of the manufacturing process and delays within the chip. There is also a relationship between the number of transistors and MIPS. For example, the 8088 clocked at 5 MHz but only executed at 0.33 MIPS (about one instruction per 15 clock cycles). Modern processors can often execute at a rate of two instructions per clock cycle. That improvement is directly related to the number of transistors on the chip and will make more sense in the next section.

6.3 Inside a Microprocessor

To understand how a microprocessor works, it is helpful to look inside and learn about the logic used to create one. In the process you can also learn about assembly language-the native language of a microprocessor-and many of the things that engineers can do to boost the speed of a processor.

A microprocessor executes a collection of machine instructions that tell the processor what to do. Based on the instructions, a microprocessor does three basic things:
- Using its ALU (Arithmetic/Logic Unit), a microprocessor can perform mathematical operations like addition, subtraction, multiplication and division. Modern microprocessors contain complete floating point processors that can perform extremely sophisticated operations on large floating point numbers.
- A microprocessor can move data from one memory location to another.

- A microprocessor can make decisions and jump to a new set of instructions based on those decisions.

There may be very sophisticated things that a microprocessor does, but those are its three basic activities. Figure 6.4 shows an extremely simple microprocessor capable of doing those three things.

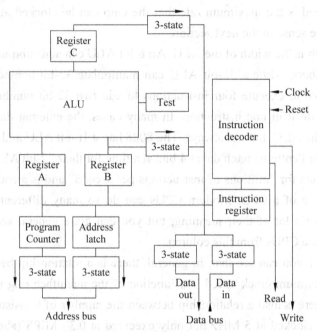

Figure 6.4   A simple microprocessor

This is about as simple as a microprocessor gets. This microprocessor has:
- An address bus (that may be 8, 16 or 32 bits wide) that sends an address to memory.
- A data bus (that may be 8, 16 or 32 bits wide) that can send data to memory or receive data from memory.
- An RD (read) and WR (write) line to tell the memory whether it wants to set or get the addressed location.
- A clock line that lets a clock pulse sequence the processor.
- A reset line that resets the program counter to zero (or whatever) and restarts execution.

Let's assume that both the address and data buses are 8 bits wide in this example.

Here are the components of this simple microprocessor:
- Registers A, B and C are simply latches made out of flip-flops.
- The address latch is just like registers A, B and C.
- The program counter is a latch with the extra ability to increment by 1 when told to do so, and also to reset to zero when told to do so.
- The ALU could be as simple as an 8-bit adder, or it might be able to add, subtract,

multiply and divide 8-bit values. Let's assume the latter here.
- The test register is a special latch that can hold values from comparisons performed in the ALU. An ALU can normally compare two numbers and determine if they are equal, if one is greater than the other, etc. The test register can also normally hold a carry bit from the last stage of the adder. It stores these values in flip-flops and then the instruction decoder can use the values to make decisions.
- There are six boxes marked "3-State" in the diagram. These are tri-state buffers. A tri-state buffer can pass a 1, a 0 or it can essentially disconnect its output (imagine a switch that totally disconnects the output line from the wire that the output is heading toward). A tri-state buffer allows multiple outputs to connect to a wire, but only one of them to actually drive a 1 or a 0 onto the line.
- The instruction register and instruction decoder are responsible for controlling all of the other components.

Although they are not shown in Figure 6.4, there would be control lines from the instruction decoder that would:
- Tell the A register to latch the value currently on the data bus.
- Tell the B register to latch the value currently on the data bus.
- Tell the C register to latch the value currently on the data bus.
- Tell the program counter register to latch the value currently on the data bus.
- Tell the address register to latch the value currently on the data bus.
- Tell the instruction register to latch the value currently on the data bus.
- Tell the program counter to increment.
- Tell the program counter to reset to zero.
- Activate any of the six tri-state buffers (six separate lines).
- Tell the ALU what operation to perform.
- Tell the test register to latch the ALU's test bits.
- Activate the RD line.
- Activate the WR line.

Coming into the instruction decoder are the bits from the test register and the clock line, as well as the bits from the instruction register.

6.4 Microprocessor Instructions

Even the incredibly simple microprocessor shown in the previous example will have a fairly large set of instructions that it can perform. The collection of instructions is implemented as bit patterns, each one of which has a different meaning when loaded into the instruction register. Humans are not particularly good at remembering bit patterns, so a set of short words are defined to represent the different bit patterns. This collection of words is called the assembly language of the processor. An assembler can translate the words into their bit patterns very easily, and then the output of the assembler is placed in memory for the microprocessor to execute.

Here's the set of assembly language instructions that the designer might create for the simple microprocessor in our example:

- LOADA mem—Load register A from memory address.
- LOADB mem—Load register B from memory address.
- CONB con—Load a constant value into register B.
- SAVEB mem—Save register B to memory address.
- SAVEC mem—Save register C to memory address.
- ADD—Add A and B and store the result in C.
- SUB—Subtract A and B and store the result in C.
- MUL—Multiply A and B and store the result in C.
- DIV—Divide A and B and store the result in C.
- COM—Compare A and B and store the result in test.
- JUMP addr —Jump to an address.
- JEQ addr —Jump, if equal, to address.
- JNEQ addr —Jump, if not equal, to address.
- JG addr —Jump, if greater than, to address.
- JGE addr —Jump, if greater than or equal, to address.
- JL addr —Jump, if less than, to address.
- JLE addr —Jump, if less than or equal, to address.
- STOP—Stop execution.

This simple piece of C code will calculate the factorial of 5 (where the factorial of 5=5!=5*4*3 *2*1=120):

```
a=1;
f=1;
while (a <=5)
{
    f=f*a;
    a=a+1;
}
```

At the end of the program's execution, the variable f contains the factorial of 5.

A C compiler translates this C code into assembly language. Assuming that RAM starts at address 128 in this processor, and ROM (which contains the assembly language program) starts at address 0, then for our simple microprocessor the assembly language might look like this:

```
//Assume a is at address 128
//Assume F is at address 129
0  CONB 1    //a=1;
1  SAVEB 128
2  CONB 1    //f=1;
3  SAVEB 129
4  LOADA 128 //if a > 5 the jump to 17
5  CONB 5
6  COM
7  JG 17
8  LOADA 129 //f=f*a;
9  LOADB 128
10 MUL
11 SAVEC 129
12 LOADA 128 //a=a+1;
13 CONB 1
14 ADD
15 SAVEC 128
16 JUMP 4    //loop back to if
17 STOP
```

So now the question is, "How do all of these instructions look in ROM?" Each of these assembly language instructions must be represented by a binary number. For the sake of simplicity, let's assume each assembly language instruction is given a unique number, like this:

```
LOADA-1
LOADB-2
CONB-3
SAVEB-4
SAVEC mem-5
ADD-6
SUB-7
MUL-8
DIV-9
COM-10
JUMP addr-11
JEQ addr-12
JNEQ addr-13
JG addr-14
JGE addr-15
JL addr-16
JLE addr-17
STOP-18
```

The numbers are known as opcodes. In ROM, our little program would look like this:

```
//Assume a is at address 128
//Assume F is at address 129
Addr opcode/value
0    3      //CONB 1
1    1
2    4      //SAVEB 128
3    128
4    3      //CONB 1
5    1
6    4      //SAVEB 129
7    129
8    1      //LOADA 128
9    128
10   3      //CONB 5
11   5
12   10     //COM
13   14     //JG 17
14   31
15   1      //LOADA 129
16   129
17   2      //LOADB 128
18   128
19   8      //MUL
20   5      //SAVEC 129
21   129
22   1      //LOADA 128
23   128
24   3      //CONB 1
25   1
26   6      //ADD
27   5      //SAVEC 128
28   128
29   11     //JUMP 4
30   8
31   18     //STOP
```

You can see that seven lines of C code became 17 lines of assembly language, and that became 31 bytes in ROM.

6.5 Decoding Microprocessor Instructions

The instruction decoder needs to turn each of the opcodes into a set of signals that drive the different components inside the microprocessor. Let's take the ADD instruction as an example and look at what it needs to do:

(1) During the first clock cycle, we need to actually load the instruction. Therefore the instruction decoder needs to:
- activate the tri-state buffer for the program counter.
- activate the RD line.
- activate the data-in tri-state buffer.
- latch the instruction into the instruction register.

(2) During the second clock cycle, the ADD instruction is decoded. It needs to do very little:
- set the operation of the ALU to addition.
- latch the output of the ALU into the C register.

(3) During the third clock cycle, the program counter is incremented (in theory this could be overlapped into the second clock cycle).

Every instruction can be broken down as a set of sequenced operations like these that manipulate the components of the microprocessor in the proper order. Some instructions, like this ADD instruction, might take two or three clock cycles. Others might take five or six clock cycles.

6.6 Microprocessor Performance

The number of transistors available has a huge effect on the performance of a processor. As seen earlier, a typical instruction in a processor like an 8088 took 15 clock cycles to execute. Because of the design of the multiplier, it took approximately 80 cycles just to do one 16-bit multiplication on the 8088. With more transistors, much more powerful multipliers capable of single-cycle speeds become possible.

More transistors also allow for a technology called pipelining. In a pipelined architecture, instruction execution overlaps. So even though it might take five clock cycles to execute each instruction, there can be five instructions in various stages of execution simultaneously. That way it looks like one instruction completes every clock cycle.

Many modern processors have multiple instruction decoders, each with its own pipeline. This allows for multiple instruction streams, which means that more than one instruction can complete during each clock cycle. This technique can be quite complex to implement, so it takes lots of transistors.

6.7 Microprocessor Trends

The trend in processor design has primarily been toward full 32-bit ALUs with fast floating point processors built in and pipelined execution with multiple instruction streams.

The newest thing in processor design is 64-bit ALUs. There has also been a tendency toward special instructions (like the MMX instructions) that make certain operations particularly efficient, and the addition of hardware virtual memory support and L1 caching on the processor chip. All of these trends push up the transistor count, leading to the multi-million transistor powerhouses available today. These processors can execute about one billion instructions per second!

6.8 64-bit Processors

Sixty-four-bit processors have been with us since 1992, and in the 21st century they have started to become mainstream. Both Intel and AMD (as shown in Figure 6.5) have introduced 64-bit chips, and the Mac G5 sports a 64-bit processor. Sixty-four-bit processors have 64-bit ALUs, 64-bit registers, 64-bit buses and so on.

Figure 6.5 An AMD chip

One reason why the world needs 64-bit processors is because of their enlarged address spaces. Thirty-two-bit chips are often constrained to a maximum of 2 GB or 4 GB of RAM access. That sounds like a lot, given that most home computers currently use only 256 MB to 512 MB of RAM. However, a 4-GB limit can be a severe problem for server machines and machines running large databases. And even home machines will start bumping up against the 2 GB or 4 GB limit pretty soon if current trends continue. A 64-bit chip has none of these constraints because a 64-bit RAM address space is essentially infinite for the foreseeable future—2^{64} bytes of RAM is something on the order of a quadrillion gigabytes of RAM.

With a 64-bit address bus and wide, high-speed data buses on the motherboard, 64-bit machines also offer faster I/O (input/output) speeds to things like hard disk drives and video cards. These features can greatly increase system performance.

Keywords

microprocessor	微处理器
fabricate	制造
splash	喷溅
pentium	奔腾

re-introduce	重新推出
microns	微米
MIPS	每秒百万指令数
mathematical operations	数学运算
sophisticated	复杂的
Registers	寄存器
sequence	序列化
increment	增量
decoder	解码器
assembler	汇编
factorial	阶乘
buffer	缓冲器
latch	锁上
multiplier	乘法器
pipelining	流水线
architecture	体系结构
mainstream	主流
tangible	明显的

6.9 Exercise 6

Multiple or single choices.

1. The first microprocessor was the Intel 4004, introduced in 1971. The mathematical operations it could do was____.
    A. add
    B. subtract
    C. multiplication
    D. division

2. Comparing the differences between the different processors, there is a relationship between ____.
    A. transistors
    B. clock speed
    C. microns
    D. MIPS

3. A microprocessor can do the following basic things include____.

A. perform mathematical operations
   B. create a new instruction
   C. move data from one memory location to another
   D. make decisions and jump to a new set of instructions

4. A simple microprocessor has_____.
   A. An address bus
   B. A data bus
   C. A clock line
   D. An RD (read) and WR (write) line

True or False.

1. The microprocessor is the heart of any normal computer, whether it is a desktop machine, a server or a laptop.

2. MIPS stands for "millions of instructions per second" and is a rough measure of the performance of a CPU.

3. A microprocessor can move data from one memory location to another.

4. The instruction decoder needs to turn each of the opcodes into a set of signals.

5. 64-bit machines also offer faster I/O (input/output) speeds to things like hard disk drives and video cards which greatly increase system performance.

6.10 Further Reading: E-commerce

Unless you have been living under a rock for the last two years, you have heard about E-commerce! And you have heard about it from several different angles. For example:
- You have heard about all of the companies that offer E-commerce because you have been bombarded by their TV and radio ads.
- You have read all of the news stories about the shift to E-commerce and the hype that has developed around E-commerce companies.
- You have seen the huge valuations that web companies get in the stock market, even when they don't make a profit.
- And you may have actually purchased something on the web, so you have direct personal experience with E-commerce.

Still, you may feel like you don't understand E-commerce at all. What is all the hype about? Why the huge valuations? And most importantly, is there a way for you to participate? If you have an E-commerce idea, how might you get started implementing it? If you have had questions like these, then this article will help out by exposing you to the entire E-commerce space. Let's have a look!

6.10.1 Commerce

Before we get into a complete discussion of E-commerce, it is helpful to have a good mental image of plain old commerce first. If you understand commerce, then E-commerce is an easy extension.

Merriam-Webster's Collegiate Dictionary defines commerce as follows:

(1) Social intercourse: interchange of ideas, opinions, or sentiments.

(2) The exchange or buying and selling of commodities on a large scale involving transportation from place to place.

(3) Sexual intercourse.

We tend to be interested in the second definition, but that third one is interesting and unexpected-maybe that's what all of the hype is about!

So commerce is, quite simply, the exchange of goods and services, usually for money. We see commerce all around us in millions of different forms. When you buy something at a grocery store or at Wal-mart you are participating in commerce. In the same way, if you cart half of your possessions onto your front lawn for a yard sale, you are participating in commerce from a different angle. If you go to work each day for a company that produces a product, that is yet another link in the chain of commerce. When you think about commerce in these different ways, you instinctively recognize several different roles:

- Buyers—these are people with money who want to purchase a good or service.
- Sellers—these are the people who offer goods and services to buyers. Sellers are generally recognized in two different forms: retailers who sell directly to consumers and wholesalers or distributors who sell to retailers and other businesses.
- Producers—these are the people who create the products and services that sellers offer to buyers. A producer is always, by necessity, a seller as well. The producer sells the products produced to wholesalers, retailers or directly to the consumer.

You can see that at this high level, commerce is a fairly simple concept! Whether it is something as simple as a person making and selling popcorn on a street corner or as complex as a contractor delivering a space shuttle to NASA, all of commerce at its simplest level relies on buyers, sellers and producers.

6.10.2 The Elements of Commerce

When you get down to the actual elements of commerce and commercial transactions, things get slightly more complicated because you have to deal with the details. However, these details boil down to a finite number of steps. The following list highlights all of the elements of a typical commerce activity. In this case, the activity is the sale of some product by a retailer to a customer:

- If you would like to sell something to a customer, at the very core of the matter is the something itself. You must have a product or service to offer. The product can be anything from ball bearings to back rubs. You may get your products directly from a producer, or you might go through a distributor to get them, or you may produce the products yourself.
- You must also have a place from which to sell your products. Place can sometimes be very ephemeral—for example a phone number might be the place. If you are a customer in need of a back rub, if you call "Judy's Backrubs, Inc." on the telephone to order a back rub, and if Judy shows up at your office to give you a backrub, then the phone number is the place where you purchased this service. For most physical products we tend to think of the place as a store or shop of some sort. But if you think about it a bit more you realize that the place for any traditional mail order company is the combination of an ad or a catalog and a phone number or a mail box.
- You need to figure out a way to get people to come to your place. This process is known as marketing. If no one knows that your place exists, you will never sell anything. Locating your place in a busy shopping center is one way to get traffic. Sending out a mail order catalog is another. There is also advertising, word of mouth and even the guy in a chicken suit who stands by the road waving at passing cars!
- You need a way to accept orders. At Wal-mart this is handled by the check out line. In a mail order company the orders come in by mail or phone and are processed by employees of the company.
- You also need a way to accept money. If you are at Wal-mart you know that you can use cash, check or credit cards to pay for products. Business-to-business transactions often use purchase orders. Many businesses do not require you to pay for the product or service at the time of delivery, and some products and services are delivered continuously (water, power, phone and pagers are like this). That gets into the whole area of billing and collections.
- You need a way to deliver the product or service, often known as fulfillment. At a store like Wal-mart fulfillment is automatic. The customer picks up the item of desire, pays for it and walks out the door. In mail-order businesses the item is packaged and mailed. Large items must be loaded onto trucks or trains and shipped.

- Sometimes customers do not like what they buy, so you need a way to accept returns. You may or may not charge certain fees for returns, and you may or may not require the customer to get authorization before returning anything.
- Sometimes a product breaks, so you need a way to honor warranty claims. For retailers this part of the transaction is often handled by the producer.
- Many products today are so complicated that they require customer service and technical support departments to help customers use them. Computers are a good example of this sort of product. On-going products like cell phone service may also require on-going customer service because customers want to change the service they receive over time. Traditional items (for example, a head of lettuce), generally require less support that modern electronic items.

You find all of these elements in any traditional mail order company. Whether the company is selling books, consumer products, information in the form of reports and papers, or services, all of these elements come into play.

In an E-commerce sales channel you find all of these elements as well, but they change slightly. You must have the following elements to conduct E-commerce:
- A product.
- A place to sell the product-in the E-commerce case a web site displays the products in some way and acts as the place.
- A way to get people to come to your web site.
- A way to accept orders—normally an on-line form of some sort.
- A way to accept money—normally a merchant account handling credit card payments. This piece requires a secure ordering page and a connection to a bank. Or you may use more traditional billing techniques either on-line or through the mail.
- A fulfillment facility to ship products to customers (often outsource-able). In the case of software and information, however, fulfillment can occur over the Web through a file download mechanism.
- A way to accept returns.
- A way to handle warrantee claims if necessary.
- A way to provide customer service (often through e-mail, on-line forms, on-line knowledge bases and FAQs, etc.).

In addition, there is often a strong desire to integrate other business functions or practices into the E-commerce offering. An extremely simple example—you might want to be able to show the customer the exact status of an order.

6.10.3 Why the Hype

There is a huge amount of hype that surrounds E-commerce. Given the similarities with mail order commerce, you may be wondering why the hype is so common. For example:

- "On the retail side alone, Forrester projects $17 billion in sales to consumers over the Internet by the year 2001. Some segments are really starting to take off." —Forrester Research, "Content and Context," DMA Insider, Spring 1998.
- "Worldwide business access to the Web is expected to grow at an even faster rate than the US market-from 1.3 million in 1996 to 8 million by 2001." —O'Reilly & Associates.
- "Home continues to be the most popular access location, with nearly 70% of users accessing from their homes...almost 60% shop online. The most popular activities include finding information about a product's price or features, checking on product selection and determining where to purchase a product." —IntelliQuest Information Group, Inc., WWITS Survey.
- "In general, the more difficult and time-consuming a purchase category is, the more likely consumers will prefer to use the internet versus standard physical means." eMarketer.

This sort of hype applies to a wide range of products. According to eMarketer the biggest product categories include:
- Computer products (hardware, software, accessories).
- Books.
- Music.
- Financial Services.
- Entertainment.
- Home Electronics.
- Apparel.
- Gifts and flowers.
- Travel services.
- Toys.
- Tickets.
- Information.

6.10.4 The Dell Example

But this doesn't explain the frantic rush by companies, both large and small, to get to the web. Nor does it justify a small business making a big expenditure on an E-commerce facility. What is driving this sort of frenzy? To understand it a bit, let's take a look at one of the most successful E-commerce companies: Dell.

Dell is a straightforward company that, like Gateway 2000, Micron and a host of others, sells custom-configured PCs to consumers and businesses. Dell started as a mail-order company that advertised in the back of magazines and sold their computers over the phone. Dell's E-commerce presence is widely publicized these days because Dell is able to sell so much merchandise over the web. According to this page from IDG, Dell currently sells something like $14,000,000 in equipment every day. 25% of Dell's sales is over the web.

Does this matter? Dell has been selling computers by mail over the phone for more than a decade. Mail order sales is a standard way of doing things that has been around for over a century (Sears, after all, was a mail order company originally). So if 25% of Dell's sales move over to the web instead of using the telephone, is that a big deal? The answer could be YES for three reasons:

- If Dell were to lose 25% of its phone sales to achieve its 25% of sales over the web, then it is not clear that E-commerce has any advantage. Dell would be selling no more computers. But what if the sales conducted over the web cost the company less (for example, because the company does not have to hire someone to answer the phone)? Or what if people purchasing over the web tend to purchase more accessories? If the transaction cost on the web is lower, or if the presentation of merchandise on the web is more inviting and encourages larger transactions, then moving to the web is productive for Dell.
- What if, in the process of selling merchandise over the Web, Dell lost no sales through its traditional phone channel? That is, what if there just happens to be a percentage of the population that prefers to buy things over the Web (perhaps because there is more time to think, or because you can try lots of different options to see what happens to the final price, or because you can compare multiple vendors easily, or whatever). In building its web site to attract these buyers, Dell may be able to lure away customers from other vendors who do not offer such a service. This gives Dell a competitive advantage that lets it increase its market share.
- There is also a widely held belief that once a customer starts working with a vendor, it is much easier to keep that customer than it is to bring in new customers. So if you can build brand loyalty for a web site early, it gives you an advantage over other vendors who try to enter the market later. Dell implemented its Web site very early, and that presumably gives it an advantage over the competition.

These three trends are the main drivers behind the E-commerce buzz. There are other factors as well.

6.10.5 The Lure of E-commerce

The following list summarizes what might be called the "lure of E-commerce":

- Lower transaction costs—if an E-commerce site is implemented well, the web can significantly lower both order—taking costs up front and customer service costs after the sale by automating processes.
- Larger purchases per transaction—Amazon offers a feature that no normal store offers. When you read the description of a book, you also can see "what other people

who ordered this book also purchased". That is, you can see the related books that people are actually buying. Because of features like these it is common for people to buy more books that they might buy at a normal bookstore.
- Integration into the business cycle—A Web site that is well-integrated into the business cycle can offer customers more information than previously available. For example, if Dell tracks each computer through the manufacturing and shipping process, customers can see exactly where their order is at any time. This is what FedEx did when they introduced on-line package tracking—FedEx made far more information available to the customer.
- People can shop in different ways. Traditional mail order companies introduced the concept of shopping from home in your pajamas, and E-commerce offers this same luxury. New features that web sites offer include:
  - The ability to build an order over several days.
  - The ability to configure products and see actual prices.
  - The ability to easily build complicated custom orders.
  - The ability to compare prices between multiple vendors easily.
  - The ability to search large catalogs easily.
- Larger catalogs—A company can build a catalog on the web that would never fit in an ordinary mailbox. For example, Amazon sells 3,000,000 books. Imagine trying to fit all of the information available in Amazon's database into a paper catalog!
- Improved customer interactions—With automated tools it is possible to interact with a customer in richer ways at virtually no cost. For example, the customer might get an e-mail when the order is confirmed, when the order is shipped and after the order arrives. A happy customer is more likely to purchase something else from the company.

It is these sorts of advantages that create the buzz that surrounds E-commerce right now.

There is one final point for E-commerce that needs to be made. E-commerce allows people to create completely new business models. In a mail order company there is a high cost to printing and mailing catalogs that often end up in the trash. There is also a high cost in staffing the order-taking department that answers the phone. In E-commerce both the catalog distribution cost and the order taking cost fall toward zero. That means that it may be possible to offer products at a lower price, or to offer products that could not be offered before because of the change in cost dynamics.

However, it is important to point out that the impact of E-commerce only goes so far. Mail order sales channels offer many of these same advantages, but that does not stop your town from having a mall. The mall has social and entertainment aspects that attract people, and at the mall you can touch the product and take delivery instantly. E-commerce cannot offer any of these features. The mall is not going to go away anytime soon...

6.10.6 Easy and Hard Aspects of E-commerce

The things that are hard about E-commerce include:
- Getting traffic to come to your web site.
- Getting traffic to return to your web site a second time.
- Differentiating yourself from the competition.
- Getting people to buy something from your web site. Having people look at your site is one thing. Getting them to actually type in their credit card numbers is another.
- Integrating an E-commerce web site with existing business data (if applicable).

There are so many web sites, and it is so easy to create a new E-commerce web site, that getting people to look at yours is the biggest problem.

The things that are easy about E-commerce, especially for small businesses and individuals, include:
- Creating the web site.
- Taking the orders.
- Accepting payment.

There are inumerable companies that will help you build and put up your electronic store. We'll discuss some options in the next section.

6.10.7 Building an E-commerce Site

The things you need to keep in mind when thinking about building an E-commerce site include:
- Suppliers—this is no different from the concern that any normal store or mail order company has. Without good suppliers you cannot offer products.
- Your price point—a big part of E-commerce is the fact that price comparisons are extremely easy for the consumer. Your price point is important in a transparent market.
- Customer relations—E-commerce offers a variety of different ways to relate to your customer. E-mail, FAQs, knowledge bases, forums, chat rooms... Integrating these features into your E-commerce offering helps you differentiate yourself from the competition.
- The back end: fulfillment, returns, customer service—These processes make or break any retail establishment. They define, in a big way, your relationship with your customer.

When you think about E-commerce, you may also want to consider these other desirable capabilities:
- Gift-sending.

- Affiliate programs.
- Special Discounts.
- Repeat buyer programs.
- Seasonal or periodic sales.

The reason why you want to keep these things in mind is because they are all difficult unless your E-commerce software supports them. If the software does support them, they are trivial.

6.10.8 Affiliate Programs

A big part of today's E-commerce landscape is the affiliate program (also known as associate programs). This area was pioneered by Amazon. Amazon allows anyone to set up a specialty book store. When people buy books from the specialty store, the person who owns the specialty bookstore gets a commission (up to 15% of the book's list price) from Amazon. The affiliate program gives Amazon great exposure because hundreds of thousands of specialty bookstores popped up all over the web. Therefore this model is now copied by thousands of E-commerce sites. If you are setting up an E-commerce site you will want to consider an affiliate program as one way to get exposure. BeFree and Link Share are two companies that help E-commerce sites set up affiliate programs.

A relatively new twist on affiliate programs is the CPC Link (CPC=Cost Per Click), also known as affiliate links or click-thru links. You put a link on your site and the company pays you when someone clicks on the link. A typical payment ranges from 5 cents to 20 cents per click. Affiliate links represent the middle ground between banner ads and commission-based affiliate programs. With banner ads, the advertiser takes all the risk-if no one clicks on the banner then the advertiser wastes money. Commission-based affiliate programs place all the risk on the web-site. If the web site sends a bunch of people to the affiliate E-commerce site but no one buys anything, then it has no value for the web site. In CPC links, both sides share risks and rewards equally. You may want to consider setting up this sort of affiliate program to gain exposure for your E-commerce site.

Similar companies:
- AFFILIATIONSPLUS.
- Business Sources & Resources-Business Opportunities.
- E-Commerce Times.
- CashPile Directory Revenue Sharing Web Marketing.

6.10.9 Implementing an E-commerce Site

Let's say that you would like to create an E-commerce site. There are three general ways to implement the site with all sorts of variations in between. The three general ways are:

- Enterprise computing.
- Virtual hosting services.
- Simplified E-commerce.

These are in order of decreasing flexibility and increasing simplicity.

Enterprise computing means that you purchase hardware and software and hire a staff of developers to create your E-commerce web site. Amazon, Dell and all of the other big players participate in E-commerce at the enterprise level. You might need to consider enterprise computing solutions if:

- You have immensely high traffic-millions of visitors per month.
- You have a large database that holds your catalog of products (especially if the catalog is changing constantly).
- You have a complicated sales cycle that requires lots of customized forms, pricing tables, etc.
- You have other business processes already in place and you want your E-commerce offering to integrate into them.

Virtual hosting services give you some of the flexibility of enterprise computing, but what you get depends on the vendor. In general the vendor maintains the equipment and software and sells them in standardized packages. Part of the package includes security, and almost always a merchant account is also an option. Database access is sometimes a part of the package. You provide the web designers and developers to create and maintain your site.

Simplified E-commerce is what most small businesses and individuals are using to get into E-commerce. In this option the vendor provides a simplified system for creating your store. The system usually involves a set of forms that you fill out online. The vendor's software then generates all of the web pages for the store for you. Two good examples of this sort of offering include Yahoo Stores and Verio Stores (if you'd like to speak with someone at Verio, Gregorio Gonzalez, 877-273-3190 ext. 4672 has been helpful). You pay by the month for these services.

Chapter 7
Application Software

7.1 What is Software

Let's start with a definition. A program is a series of instructions that guides a computer through a process. Each instruction tells the machine to perform one of its basic functions: add, subtract multiply, divide, compare copy, request input, or request output. We learned that the processor fetches and executes a single instruction during each machine cycle. A typical instruction contains an operation code that specifies the locations or registers holding the data to be manipulated.

For example, the instruction
   ADD 3, 4
tells a hypothetical computer to add registers 3 and 4.

Because a computer's instruction set is so limited, even simple logical operations call for several instructions. For example, imagine two data values stored in main memory. To add them on many computers, both values are first loaded (of copied) into registers, the registers are added, and then the answer is stored (or copied) back into main memory. That's four instructions: LOAD, LOAD, ADD, and STORE. If four instructions are needed to add two numbers, imagine the number of instructions in a complete program. A computer is controlled by a program stored in its own main memory. Because main memory stores bits, the program must exist in binary form. If programmers had to write in machine language, there would be very few programmers.

7.2 Programming Languages

7.2.1 Assemblers

An option is writing instructions in an assembler language. The programmer writes one mnemonic (memory-aiding) instruction for each machine-level instruction. AR (for add registers) is much easier to remember than the equivalent binary operation code: 00011010.L (for load) is much easier to remember than 01011000. The operands use labels, such as A, B, and C, instead of numbers to represent main memory addresses, and that simplifies the code, too.

Unfortunately, there are no computers that can directly execute assembler language instructions. Writing mnemonic codes may simplify the programmer's job, but computers are still binary machines and require binary instructions. Thus, translation is necessary. An assembler program reads a programmer's source code, translates the source statements to binary, and produces an object module. Because the object module is a machine level version of the programmer's code, it can be loaded into memory and executed.

An assembler language programmer writes one mnemonic instruction for each machine-level instruction. Because of the one-to-one relationship between the language and the machine, assemblers are machine dependent, and a program written for one type of computer won't run on another. On a given machine, assembler language generates the most efficient programs possible, and thus is often used to write operating systems and other system software.

However, when it comes to application programs, machine dependency is a high price to pay for efficiency, so application programs are rarely written in assembler.

7.2.2 Compilers and Interpreters

A computer needs four machine-level instructions to add two numbers, because that's the way a computer works. Human beings shouldn't have to think like computers. Why not simply allow the programmer to indicate addition and assume the other instructions? For example, one way to view addition is as an algebraic expression:

$$C=A+B$$

Why not allow a programmer to write statements in a form similar to algebraic expressions, read those source statements into a program, and let the program generate the necessary machine-level code. That's exactly what happens with a compiler.

Many compiler languages, including RFORTRAN, BASIC, Pascal, PL/1, and ALGOL, are algebraically based. The most popular business-oriented language, COBOL, calls fir statements that resemble brief English-language sentences. Note, however, that no matter what language is used, the objective is the same. The programmer writes source code. An assembler program accepts mnemonic source code and generates a machine-level object module. A FORTRAN compiler accepts FORTRAN source code and generates a machine-level object module. A COBOL compiler accepts COBOL source code and generates a machine-level object module.

What's the difference between an assembler and compiler? With an assembler, each source statement is converted to a single machine-level instruction. With a compiler, a given source statement may be converted to any number of machine-level instructions. An option is using an interpreter. An assembler or a compiler reads a complete source program and generates a compete object module. An interpreter, on the other hand, works with one source statement at a time, reading it, translating it to machine-level, executing the resulting binary instructions, and then moving on to the next source statement. Both compilers and interpreters generate machine-level instructions, but the process is different.

Each language has its own syntax, punctuation, and spelling rules; for example, a Pascal source program is meaningless to a COBOL compiler or a BASIC interpreter. They all support writing programs, however. No matter what language is used, the programmer's objective is the same: defining a series of steps to guide the computer through some process.

7.2.3 Nonprocedural Languages

With traditional assemblers, compilers, and interpreters, the programmer defines a procedure telling the computer exactly how to solve a problem.

However, with a modern, nonprocedural language (sometimes called a 4th-generation or declarative language), the programmer simply defines the logical structure of the problem and lets the language translator figure out how to solve it. Examples of commercially available nonprocedural languages include Prolog, Focus, Lotus 1-2-3, and many others.

7.3 Libraries

Picture a programmer writing a large program. As source statements are typed, they are manipulated by an editor program and stored on disk. Because large programs are rarely written in a single session, the programmer will eventually stop working and remove the disk from the drive. Later, when work resumes, the disk is reinserted, and new source statements are added to the old ones. That same disk might hold other source programs and even routines written by other programmers, It's good example of a source statement library.

Eventually, the source program is completed and compiled. The resulting object module might be loaded directly into main memory, but more often, it is stored on an object module library. Because object modules are binary, machine-level routines, there is no inherent difference between one produced by an assembler and one produced by a FORTRAN compiler (or any other compiler for that matter). Thus, object modules generated by different source languages can be stored on the same library.

Some object modules can be loaded onto memory and executed. Others however, include references to subroutines that are not part of the object module. For example, imagine a program that allows a computer to simulate a game of cards. If, some time ago, another programmer wrote an excellent subroutine to deal cards, it would make sense to reuse that logic. Picture the new program after it is has been written, compiled, and stored on the object module library. The subroutine that deals cards is stored on the same library. Before the program is loaded, the two routines must be combined to form a load module. An object module is a machine-language translation of a source module, and may include references to other subroutines. A load module is a complete, ready to execute program with all subroutines in place. Combining object modules to form a load module is the job of the linkage editor or loader.

Video games, spreadsheet programs, word processors, database programs, accounting routines, and other commercial software packages are generally purchased on disk in load module form. Given a choice between source code, object modules, and load modules, most people would find the load module easier to use simply because the computer can execute it without translation. Load modules are difficult to change, however. If a programmer plans to modify or customize a software package, the source code is essential.

7.4 The Program Development Process

How does a programmer go about modifying a software package? More generally, how does a professional programmer go about the task of writing an original program? Programming is not quite a science; there is a touch of art involved. Thus it is not surprising that different programmers work in different ways. Most, however, begin by carefully difining the problem, and then planning their solution in detail before writing the code. Let's briefly investigate the program development process.

7.4.1 Problem Definition

The first step is defining the problem. That seems like common sense, but all too often, programs are written without a clear idea of why they are needed. A solution, even a great solution, to the wrong problem is useless.

Programs are written because people need information; thus, the programmer begins by identifying the desired information. Next, the algorithms, or rules, for generating that information are specified. Given the disired information (output) and the algorithms, the necessary input data can be defined. The result is a clear problem definition that gives the programmer a good idea of what the program must accomplish.

Incidentally, programs are usually defined in the context of a system.

7.4.2 Planning

The algorithms define what must be done; the next task is deciding how to do it. The objective is to state a problem solution in terms the computer can understand. A computer can perform arithmetic, compare, copy, and request input or output; thus, the programmer is limited to these basic operations. A good starting point is solving a small version of the problem; by actually solving the algorithm, even on a limited scale, the programmer can gain a good sense of the steps required to program it.

Programmers use a number of tools to help convert a problem solution to computer terms. For example, flowcharts can programmer can "draft" the logic before converting it to source code. More complex programs are often written by two or more programmers, or involve a great deal of logic. Such programs are typically broken down into smaller, single-function modules that can be independently coded. A good programmer plans the contents of each module and carefully defines the relationships between the modules before starting to write the source code. Just as a contractor prepares detailed blueprints before starting to build a house, a programmer develops a detailed program plan before starting to write code.

7.4.3 Writing the Program

During implementation, the programmer translates the problem solution into a series of source statements written in some programming language. While each programming language has its own syntax, punctuation, and spelling rules, and learning a new language takes time, writing instructions is basically a mechanical task. The real secret to programming isn't simply coding instructions; it's knowing what instruction to code next. That requires logic. Fortunately, knowing how to program is not a prerequisite for using a computer.

7.4.4 Debug and Documentation

Once the program is coded, the programmer must debug it. Often, the first step is correcting mechanical errors such as incorrect punctuation or spelling; the compiler or interprete usually spots them. Much more difficult is finding and correcting logical errors, or bugs. That result from coding the wrong instruction. Valid instructions are not enough; they must be the right instructions in the right order. Once again, careful planning is the key; good planning simplifies program debug.

Program documentation consists of diagrams, comments, and other descriptive materials that explain or clarify the code. Documentation is invaluable during program debug, and essential for efficient program maintenance. Most useful are comments that appear in a program listing and explain the logic.

7.4.5 Maintenance

Once a program is completed, maintenance begins. Since it is impossible to test many large programs exhaustively, bugs can slip through the debug stage, only to show up months and even years later. Fixing such bugs is an important maintenance task. More significant is the need to update a program to keep it current; for example, because income tax rates change frequently, a payroll program must be constantly updated. The keys to maintenance are careful planning, good documentation, and good program design.

7.5 Writing your Own Programs

Playing an instrument is not essential to enjoying music. Likewise, knowing how to program is not a prerequisite to using a computer. Most computer users can't program. Still, just as a rudimentary knowledge of an instrument increases your appreciation for music knowing how to program can make you more effective computer user. Of course, it goes without saying that if you hope to earn your living as a computer professional, a knowledge of programming is essential. Some people find programming easy. Others find it extremely difficult. The key is practice. The only way to learn how to program is to program.

Keywords

mnemonic	助记符
simplify	简化
module	组件
dependency	依赖
algebraic	代数
syntax	句法
punctuation	标点
nonprocedural	非过程
inherent	固有
subroutines	子程序
customize	定制
algorithms	算法
flowcharts	流程图
contractor	承包商
blueprints	蓝图
prerequisite	条件
debug	调试
diagrams	图表
maintenance	维护
exhaustively	详尽
update	更新
rudimentary	初步

7.6 Exercise 7

Multiple or single choices.

1. A program written for computer in windows system could run on another computer in ____ system.
   A. IOS
   B. Android
   C. Windows
   D. Linux

2. The following compiler languages are algebraically based languages ____.
   A. BASIC
   B. Pascal
   C. Python
   D. Algol

3. Each language has its own ____.
   A. syntax
   B. flowcharts
   C. punctuation
   D. spelling rules

4. The program development process usually includes ____.
   A. problem definition
   B. algorithms research
   C. write the program
   D. debug

5. When debugging a program, you may come to the problems such as ____.
   A. incorrect punctuation or spelling
   B. logical errors
   C. inconsistent capitalization
   D. Null pointer

True or False.

1. With an assembler, each source statement is converted to any number of machine-level

instructions.

2. If a programmer plans to modify or customize a software package, the source code is essential.

3. The first step of program development is defining the problem. Figure out the disired information (output) and the algorithms.

4. Flowcharts can help programmer "draft" the logic before converting it to source code.

5. The most useful mean to debug your code is make comments that appear in a program listing and explain the logic.

7.7 Further Reading: Computer Software

Computer software (or simply software) is essentially a computer program encoded in such a fashion that the program (the instruction set) contents can be changed with minimal effort. Computer software can have various functions such as controlling hardware, performing computations, communication with other software, human interaction, etc; all of which are prescribed in the program.

The term "software" was first used in this sense by John W. Tukey in 1957; computer science and software engineering, computer software is all information processed by computer system, programs and data.

7.7.1 Relationship to Hardware

Computer software is so called in contrast to computer hardware, which is the physical substrate which stores and executes (or "runs") the software.

Software has historically been considered an intermediary between electronic hardware and data, which the hardware processes in some manner, according to instructions defined by the software. More specifically it has been considered to be a conceptual interface composed of a binary representation of electronics-readable code or logic. The purpose of software is to cause a task, process, or computation to be performed. A task can include the retrieval, storage, or display of information, or the transformation of data from one form to another.

As computational science becomes increasingly complex, the distinction between software and data becomes less precise. Data has generally been considered to be either the output of or input for software (n.b. that "data" is not the only possible output or input; for example, configuration information can also be considered input, though not necessarily considered to be data). The output of a particular piece of software may be the input for another piece of software. Therefore, software may be considered to be an interface between hardware, data, or software.

It is generally accepted that software interfaces with electronic devices, or electronics. The terms electronics recently can be defined to include devices which have biological components or biological interfaces. Instructions processed by an electronic device which cause a muscle to contract, for example, may be considered software. The instruction from the electronic device to the muscle may also be considered software because it is the output, a task, of electronics readable code or logic.

7.7.2 System and Application Software

Practical computer systems divide software into two major classes: system software and application software, although the distinction is somewhat arbitrary, and often blurred.

System software helps run the computer hardware and computer system. It includes operating systems, device drivers, programming tools, servers, windowing systems, utilities and more.

Application software allows a user to accomplish one or more specific tasks. Typical applications include office suites, business software, educational software, databases and computer games. Most application software has a graphical user interface (GUI).

Software libraries, which are software components which are used by stand-alone programs, but which cannot be executed on their own, can be considered either system or application software, or neither.

7.7.3 Users See Three Layers of Software

Users often see things differently than programmers. People who use modern general purpose computers (as opposed to embedded systems) usually see three layers of software performing a variety of tasks: platform, application, and user software.

1. Platform software

Platform includes the basic input-output system (often described as firmware rather than software), device drivers, an operating system, and typically a graphical user interface which, in total, allow a user to interact with the computer and its peripherals (associated equipment). Platform software often comes bundled with the computer, and users may not realize that it exists or that they have a choice to use different platform software.

2. Application software

Applications are what most people think of when they think of software. Typical examples include office suites and video games. Application software is often purchased separately from computer hardware. Sometimes applications are bundled with the computer, but that does not change the fact that they run as independent applications. Applications are almost always independent programs from the operating system, though they are often tailored for specific platforms. Most users think of compilers, databases, and other "system software" as applications.

3. User-written software

User software tailors systems to meet the users specific needs. User software include spreadsheet templates, word processor macros, scientific simulations, graphics and animation scripts. Even e-mail filters are a kind of user software. Users create this software themselves and often overlook how important it is.

7.7.4 Software Creation

Software is created with programming languages and related utilities, which may come in several of the above forms: single programs like script interpreters, packages containing a compiler, linker, and other tools; and large suites (often called Integrated Development Environments) that include editors, debuggers, and other tools for multiple languages.

7.7.5 Software in Operation

Computer software has to be "loaded" into the computer's storage (also known as memory and RAM).

Once the software is loaded, the computer is able to operate the software. Computers operate by executing the computer program. This involves passing instructions from the application software, through the system software, to the hardware which ultimately receives the instruction as machine code. Each instruction causes the computer to carry out an operation-moving data, carrying out a computation, or altering the flow of instructions.

Kinds of software by operation: computer program as executable, source code or script, configuration.

7.7.6 Software Reliability

Software reliability considers the errors, faults, and failures related to the creation and operation of software.

7.7.7 Software Patents

The issue of software patents is very controversial, since while patents protect the ideas of "inventors", they are widely believed to hinder software development.

7.7.8 System Software

System software is a generic term referring to any computer software whose purpose is to help run the computer system. Most of it is responsible directly for controlling, integrating, and managing the individual hardware components of a computer system.

System software is opposed to application software that helps solve user problems directly.

System software performs tasks like transferring data from memory to disk, or rendering text onto a display. Specific kinds of system software include loading programs, operating systems, device drivers, programming tools, compilers, assemblers, linkers, and utilities.

Software libraries that perform generic functions also tend to be regarded as system software, although the dividing line is fuzzy; while a C runtime library is generally agreed to be part of the system, an OpenGL or database library is less obviously so.

System software can be stored on non-volatile storage on integrated circuits that is usually termed firmware.

Chapter 8

Compiler

A compiler is a computer program that translates a computer program written in one computer language (called the source language) into an equivalent program written in another computer language (called the output or the target language).

8.1 Introduction and History

Most compilers translate source code written in a high level language to object code or machine language that may be directly executed by a computer or a virtual machine. However, translation from a low level language to a high level one is also possible; this is normally known as a decompiler if it is reconstructing a high level language program which (could have) generated the low level language program. Compilers also exist which translate from one high level language to another (cross compilers), or sometimes to an intermediate language that still needs further processing; these are sometimes known as cascaders.

Typical compilers output so-called objects that basically contain machine code augmented by information about the name and location of entry points and external calls (to functions not contained in the object). A set of object files, which need not have all come from a single compiler provided that the compilers used share a common output format, may then be linked together to create the final executable which can be run directly by a user.

Several experimental compilers were developed in the 1950s, but the FORTRAN team led by John Backus at IBM is generally credited as having introduced the first complete compiler, in 1957. COBOL was an early language to be compiled on multiple architectures, in 1960. The idea of compilation quickly caught on, and most of the principles of compiler design were developed during the 1960s.

A compiler is itself a computer program written in some implementation language. Early compilers were written in assembly language. The first self-hosting compiler—capable of compiling its own source code in a high-level language—was created for Lisp by Hart and Levin at MIT in 1962. The use of high-level languages for writing compilers gained added impetus in the early 1970s when Pascal and C compilers were written in their own languages. Building a self-hosting compiler is a bootstrapping problem—the first such compiler for a language must be compiled either by a compiler written in a different language, or (as in Hart and Levin's Lisp compiler) compiled by running the compiler in an interpreter.

During the 1990s a large number of free compilers and compiler development tools were developed for all kinds of languages, both as part of the GNU project and other open-source initiatives. Some of them are considered to be of high quality and their free source code makes a nice read for anyone interested in modern compiler concepts.

8.2 Types of Compilers

A compiler may produce code intended to run on the same type of computer and operating system ("platform") as the compiler itself runs on. This is sometimes called a native-code compiler. Alternatively, it might produce code designed to run on a different platform. This is known as a cross compiler. Cross compilers are very useful when bringing up a new hardware platform for the first time. A "source to source compiler" is a type of compiler that takes a high level language as its input and outputs a high level language. For example, an automatic parallelizing compiler will frequently take in a high level language program as an input and then transform the code and annotate it with parallel code annotations (e.g. OpenMP) or language constructs (e.g. Fortran's DOALL statements).

(1) One-pass compiler, like early compilers for Pascal The compilation is done in one pass, hence it is very fast.

(2) Threaded code compiler (or interpreter), like most implementations of forth. This kind of compiler can be thought of as a database lookup program. It just replaces given strings in the source with given binary code. The level of this binary code can vary; in fact, some FORTH compilers can compile programs that don't even need an operating system.

(3) Incremental compiler, like many Lisp systems Individual functions can be compiled in a run-time environment that also includes interpreted functions. Incremental compilation dates back to 1962 and the first Lisp compiler, and is still used in Common Lisp systems.

(4) Stage compiler that compiles to assembly language of a theoretical machine, like some Prolog implementations This Prolog machine is also known as the Warren abstract machine (or WAM). Byte-code compilers for Java, Python (and many more) are also a subtype of this.

(5) Just-in-time compiler, used by Smalltalk and Java systems Applications are delivered in byte code, which is compiled to native machine code just prior to execution.

(6) A retargetable compiler is a compiler that can relatively easily be modified to generate code for different CPU architectures. The object code produced by these is frequently of lesser quality than that produced by a compiler developed specifically for a processor. Retargetable compilers are often also cross compilers. GCC is an example of a retargetable compiler.

(7) A parallelizing compiler converts a serial input program into a form suitable for efficient execution on a parallel computer architecture.

8.3 Compiled vs. Interpreted Languages

Many people divide higher-level programming languages into compiled languages and interpreted languages. However, there is rarely anything about a language that requires it to be compiled or interpreted. Compilers and interpreters are implementations of languages, not languages themselves. The categorization usually reflects the most popular or widespread implementations of a language—for instance, BASIC is thought of as an interpreted language, and C a compiled one, despite the existence of BASIC compilers and C interpreters. There are exceptions, however; some language specifications assume the use of a compiler (as with C), or spell out that implementations must include a compilation facility (as with Common Lisp).

8.4 Compiler Design

In the past, compilers were divided into many passes to save space. A pass in this context is a run of the compiler through the source code of the program to be compiled, resulting in the building up of the internal data of the compiler (such as the evolving symbols table and other assisting data). When each pass is finished, the compiler can free the internal data space needed during that pass. This "multipass" method of compiling was the common compiler technology at the time, but was also due to the small main memories of host computers relative to the source code and data.

Many modern compilers share a common "two stage" design. The front end translates the source language into an intermediate representation. The second stage is the back end, which works with the internal representation to produce code in the output language. The front end and back end may operate as separate passes, or the front end may call the back end as a subroutine, passing it the intermediate representation.

This approach mitigates complexity separating the concerns of the front end, which typically revolve around language semantics, error checking, and the like, from the concerns of the back end, which concentrates on producing output that is both efficient and correct. It also has the advantage of allowing the use of a single back end for multiple source languages, and similarly allows the use of different back ends for different targets.

Often, optimizers and error checkers can be shared by both front ends and back ends if they are designed to operate on the intermediate language that a front-end passes to a back end. This can let many compilers (combinations of front and back ends) reuse the large amounts of work that often go into code analyzers and optimizers.

Certain languages, due to the design of the language and certain rules placed on the declaration of variables and other objects used, and the predeclaration of executable procedures prior to reference or use, are capable of being compiled in a single pass. The Pascal programming language is well known for this capability, and in fact many Pascal compilers are themselves written in the Pascal language because of the rigid specification of the language and the capability to use a single pass to compile Pascal language programs.

8.5 Compiler Front End

The compiler front end consists of multiple phases itself, each informed by formal language theory:

(1) Lexical analysis—breaking the source code text into small pieces ("tokens" or "terminals"), each representing a single atomic unit of the language, for instance a keyword, identifier or symbol names. The token language is typically a regular language, so a finite state automaton constructed from a regular expression can be used to recognize it. This phase is also called lexing or scanning.

(2) Syntax analysis—Identifying syntactic structures of source code. It only focuses on the structure. In other words, it identifies the order of tokens and understand hierarchical structures in code. This phase is also called parsing.

(3) Semantic analysis is to recognize the meaning of program code and start to prepare for output. In that phase, type checking is done and most of compiler errors show up.

(4) Intermediate language generation—an equivalent to the original program is created in an intermediate language.

8.6 Compiler Back End

While there are applications where only the compiler front end is necessary, such as static language verification tools, a real compiler hands the intermediate representation generated by the front end to the back end, which produces a functional equivalent program in the output language. This is done in multiple steps:

(1) Compiler Analysis—This is the process to gather program information from the intermediate representation of the input source files. Typical analysis are variable define-use and use-define chain, data dependence analysis, alias analysis etc. Accurate analysis is the base for any compiler optimizations. The call graph and control flow graph are usually also built during the analysis phase.

(2) Optimization—the intermediate language representation is transformed into functionally equivalent but faster (or smaller) forms. Popular optimizations are in-line expansion, dead code elimination, constant propagation, loop transformation, register allocation or even auto parallelization.

(3) Code generation—the transformed intermediate language is translated into the output language, usually the native machine language of the system. This involves resource and

storage decisions, such as deciding which variables to fit into registers and memory and the selection and scheduling of appropriate machine instructions along with their associated addressing modes.

Notes

A pass has also been known as a parse in some textbooks. The idea is that the source code is parsed by gradual, iterative refinement to produce the completely translated object code at the end of the process. There is, however, some dispute over the general use of parse for all those phases (passes), since some of them, e.g. object code generation, are arguably not regarded to be parsing as such.

References

(1) Compilers: Principles, Techniques and Tools by Alfred V. Aho, Ravi Sethi, and Jeffrey D. Ullman (ISBN 0201100886) is considered to be the standard authority on compiler basics, and makes a good primer for the techniques mentioned above. (It is often called the Dragon Book because of the picture on its cover showing a Knight of Programming fighting the Dragon of Compiler Design.)

(2) Understanding and Writing Compilers: A Do It Yourself Guide (ISBN 0333217322) by Richard Bornat is an unusually helpful book, being one of the few that adequately explains the recursive generation of machine instructions from a parse-tree. Having learnt his subject in the early days of mainframes and minicomputers, the author has many useful insights that more recent books often fail to convey.

Keywords

virtual	虚拟
decompiler	反编译
cascaders	层译器
platform	平台
native-code compiler	本地编码编译器
cross compiler	交叉编译器
compilation	汇编
Threaded code compiler	线性代码编译器
Incremental compiler	增量编译器
subtype	子类型
retargetable	可重定向
parallelizing	并行化
reflect	反映

specification   规格
evolving symbol 扩展符
semantics   语义
optimizers   优化
rigid   严格的
tokens   象征符
hierarchical   阶级式
verification   验证
propagation   传递
iterative   迭代
dispute   争议

8.7   Exercise 8

Multiple or single choices.

1. Translation ____ is possible.
   A. from a low level language to a low level one
   B. from a low level language to a high level one
   C. from a high level language to a low level one
   D. from a high level language to a high level one

2. A compiler may produce code intended to run on the same type of computer and operating system ("platform") as the compiler itself runs on, which is called ____.
   A. one-pass compiler
   B. native-code compiler
   C. threaded code compiler
   D. Incremental compiler

3. Compilers were divided into many passes ____.
   A. to save space
   B. to run fast
   C. to separate data
   D. to maintenance data

4. The front end of a compiler typically revolve around ____.
   A. language semantics
   B. producing output

C. analysis data

D. error checking

5. Compiler back end mainly due to ____.

A. optimization

B. semantic analysis

C. code generation

D. syntax analysis

True or False.

1. Translation from a low level language to a high level one is also possible.

2. A native-code compiler might produce code designed to run on a different platform.

3. Compilers were divided into many passes to save space earlier.

4. Many modern compilers share a common two stage design which the front end translates the source language into the output language.

5. Semantic analysis is to recognize the meaning of program code and start to prepare for output.

8.8 Further Reading: Assembly Language

An assembler is a computer program for translating assembly language—essentially, a mnemonic representation of machine language—into object code. A cross assembler produces code for one type of processor, but runs on another.

As well as translating assembly instruction mnemonics into opcodes, assemblers provide the ability to use symbolic names for memory locations (saving tedious calculations and manually updating addresses when a program is slightly modified), and macro facilities for performing textual substitution-typically used to encode common short sequences of instructions to run inline instead of in a subroutine.

Assemblers are far simpler to write than compilers for high-level languages, and have been available since the 1950s. Modern assemblers, especially for RISC based architectures, such as MIPS, Sun SPARC and HP PA-RISC, optimize instruction scheduling to exploit the CPU pipeline efficiently.

High-level assemblers provide high-level-language abstractions such as advanced control structures, high-level procedure/function declarations and invocations, and high-level abstract data types including structures/records, unions, classes, and sets.

8.8.1 Assemblers

Hundreds of assemblers have been written; some notable examples are:
- ASEM-51—for Intel MCS-51 microcontrollers family.
- A56—for Motorola DSP56000 DSPs (DSP56k series).
- AKI (AvtoKod Ingenera, or "engineer's autocode")—for Minsk family of computers was a half-step away from assembly languages.
- ASCENT (ASsembler for CENTral Processor Unit)—for Control Data Corporation computer systems pre-COMPASS.
- ASPER (ASsembler for PERipheral Processor Units)—for Control Data Corporation computer systems pre-COMPASS.
- C—name used by a few languages that bring C language closer to Assembly.
- COMPASS (COMPrehensive ASSembler)—macro assembler for Control Data Corporation 3000 series minicomputers, 6400/6500/6600, 7600 and Cyber series supercomputers.
- Emu8086—x86 assembler and Intel's 8086 microprocessor emulator.
- FAP (FORTRAN Assembly Program)—for mainframes IBM 709, 7090, 7094.
- FASM (Flat Assembler)—for IA-32, IA-64, open source.
- GAS (GNU Assembler)—for many processors, open source.
- HLA (High Level Assembler)—for x86, public domain.
- HLASM (High Level Assembler)—for mainframes.
- Linoleum—for cross platform use.
- MACRO-11—for DEC PDP-11.
- MACRO-32—for DEC VAX.
- MASM (Macro/MS Assembler)—for x86, from Microsoft.
- MI (Machine Interface)—for AS/400, compile-time intermediate language has many features normally found in high-level languages.
- NASM (Netwide Assembler)—for x86, open source.
- PAL-III—for DEC PDP-8.
- RosASM—32 bit Assembler; The Bottom Up Assembler, open source GPL.
- Sphinx C—mix of Assembly and C, allows combining Assembly commands with C-like structures.
- SSK (Sistema Simvolicheskogo Kodirovaniya, or "System of symbolic coding")—for Minsk family of computers.
- TASM (Turbo Assembler)—for x86 from Borland.
- x86 Assembler Chart—tries to be fairly complete, shows general lineage.

- AS Macro Assembler—Cross assembler for a large variety of processors running on DOS, Win32, and OS/2.

On UNIX systems, the assembler is traditionally called as, although it is not a single body of code, being typically written anew for each port. A number of UNIX variants use GAS.

Within processor groups, each assembler has its own dialect. Sometimes, some assemblers can read another assembler's dialect, for example, TASM can read old MASM code, but not the reverse. FASM and NASM have similar syntax, but each support different macros that could make them difficult to translate to each other. The basics are all the same, but the advanced features will differ.

Also, assembly can sometimes be portable across different operating systems on the same type of CPU. Calling conventions between operating systems often differ slightly to none at all, and with care it is possible to gain some portability in assembly language, usually by linking with a C library that does not change between operating systems. However, it is not possible to link portably with C libraries that require the caller to use preprocessor macros that may change between operating systems. For example, many things in libc depend on the preprocessor to do OS-specific, C-specific things to the program before compiling. In fact, some functions and symbols are not even guaranteed to exist outside of the preprocessor. Worse, the size and field order of structs, as well as the size of certain typedefs such as off_t, are entirely unavailable in assembly language, and do differ even between versions of Linux, making it impossible to portably call functions in libc other than ones that only take simple integers/pointers as parameters.

Many people use an emulator to debug assembly-language programs.

8.8.2 Assembly Language

Assembly language or simply assembly is a human-readable notation for the machine language that a specific computer architecture uses. Machine language, a pattern of bits encoding machine operations, is made readable by replacing the raw values with symbols called mnemonics.

For example, a computer with the appropriate processor will understand this x86/IA-32 machine instruction:

10110000 01100001

For programmers, however, it is easier to remember the equivalent assembly language representation:

mov al, 0x61

which means to move the hexadecimal value 61 (97 decimal) into the processor register with the name "al". The mnemonic "mov" is short for "move", and a comma-separated list of arguments or parameters follows it; this is a typical assembly language statement.

Transforming assembly into machine language is accomplished by an assembler, and the reverse by a disassembler. Unlike in high-level languages, there is usually a 1-to-1 correspondence between simple assembly statements and machine language instructions. However, in some cases an assembler may provide pseudoinstructions which expand into several matching language instructions to provide commonly needed functionality. For example, for a machine that lacks a "branch if greater or equal" instruction, an assembler may provide a pseudoinstruction that expands to the machine's "set if less than" and "branch if zero (on the result of the set instruction)".

Every computer architecture has its own machine language, and therefore its own assembly language. Computers differ by the number and type of operations that they support. They may also have different sizes and numbers of registers, and different representations of data types in storage. While all general-purpose computers are able to carry out essentially the same functionality, the way they do it differs, and the corresponding assembly language must reflect these differences.

In addition, multiple sets of mnemonics or assembly-language syntax may exist for a single instruction set. In these cases, the most popular one is usually that used by the manufacturer in their documentation.

8.8.3 Machine Instructions

Instructions in assembly language are generally very simple, unlike in a high-level language. Any instruction that references memory (for data or as a jump target) will also have an addressing mode to determine how to calculate the required memory address. More complex operations must be built up out of these simple operations. Some operations available in most instruction sets include:

- movings.
  a. set a register (a temporary "scratchpad" location in the CPU itself) to a fixed constant value.
  b. move data from a memory location to a register, or vice versa. This is done to obtain the data to perform a computation on it later, or to store the result of a computation.
  c. read and write data from hardware devices.
- computing.
  a. add, subtract, multiply, or divide the values of two registers, placing the result in a register.
  b. perform bitwise operations, taking the conjunction/disjunction (and/or) of corresponding bits in a pair of registers, or the negation (not) of each bit in a register.
  c. compare two values in registers (for example, to see if one is less, or if they are equal).
- affecting program flow.
  a. jump to another location in the program and execute instructions there.
  b. jump to another location if a certain condition holds.

c. jump to another location, but save the location of the next instruction as a point to return to (a call).

Some computers include one or more "complex" instructions in their instruction set. A single "complex" instruction does something that may take many instructions on other computers. Such instructions are typified by instructions that take multiple steps, may issue to multiple functional units, or otherwise appear to be a design exception to the simplest instructions which are implemented for the given processor. Some examples of such instruction include:

- saving many registers on the stack at once.
- moving large blocks of memory.
- complex and/or floating-point arithmetic (sine, cosine, square root, etc.).
- performing an atomic test and set instruction.
- instructions that combine ALU with an operand from memory rather than a register.

A form of complex instructions that has become particularly popular recently are SIMD operations that perform the same arithmetic operation to multiple pieces of data at the same time, which have appeared under various trade names beginning with MMX and AltiVec.

The design of instruction sets is a complex issue, with a simpler instruction set (generally grouped under the concept RISC) perhaps offering the potential for higher speeds, while a more complex one (traditionally called CISC) may offer particularly fast implementations of common performance-demanding tasks, may use memory (and thus cache) more efficiently), and be somewhat easier to program directly in assembler. See instruction set for a fuller discussion of this point.

8.8.4 Assembly Language Directives

In addition to codes for machine instructions, assembly languages have extra directives for assembling blocks of data, and assigning address locations for instructions or code.

They usually have a simple symbolic capability for defining values as symbolic expressions which are evaluated at assembly time, making it possible to write code that is easier to read and understand.

Like most computer languages, comments can be added to the source code; these often provide useful additional information to human readers of the code but are ignored by the assembler and so may be used freely.

They also usually have an embedded macro language to make it easier to generate complex pieces of code or data.

In practice, the absence of comments and the replacement of symbols with actual numbers makes the human interpretation of disassembled code considerably more difficult than the original source would be.

8.8.5 Usage of Assembly Language

There is some debate over the usefulness of assembly language. It is often said that modern compilers can render higher-level languages into codes that run as fast as hand-written assembly, but counter-examples can be made, and there is no clear consensus on this topic. It is reasonably certain that, given the increase in complexity of modern processors, effective hand-optimization is increasingly difficult and requires a great deal of knowledge.

However, some discrete calculations can still be rendered into faster running code with assembly, and some low-level programming is simply easier to do with assembly. Some system-dependent tasks performed by operating systems simply cannot be expressed in high-level languages. In particular, assembly is often used in writing the low level interaction between the operating system and the hardware, for instance in device drivers. Many compilers also render high-level languages into assembly first before fully compiling, allowing the assembly code to be viewed for debugging and optimization purposes.

It's also common, especially in relatively low-level languages such as C, to be able to embed assembly language into the source code with special syntax. Programs using such facilities, such as the Linux kernel, often construct abstractions where different assembly language is used on each platform the program supports, but it is called by portable code through a uniform interface.

Many embedded systems are also programmed in assembly to obtain the absolute maximum functionality out of what is often very limited computational resources, though this is gradually changing in some areas as more powerful chips become available for the same minimal cost.

Another common area of assembly language use is in the system BIOS of a computer. This low-level code is used to initialize and test the system hardware prior to booting the OS and is stored in ROM. Once a certain level of hardware initialization has taken place, code written in higher level languages can be used, but almost always the code running immediately after power is applied is written in assembly language. This is usually due to the fact that system RAM may not yet be initialized at power-up and assembly language can execute without explicit use of memory, especially in the form of a stack.

Computer systems vendors may charge high prices for compiler language runtime libraries, thereby virtually assuring not every installation supports applications that are written in a particular language, except assembly language. Under this premise, assembly language is forced on Independent Software Vendors to keep the prospective buyer's costs down. What is good from a software engineering viewpoint is bad for business.

Assembly language is also valuable in reverse engineering, since many programs are distributed only in machine code form, and machine code is usually easy to translate into assembly language and carefully examine in this form, but very difficult to translate into a higher-level language. Tools such as the Interactive Disassembler make extensive use of disassembly for such a purpose.

8.8.6 Cross Compiler

A cross compiler is a compiler capable of creating executable code for another platform than the one on which the cross compiler is run. Such a tool is handy when you want to compile code for a platform that you don't have access to, or because it is inconvenient or impossible to compile on that platform (as is the case with embedded systems).

8.8.7 Compiling a Gcc Cross Compiler

Gcc is a free cross compiler that supports dozens of platforms and a handful of languages. However, due to limited volunteer time and the huge amount of work it takes to maintain working cross compilers, in many releases some of the cross compilers are broken.

Gcc relies upon the binaries of binutils for the targeted platform to be available. Especially important is the GNU Assembler. Therefore, binutils first has to be compiled correctly with the switch-target=some-target sent to the configure script. Gcc also has to be configured with the same-target option. Then gcc can be compiled as normal provided that the tools binutils creates are available in the path. On UNIX-like OS:es with bash, that can be accomplished with the following:

PATH=/path/to/binutils/bin:$PATH make

Cross compiling gcc requires that a portion of the C standard library is available for the targetted platform on the host platform. At least the crt0, ... components of the library has to be available in some way. You can choose to compile the full C library, but that can be too large for many platforms. The alternative is to use newlib, which is a small C library containing only the most essential components required to get C source code compiled. To configure gcc with newlib, use the switch-with-newlib to the configure script.

Chapter 9
How Java Works

Have you ever wondered how computer programs work? Have you ever wanted to learn how to write your own computer programs? Whether you are 14 years old and hoping to learn how to write your first game, or you are 70 years old and have been curious about computer programming for 20 years, this article is for you. In this article, I'm going to teach you how computer programs work by teaching you how to program in the Java programming language.

In order to teach you about computer programming, I am going to make several assumptions from the start:

- I am going to assume that you know nothing about computer programming now. If you already know something then the first part of this article will seem elementary to you. Please feel free to skip forward until you get to something you don't know.
- I am going to assume you do know something about the computer you are using. That is, I am going to assume you already know how to edit a file, copy and delete files, rename files, find information on your system, etc.
- For simplicity, I am going to assume that you are using a machine running Windows 95, 98, 2000, NT or XP. It should be relatively straightforward for people running other operating systems to map the concepts over to those.
- I am going to assume that you have a desire to learn.

All of the tools you need to start programming in Java are widely available on the Web for free. There is also a huge amount of educational material for Java available on the Web, so once you finish this article you can easily go learn more to advance your skills. You can learn Java programming here without spending any money on compilers, development environments, reading materials, etc. Once you learn Java it is easy to learn other languages, so this is a good place to start.

Having said these things, we are ready to go. Let's get started!

9.1 A Little Terminology

Keep in mind that I am assuming that you know nothing about programming. Here are several vocabulary terms that will make things understandable:

- Computer program—A computer program is a set of instructions that tell a computer

exactly what to do. The instructions might tell the computer to add up a set of numbers, or compare two numbers and make a decision based on the result, or whatever. But a computer program is simply a set of instructions for the computer, like a recipe is a set of instructions for a cook or musical notes are a set of instructions for a musician. The computer follows your instructions exactly and in the process does something useful—like balancing a checkbook or displaying a game on the screen or implementing a word processor.
- Programming language—In order for a computer to recognize the instructions you give it, those instructions need to be written in a language the computer understands—a programming language. There are many computer programming languages—Fortran, Cobol, Basic, Pascal, C, C++, Java, Perl—just like there are many spoken languages. They all express approximately the same concepts in different ways.
- Compiler—A compiler translates a computer program written in a human-readable computer language (like Java) into a form that a computer can execute. You have probably seen EXE files on your computer. These EXE files are the output of compilers. They contain executables-machine-readable programs translated from human-readable programs.

In order for you to start writing computer programs in a programming language called Java, you need a compiler for the Java language. The next section guides you through the process of downloading and installing a compiler. Once you have a compiler, we can get started. This process is going to take several hours, much of that time being download time for several large files. You are also going to need about 40 megabytes of free disk space (make sure you have the space available before you get started).

9.2 Downloading the Java Compiler

In order to get a Java development environment set up on your machine—you "develop" (write) computer programs using a "development environment"—you will have to complete the following steps:

Download a large file containing the Java development environment (the compiler and other tools).

(1) Download a large file containing the Java documentation.

(2) If you do not already have WinZip (or an equivalent) on your machine, you will need to download a large file containing WinZip and install it.

(3) Install the Java development environment.

(4) Install the documentation.

(5) Adjust several environment variables.

(6) Test everything out.

Before getting started, it would make things easier if you create a new directory in your temp directory to hold the files we are about to download. We will call this the download directory.

Step 1—Download the Java development environment

Go to the page http://java.sun.com/j2se/1.4.2/download.html. Download the SDK software by clicking on the "Download J2SE SDK" link. You will be shown a licensing agreement. Click Accept. Select your operating system and download the file to your download directory. This is a huge file, and it will take several hours to download over a normal phone-line modem. The next two files are also large.

Step 2—Download the Java documentation

Download the documentation by selecting your operating system and clicking the SDK 1.4.1 documentation link.

Step 3—Download and install WinZip

If you do not have a version of WinZip or an equivalent on your machine, go to the page http://www.winzip.com/ and download an evaluation copy of WinZip. Run the EXE you get to install it. We will use it in a moment to install the documentation.

Step 4—Install the development kit

Run the j2sdk-1_4_1-.exe file that you downloaded in step 1. It will unpack and install the development kit automatically.

Step 5—Install the documentation

Read the installation instructions for the documentation. They will instruct you to move the documentation file to same directory as that containing the development kit you just installed. Unzip the documentation and it will drop into the proper place.

Step 6—Adjust your environment

As instructed on this page, you need to change your path variable. This is most easily done by opening an MS-DOS prompt and typing PATH to see what the path is set to currently. Then open autoexec.bat in Notepad and make the changes to PATH specified in the instructions.

Step 7—Test

Now you should be able to open another MS-DOS window and type javac. If everything is set up properly, then you should see a two-line blob of text come out that tells you how to use javac. That means you are ready to go. If you see the message "Bad Command or File Name" it means you are not ready to go. Figure out what you did wrong by rereading the installation instructions. Make sure the PATH is set properly and working. Go back and reread the Programmer's Creed above and be persistent until the problem is resolved.

You are now the proud owner of a machine that can compile Java programs. You are ready to start writing software!

By the way, one of the things you just unpacked is a demo directory full of neat examples.

All of the examples are ready to run, so you might want to find the directory and play with some of the samples. Many of them make sounds, so be sure to turn on your speakers. To run the examples, find pages with names like example1.html and load them into your usual Web browser.

9.3 Your First Program

Your first program will be short and sweet. It is going to create a drawing area and draw a diagonal line across it. To create this program you will need to:

Open Notepad and type in (or cut and paste) the program.

(1) Save the program.

(2) Compile the program with the Java compiler to create a Java applet.

(3) Fix any problems.

(4) Create an HTML web page to "hold" the Java Applet you created.

(5) Run the Java applet.

Here is the program we will use for this demonstration:

```
import java.awt.Graphics;
public class FirstApplet extends java.applet.Applet
{
  public void paint(Graphics g)
  {
    g.drawLine(0, 0, 200, 200);
  }
}
```

Step 1—Type in the program

Create a new directory to hold your program. Open up Notepad (or any other text editor that can create TXT files). Type or cut and paste the program into the Notepad window. This is important: When you type the program in, case matters. That means that you must type the uppercase and lowercase characters exactly as they appear in the program. Review the programmer's creed above. If you do not type it EXACTLY as shown, it is not going to work.

Step 2—Save the file

Save the file to the filename FirstApplet.java in the directory that you created in step 1. Case matters in the filename. Make sure the "F" and "A" are uppercase and all other characters are lowercase, as shown.

Step 3—Compile the program

Open an MS-DOS window. Change directory ("cd") to the directory containing FirstApplet.java. Type:

javac FirstApplet.java

Case matters! Either it will work, in which case nothing will be printed to the window, or there will be errors. If there are no errors, a file named FirstApplet.class will be created in the directory right next to FirstApplet.java.

Step 4—Fix any problems

If there are errors, fix them. Compare your program to the program above and get them to match exactly. Keep recompiling until you see no errors. If javac seems to not be working, look back at the previous section and fix your installation.

Step 5—Create an HTML Page

Create an HTML page to hold the applet. Open another Notepad window. Type into it the following:

```
<html>
<body>
<applet code=FirstApplet.class width=200 height=200>
</applet>
</body>
</html>
```

Save this file in the same directory with the name applet.htm.

Step 6—Run the Applet

In your MS-DOS window, type:

appletviewer applet.htm

You should see a diagonal line running from the upper left corner to the lower right corner, as shown in Figure 9.1.

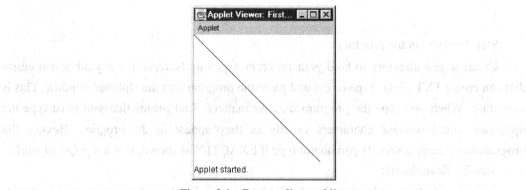

Figure 9.1 Draw a diagonal line

Pull the applet viewer a little bigger to see the whole line. You should also be able to load the HTML page into any modern browser like Netscape Navigator or Microsoft Internet Explorer and see approximately the same thing.

You have successfully created your first program!!!

9.4 Understanding What Just Happened

So what just happened? First, you wrote a piece of code for an extremely simple Java applet. An applet is a Java program that can run within a Web browser, as opposed to a Java application, which is a stand-alone program that runs on your local machine (Java applications are slightly more complicated and somewhat less popular, so we will start with applets). We compiled the applet using javac. We then created an extremely simple Web page to "hold" the applet. We ran the applet using appletviewer, but you can just as easily run it in a browser.

The program itself is about 8 lines long:

```
import java.awt.Graphics;
public class FirstApplet extends java.applet.Applet
{
    public void paint(Graphics g)
    {
       g.drawLine(0, 0, 200, 200);
    }
}
```

This is about the simplest Java applet you can create. To fully understand it you will have to learn a fair amount, particularly in the area of object oriented programming techniques. Since I am assuming that you have zero programming experience, what I would like you to do is focus your attention on just one line in this program for the moment:

```
g.drawLine(0, 0, 200, 200);
```

This is the line in this program that does the work. It draws the diagonal line. The rest of the program is scaffolding that supports that one line, and we can ignore the scaffolding for the moment. What happened here was that we told the computer to draw one line from the upper left hand corner (0,0) to the bottom right hand corner (200, 200). The computer drew it just like we told it to. That is the essence of computer programming!

(Note also that in the HTML page, we set the size of the applet's window in step 5 above to have a width of 200 and a height of 200.)

In this program, we called a method (a.k.a. function) called drawLine and we passed it four parameters (0, 0, 200, 200). The line ends in a semicolon. The semicolon acts like the period at the end of the sentence. The line begins with g., signifying that we want to call the method named drawLine on the specific object named g (which you can see one line up is of the class Graphics—we will get into classes and methods of classes in much more detail later in this article).

A method is simply a command—it tells the computer to do something. In this case, drawLine tells the computer to draw a line between the points specified as shown in Figure 9.2: (0, 0) and (200, 200). You can think of the window as having its 0,0 coordinate in the upper left corner, with positive X and Y axes extending to the right and down. Each dot on the screen (each pixel) is one increment on the scale.

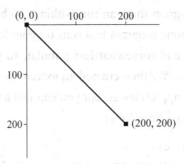

Figure 9.2   A diagonal line

Try experimenting by using different numbers for the four parameters. Change a number or two, save your changes, recompile with javac and rerun after each change in appletviewer, and see what you discover.

What other functions are available besides drawLine? You find this out by looking at the documentation for the Graphics class. When you installed the Java development kit and unpacked the documentation, one of the files unloaded in the process is called java.awt.Graphics.html, and it is on your machine. This is the file that explains the Graphics class. On my machine, the exact path to this file is D:\\jdk1.1.7\\docs\\api\\java.awt.Graphics.html. On your machine the path is likely to be slightly different, but close—it depends on exactly where you installed things. Find the file and open it. Up toward the top of this file there is a section called "Method Index." This is a list of all of the methods this class supports. The drawLine method is one of them, but you can see many others. You can draw, among other things:

- Lines.
- Arcs.
- Ovals.
- Polygons.
- Rectangles.
- Strings.
- Characters.

Read about and try experimenting with some of these different methods to discover what is possible. For example, try this code:

g.drawLine(0, 0, 200, 200);
g.drawRect(0, 0, 200, 200);
g.drawLine(200, 0, 0, 200);

It will draw a box with two diagonals (be sure to pull the window big enough to see the whole thing). Try drawing other shapes. Read about and try changing the color with the setColor method. For example:

```
import java.awt.Graphics;
import java.awt.Color;
public class FirstApplet extends java.applet.Applet
{
  public void paint(Graphics g)
  {
    g.setColor(Color.red);
    g.fillRect(0, 0, 200, 200);
    g.setColor(Color.black);
    g.drawLine(0, 0, 200, 200);
    g.drawLine(200, 0, 0, 200);
  }
}
```

Note the addition of the new import line in the second line of the program. The output of this program looks like Figure 9.3.

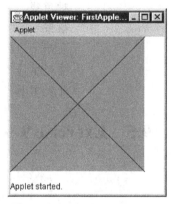

Figure 9.3　Add a second line

One thing that might be going through your head right now is, "How did he know to use Color. red rather than simply red, and how did he know to add the second import line?" You learn things like that by example. Because I just showed you an example of how to call the setColor method, you now know that whenever you want to change the color you will use Color. followed by a color name as a parameter to the setColor method, and you will add the appropriate import line at the top of the program. If you look up setColor, it has a link that will tell you about the Color class, and in it is a list of all the valid color names along with techniques for creating new (unnamed) colors. You read that information, you store it in your head and now you know how to change colors in Java. That is the essence of becoming a computer programmer—you learn techniques and remember them for the next program you write. You learn the techniques either by reading an example (as you did here) or by reading through the documentation or by looking at example code (as in the demo directory). If you have a brain that likes exploring and learning and remembering things, then you will love programming!

Keywords

elementary	初级
skip forward	快进
recipe	菜谱
licensing agreement	许可协议
prompt	提示
unpacked	解压
directory	目录
uppercase	大写
lowercase	小写
applet	小程序
opposed	对应的
diagonal	对角线
scaffolding	框架
signify	表示
recompile	重新编译
development kit	开发工具

9.5　Exercise 9

Multiple or single choices.

1. A computer program is ____.

　　A. a set of code

　　B. a set of instructions

　　C. to tell a computer how to run

　　D. to tell a computer exactly what to do

2. A program for a computer could be look like ____.

　　A. musical notes for a musician

　　B. gun for a policeman

　　C. recipe for a cooker

　　D. numbers for a teacher

3. If you want to draw a box, you may input the following code: _____.
   A. g.drawLine(200, 0, 0, 200)
   B. g.drawLine(0, 200, 0, 200)
   C. g.drawLine(200, 0, 200, 0)
   D. g.drawLine(0, 0, 200, 200)

True or False.

1. A method is simply a command—it tells the computer to do something.

2. You don't need to care about the uppercase and lowercase characters exactly as they appear in the program.

3. In order to use the setcolor method, you must add the appropriate import line at the top of the program.

9.6 Further Reading: How Perl Works

Perl is a fairly straightforward, widely known and well-respected scripting language. It is used for a variety of tasks (for example, you can use it to create the equivalent of DOS batch files or C shell scripts), but in the context of Web development it is used to develop CGI scripts.

One of the nice things about Perl is that, because it is a scripting language, people give away source code for their programs. This gives you the opportunity to learn Perl by example, and you can also download and modify thousands of Perl scripts for your own use. One of the bad things about Perl is that much of this free code is impossible to understand. Perl lends itself to an unbelievably cryptic style!

This article assumes that you already know how to program (if you know the C programming language, this will be especially easy for you). Perl is easy to use once you know the basics. In this article, we're going to start at the beginning and show you how to do the most common programming tasks using Perl. By the end of this article, you will be able to write your own Perl scripts with relative ease, and read cryptic scripts written by others with somewhat less ease, but this will be a good starting point.

9.6.1 Getting Started

To start with Perl you need the Perl interpreter. On any UNIX machine there is a 99.99-percent probability that it's already there. On a Windows machine or a Mac, you need to download the latest release of the language and install it on your machine. (See the links at the end of this article for more information.) Perl is widely available on the Web and is free.

Next, make sure you look in the DOCS directory that comes with Perl—there will be user-manual-type stuff in there. At some point, it would not hurt to read through all of the documentation, or at least scan it. Initially it will be cumbersome, but after reading this article it will make much more sense.

9.6.2 Hello World

Once you have Perl loaded, make sure you have your path properly set to include the Perl executable. Then, open a text editor and create a text file. In the file, place the following line:

```
print "Hello World!\\n";
```

Name the file "test1.pl". At the command prompt, type:

```
perl test1.pl
```

Perl will run and execute the code in the text file. You should see the words "Hello World!" printed to stdout (standard out). As you can see, it is extremely easy to create and run programs in Perl. (If you are using UNIX, you can place a comment like #! /usr/bin/perl on the first line, and then you will not have to type the word "perl" at the command line.)

The print command prints things to stdout. The \\n notation is a line feed. That would be more clear if you modified the test program to look like this (# denotes a comment):

```
# Print on two lines
print "Hello\\nWorld!\\n";
```

Note that the print command understood that it should interpret the "\\n" as a line feed and not as the literal characters. The interpretation occurred not because of the print command, but because of the use of double quotes (a practice called quoting in Perl). If you were to use single quotes instead, as in:

```
print 'Hello\\nWorld!\\n';
```

the \\n character would not be interpreted but instead would be used literally.

There is also the backquote character: '. A pair of these imply that what is inside the quotes should be interpreted as an operating system command, and that command should be executed with the output of the command being printed. If you were to place inside the backquotes a command-line operation from the operating system, it would execute. For example, on Windows NT you can say:

```
print 'cmd /c dir';
```

to run the DIR command and see a list of files from the current directory.

PERL Note: In Windows NT, you cannot say:

```
print 'dir';
```

because dir is not a separate executable—it's part of the command interpreter cmd. Type cmd /? at a DOS prompt for details.

You will also see the / character used for quoting regular expressions.

The print command understands commas as separators. For example:

```
print 'hello', "\\n", 'world!';
```

However, you will also see a period:

```
print 'hello'. "\\n". 'world!';
```

The period is actually a string concatenation operator.

There is also a printf operator for C folks.

9.6.3 Variables

Variables are interesting in Perl. You do not declare them, and you always use a $ to denote them. They come into existence at first use. For example:

```
$s="Hello\\nWorld\\n";
$t='Hello\\nWorld\\n';
print $s, "\\n", $t;Or:       $i=5;
$j=$i+5;
print $i, "\\t", $i+1, "\\t", $j; # \\t=tabOr:      $a="Hello ";
$b="World\\n";
$c=$a . $b;  # note use of . to concat strings
```

print $c;Since . is string concatenation, .=has the expected meaning in the same way that "+=" does in C. Therefore, you can say:

```
$a="Hello ";
$b="World\\n";
$a .=$b;
print $a;
```

You can also create arrays:

```
@a=('cat', 'dog', 'eel');
  print @a, "\\n";
  print $#a, "\\n";  # The value of the highest index, zero based
  print $a\, "\\n";
  print $a\, $a\, $a\, "\\n";
```

The $# notation gets the highest index in the array, equivalent to the number of elements in the array minus 1. As in C, all arrays start indexing at zero.

You can also create hashes:

```
%h=('dog', 'bark', 'cat', 'meow', 'eel', 'zap');
print "The dog says ", $h{'dog'};
```

Here, "bark" is associated with the word "dog", "meow" with "cat", and so on. A more expressive syntax for the same declaration is:

```
%h=(
    dog => 'bark',
    cat => 'meow',
    eel => 'zap'
);
```

The => operator quotes the left string and acts as a comma.

9.6.4 Loops and Ifs

You can create a simple for loop like you do in C:

```
for ($i=0; $i<10; $i++)
  {
    print $i, "\\n";
  }
```

Perl Note:

```
You must use the "begin" and "end" braces-{ and }-even for a single line.
```

While statements are easy:

```
$i=0;
while ($i<10)
{
  print $i, "\\n";
  $i++;
}
```

If statements are similarly easy:

```
for ($i=0; $i<10; $i++)
{
  if ($i !=5)
  {
    print $i, "\\n";
  }
}
```

The boolean operators work like they do in C:
- && and
- || or
- ! not

For numbers:
- == equal
- != not equal
- <, <=, >, >= (as expected)

Others:
- eq
- ne
- lt
- le
- gt
- ge

If you have an array, you can loop through it easily with foreach:

```
@a=('dog', 'cat', 'eel');
foreach $b (@a)
{
  print $b, "\\n";
}
```

9.6.5 Functions

You create a subroutine with the word sub. All variables passed to the subroutine arrive in an array called _. Therefore, the following code works:

```
show ('cat', 'dog', 'eel');
sub show
{
  for ($i=0; $i <=$#_; $i++)
  {
    print $_\, "\\n";
  }
}
```

Remember that $# returns the highest index in the array (the number of elements minus 1), so $#_ is the number of parameters minus 1. If you like that sort of obtuseness, then you will love PERL.

You can declare local variables in a subroutine with the word local, as in:

```
sub xxx
{
  local ($a, $b, $c)
  ...
}
```

You can also call a function using &, as in:

&show ('a', 'b', 'c'); The & symbol is required only when there is ambiguity, but some programmers use it all the time.

To return a value from a subroutine, use the keyword return.

9.6.6 Reading

1. Reading from STDIN

To read in data from the stdin (standard in), use the STDIN handle. For example:

```
print "Enter high number: ";
$i=<STDIN>;
for ($j=0; $j <=$i; $j++)
{
  print $j, "\\n";
}
```

As long as you enter an integer number, this program will work as expected. <STDIN> reads a line at a time. You can also use getc to read one character, as in:

```
$i=getc(STDIN);
```

Or use read:

```
read(STDIN, $i, 1);
```

The 1 in the third parameter to the read command is the length of the input to read.

2. Reading Environment Variables

Perl defines a global hash named ENV, and you can use it to retrieve the values of environment variables. For example:

```
print $ENV{'PATH'};
```

PERL Note: The name of the environment variable must be upper case.

3. Reading Command Line Arguments

Perl defines a global array ARGV, which contains any command line arguments passed to the script. $#ARGV is the number of arguments passed minus 1, $ARGV\ is the first argument passed, $ARGV\ is the second, and so on.

Chapter 10
Database & C++

10.1 Text

Between November 1992 and the end of 1995, Microsoft introduced a number of new Windows relational database products: Access 7, Visual FoxPro 3.0 for Windows, and Visual C++ 4.0. Microsoft heralded Access as "the database that anyone can use" and sold 750,000 bargain-priced copies in 90 days when Access 1.x was first released. FoxPro for Windows targeted existing FoxPro developers and prospective users of Borland's long-promised dBase for Windows. Both Access and FoxPro targeted the market for Borland's Paradox for Windows, which emerged shortly after the retail release of Access 1.0. Access (upgraded to version 7.0 with the introduction of Windows 95), Visual FoxPro, and Paradox for Windows are categorized as desktop databases.

Since its introduction, Visual Basic has had better support for database interface than Visual C++. Only with Visual C++ 4 has the C/C++ programmer had a real interface with the Microsoft Jet database engine. Desktop databases are applications that employ a proprietary (native) database file or index structure (or both) and that can be run on a PC of moderate performance (for example, an 80486 with 8MB of RAM). Desktop databases also include an application programming language that is designed specifically for use with the native database structure.

When the first edition of this book was written, Microsoft had sold more than four million copies of Access versions 1.0, 1.1, and 2.0. Between mid-June and mid-November of 1995, Microsoft released Windows 95 and 32-bit "designed for Windows 95" versions of Access (7.0), Visual FoxPro (3.0), Visual Basic (4.0), and Visual C++ (4.0), together with the 32-bit Microsoft Office 95. Microsoft wanted to make sure that early adopters of Windows 95 would have 32-bit applications to run.

Visual C++ is Microsoft's most extensive and powerful programming language. Microsoft's original objective for Visual C++ was to provide a powerful and flexible platform that programmers could use to create their own Windows applications while running under Windows. Microsoft achieved this goal with Visual C++ 1.0. Many experienced programmers abandoned DOS-based C, C++, and Pascal in favor of Visual C++ because they could develop Windows applications faster than with traditional programming languages while working with

Windows' graphical interface. Microsoft enriched Visual C++ 1.5 with improvements in the interface and extensions to the MFC C++ libraries, while Visual C++ 2.x moved programmers into the 32-bit application world. Visual C++ 4.0 moves the programmer interface, class library, and feature set to a new high. With the introduction of Visual C++ 4.0, a new set of database features has been added. Visual C++ 4.0 supports DAO (Data Access Objects) in addition to ODBC and also greatly extends other support, such as the addition of container support for OLE Custom Controls. Independent firms have created a variety of utilities, libraries, and add-on features for Visual C++, the majority of which addressed database applications. There will be, in the very near future, a plethora of new OLE Custom Controls for database programmers.

By early 1993, a Microsoft market study showed that more than 70 percent of Windows applications involved databases in one form or another. In October of 1995, Microsoft's Visual C++ product manager noted in a speech in Boston that between 40 and 60 percent of all Visual C++ applications were database oriented. Visual C++ can be expected to also be a popular database applications development tool. Even before the introduction of Visual C++ 4.0, with its data access objects (CRecordset, CDatabase, and CRecordView) that greatly enhance database functionality, C and C++ were major but unrecognized players in the Windows database market. The introduction of Visual C++ 4.0 has now pushed Visual C++ to be a strong competitor to Visual Basic in the database development platform arena. The failure of market research firms to place Visual C++ in the Windows database category caused significant distortion of the desktop database market statistics for 1993 and later.

This chapter describes Visual C++'s role in database application development and shows how Visual C++, OLE (Object Linking and Embedding) automation, ODBC (Open Database Connectivity), DAO, and MFC fit into Microsoft's strategy to maintain its domination of the Windows application marketplace. This chapter also discusses the advantages and drawbacks of using Visual C++ for database applications and gives you a preview of many of the subjects that are covered in detail in the remaining chapters of this book.

It's becoming a 32-bit, "Designed for Windows 95" world out there, so this book concentrates on 32-bit application development with Visual C++ 4.0.

Choosing Visual C++ as Your Database Development Platform

Visual C++ now includes the database connectivity and data handling features that qualify the language as a full-fledged database application development environment. The new data access features that Microsoft added to Visual C++ position the product as a direct competitor to Visual Basic and make Visual C++'s support of Access, FoxPro, and Paradox for Windows in the desktop database market even more complete. Visual C++'s primary advantages over its database competitors are simplicity, flexibility, and extensibility:

- The 32-bit Microsoft Jet 3.0 database engine offers substantially improved performance compared to 16-bit Jet 2+. Jet 3.0 is multithreaded, with a minimum of three threads of execution. (You can increase the number of available threads by an entry in the Windows 95 or Windows NT Registry.) Overall optimization and code tuning also contributes to faster execution of Jet 3.0 queries.

- Visual C++'s built-in MFC classes, along with AppWizard, let you quickly create a form to display database information with little or no Visual C++ code.
- Visual C++ is flexible because it doesn't lock you into a particular application structure, as is the case with Access's multiple document interface (MDI). Nor do you have to use DoCmd instructions to manipulate the currently open database.
- Visual C++ 4.0 database front ends require substantially fewer resources than their Access counterparts. Most 32-bit Visual C++ 4.0 database applications run fine under Windows 95 with PCs having 8M of RAM and under Windows NT 3.51+ in the 16M range. Microsoft says Access 95 requires 12MB of RAM under Windows 95, but you need 16M to achieve adequate performance of all but trivial Access 95 applications. A typical Visual C++ database front-end program would probably run with satisfactory performance on a system with as little as 12MB of RAM under Windows 95. This same program would run well under the same amount of RAM in future versions of Windows NT.
- OLE Custom Controls, not yet available in all other database development platforms, let you add new features to Visual C++ applications with minimal programming effort. Third-party developers can create custom control add-ins to expand Visual C++'s repertoire of data access controls. Custom controls can take the form of OLE Custom Controls for the 32-bit environments.

The most important benefit of selecting Visual C++ as your primary database development platform, however, isn't evident in Microsoft's feature list. There is a vast array of tools and support for ODBC and database development with Visual C++ today.

Another reason for choosing Visual C++ for database application development is its OLE compatibility. Visual C++ was the best database development environment that incorporated OLE.

OLE automation is likely to be the most significant OLE feature for the majority of Visual C++ database developers. OLE automation lets you control the operation of other OLE-compliant server applications from within your Visual C++ database application. Applications need not include Visual Basic for Applications to act as OLE automation source applications (servers); Word for Windows 6 and later supports OLE automation using the conventional Word Basic macro language syntax.

Microsoft's multipronged attack with Visual Basic, Access, Visual FoxPro, SQL Server, and Visual C++ is forcing competing publishers of desktop database managers into their defensive trenches. As a group, Microsoft's database applications for Windows, together with ancillary products such as ODBC, DAO, and the Access Jet database engine, have a breadth and depth that no other software publisher presently can match.

Keywords

Paradox	悖论
emerge	涌现
interface	接口
engine	引擎
proprietary	私人
flexible	方便的
enriched	丰富
majority	多数
plethora	过剩
distortion	流失
drawbacks	缺陷
substantially	大幅地
threads	线程
multithreaded	多线程
trivial	细小的
satisfactory	满意的
repertoire	剧目
compatibility	兼容性
syntax	句法
multipronged	多方面的
defensive trenches	抵抗侵犯
ancillary	辅助的

10.2   Exercise 10

Multiple or single choices.

1. According to the text, Visual C++ 4.0 supports the following database include ____.
   A. DAO (Data Access Objects)
   B. ADO
   C. ODBC
   D. OLE

2. Before the introduction of Visual C++ 4.0, the data access objects include ____, which have already enhance database functionality greatly.

A. CRecordset
   B. CDatabase
   C. MFC
   D. DAO

3. Visual C++'s primary advantages over its database competitors are ____.
   A. simplicity
   B. flexibility
   C. efficiency
   D. extensibility:

4. The benefit of selecting Visual C++ as your primary database development platform is ____.
   A. a vast array of tools and support for ODBC
   B. evident in Microsoft's feature list
   C. OLE compatibility
   D. its front ends require substantially fewer resources

True or False.

1. Visual C++ 4.0 supports DAO (Data Access Objects).

2. Visual C++ qualify the language as a full-fledged database application development environment.

3. Jet 3.0 is multithreaded, with a minimum of two threads of execution.

4. Visual C++ will lock you into a particular application structure.

5. OLE Custom Controls, are now available in all other database development platforms.

10.3 Further Reading: C++

C++(pronounced "see plus plus") is a general-purpose computer programming language. It is a statically typed free-form multi-paradigm language supporting procedural programming, data abstraction, object-oriented programming, and generic programming. During the 1990s, C++ became one of the most popular commercial programming languages.

Bell Labs' Bjarne Stroustrup developed C++ (originally named "C with Classes") during the 1980s as an enhancement to the C programming language. Enhancements started with the addition of classes, followed by, among many features, virtual functions, operator overloading, multiple inheritance, templates, and exception handling. The C++ programming language standard was ratified in 1998 as ISO/IEC 14882:1998, the current version of which is the 2003 version, ISO/IEC 14882:2003.

In C and C++, the expression x++ increases the value of x by 1 (called incrementing). The name "C++" is a play on this, suggesting an incremental improvement upon C.

10.3.1 Technical Overview

The 1998 C++ standard consists of two parts: the core language and the C++ standard library; the latter includes most of the Standard Template Library and a slightly modified version of the C standard library. Many C++ libraries exist which are not part of the standard, such as Boost. Also, non-standard libraries written in C can generally be used by C++ programs.

10.3.2 Features Introduced in C++

Features introduced in C++ include declarations as statements, function-like casts, new/delete, bool, reference types, const, inline functions, default arguments, function overloading, namespaces, classes (including all class-related features such as inheritance, member functions, virtual functions, abstract classes, and constructors), operator overloading, templates, the :: operator, exception handling, and run-time type identification.

C++ also performs more type checking than C in several cases.

Comments starting with two slashes ("//") were originally part of C's predecessor, BCPL, and were reintroduced in C++.

Several features of C++ were later adopted by C, including const, inline, declarations in for loops, and C++ -style comments (using the // symbol). However, C99 also introduced features that do not exist in C++, such as variadic macros and better handling of arrays as parameters.

A very common source of confusion is a subtle terminology issue: because of its derivation from C, in C++ the term object means memory area, just like in C, and not class instance, which is what it means in most other object oriented languages. For example in both C and C++ the line

```
int i;
```

defines an object of type int, that is the memory area where the value of the variable i will be stored on assignment.

10.3.3　C++ Library

The C++ standard library incorporates the C standard library with some small modifications to make it work better with the C++ language. Another large part of the C++ library is based on the Standard Template Library (STL). This provides such useful tools as containers (for example vectors and lists) and iterators (generalized pointers) to provide these containers with array-like access. Furthermore (multi)maps (associative arrays) and (multi)sets are provided, all of which export compatible interfaces. Therefore it is possible, using templates, to write generic algorithms that work with any container or on any sequence defined by iterators. As in C, the features of the library are accessed by using the #include directive to include a standard header. C++ provides sixty-nine standard headers, of which nineteen are deprecated.

The STL was originally a third-party library from HP and later SGI, before its incorporation into the C++ standard. The standard does not refer to it as "STL", as it is merely a part of the standard library, but many people still use that term to distinguish it from the rest of the library (input/output streams (IOstreams), internationalization, diagnostics, the C library subset, etc).

A project known as STLPort, based on the SGI STL, maintains an up-to-date implementation of the STL, IOStreams and strings. Other projects also make variant custom implementations of the standard library with various design goals. Every C++ compiler vendor or distributor includes some implementation of the library, since this is an important part of the standard and is expected by users.

10.3.4　Object-oriented Features of C++

C++ introduces some object-oriented features to C. It offers classes which provide the four features commonly present in OO (and some non OO) languages: abstraction, encapsulation, polymorphism, and inheritance.

1. Encapsulation

C++ implements encapsulation by allowing all members of a class to be declared as either public, private, or protected. A public member of the class will be accessible to any function. A private member will only be accessible to functions that are members of that class and to functions and classes explicitly granted access permission by the class ("friends"). A protected member will be accessible to members of classes that inherit from the class in addition to the class itself and any friends.

The OO principle is that all and only the functions that can access the internal representation of a type should be encapsulated within the type definition. C++ supports this (via member functions and friend functions), but does not enforce it: the programmer can declare parts or all of the representation of a type to be public, and is also allowed to make public entities that are not part of the representation of the type. Because of this C++ supports not just OO programming but other weaker decomposition paradigms, like modular programming.

It is generally considered good practice to make all data private, or at least protected, and to make public only those functions that are part of a minimal interface for users of the class that hides implementation details.

2. Polymorphism

Polymorphism is a widely used and abused term that is not well defined.

In the case of C++ it is often used in connection with member function names, where the function name corresponds to several implementations, and which implementation gets invoked depends either on the type of the arguments (static polymorphism) or on the type of the class instance value (dynamic polymorphism) used on which the virtual member function is invoked.

For example, a C++ program may contain something like this:

```
/   Static polymorphism   /
extern void SendJobToDevice(PrintJobText  ,DeviceLaser );
extern void SendJobToDevice(PrintJobText  ,DeviceJet );
extern void SendJobToDevice(PrintJobHTML  ,DeviceLaser );
extern void SendJobToDevice(PrintJobHTML  ,DeviceJet );
...
SendJobToDevice(printJob,device);
/   Dynamic polymorphism   /
class Device {
public:
  virtual void print(PrintJob  );
  ...
};
PrintJob   printJob;
Device   device;
...
device->print(printJob);
//Note that since C++ does not have multiple dispatch, the above
//function call is polymorphic only based on the device's type.
```

In C, (dynamic) polymorphism of a sort can be achieved using the switch statement or function pointers.

C++ provides two more sophisticated features for polymorphism: function overloading and virtual member functions. Both features allow a program to define several different implementations of a function for use with different types of objects.

Function overloading allows programs to declare multiple functions with the same name. The functions are distinguished by the number and types of their formal parameters. For example, a program might contain the following three function declarations:

```
void pageUser(int userid);
void pageUser(int userid, string message);
void pageUser(string username);
```

Three different pageUser() functions are declared. When the compiler afterwards encounters a call to pageUser(), it determines which function to call based on the number and type of the arguments provided. (The compiler considers only the parameters, not the return type.) Because the compiler determines which function to call at compile time, this is called static polymorphism. (The word static is used here in the sense of "not moving". It denotes that the determination is made based solely on static analysis of the source code: by reading it, not by running it. By the time the program executes, the decision has been made.)

Operator overloading is a form of function overloading. It is one of C++'s most controversial features. Many consider operator overloading to be widely misused, while others think it is a great tool for increasing expressiveness. An operator is one of the symbols defined in the C++ language, such as +, !=, <, or &. Much as function overloading allows the programmer to define different versions of a function for use with different argument types, operator overloading lets the programmer define different versions of an operator for use with different operand types. For example, if the class Integer contains a declaration like this:

```
Integer& operator++();
```

then the program can use the ++ operator with objects of type Integer. For example, the code

```
Integer a=2;
++a;
```

behaves exactly like this:

```
Integer a=2;
a.operator++();
```

In most cases, this would then increment the value of the variable a to 3. However, the programmer who created the Integer class can define the Integer::operator++() member function to do whatever he wants. Because operators are commonly used implicitly, it is considered bad style to declare an operator except when its meaning is obvious and unambiguous. Curiously, it can be argued that the standard libraries do not follow this convention. For example, the object cout, used for outputting text to the console, has an overloaded << operator for outputting the text. Critics argue that this use is non-obvious because << is the operator for a bit shift, which is clearly meaningless in this context.

Nevertheless, most people consider this use to be acceptable, and this particular example is certainly in C++ to stay.

C++ templates make heavy use of static polymorphism, including overloaded operators.

Virtual member functions provide a different type of polymorphism. In this case, different objects that share a common base class may all support an operation in different ways. For example, a PrintJob base class might contain a member function virtual int getPageCount (double pageWidth, double pageHeight) Each different type of print job, such as DoubleSpacedPrintJob, may then override the method with a function that can calculate the appropriate number of pages for that type of job. In contrast with function overloading, the parameters for a given member function are always exactly the same number and type. Only the type of the object for which this method is called varies.

When a virtual member function of an object is called, the compiler sometimes doesn't know the type of the object at compile time and therefore can't determine which function to call. The decision is therefore put off until runtime. The compiler generates code to examine the object's type at runtime and determine which function to call. Because this determination is made on the fly, this is called dynamic polymorphism.

The run-time determination and execution of a function is called dynamic dispatch. In C++, this is commonly done using virtual tables.

3. Inheritance

Inheritance from a base class may be declared as public, protected, or private. This access specifier determines whether unrelated and derived classes can access the inherited public and protected members of the base class. Only public inheritance corresponds to what is usually meant by "inheritance". The other two forms are much less frequently used. If the access specifier is omitted, inheritance is assumed to be private for a class base and public for a struct base. Base classes may be declared as virtual; this is called virtual inheritance. Virtual inheritance ensures that only one instance of a base class exists in the inheritance graph, avoiding some of the ambiguity problems of multiple inheritance.

Multiple inheritance is another controversial C++ feature. Multiple inheritance allows a class to derive from more than one base class; this can result in a complicated graph of inheritance relationships. For example, a "Flying Cat" class can inherit from both "Cat" and "Flying Mammal". Some other languages, such as Java, accomplish something similar by allowing inheritance of multiple interfaces while restricting the number of base classes to one (interfaces, unlike classes, provide no implementation).

10.3.5 Design of C++

In The Design and Evolution of C++ (ISBN 0-201-54330-3), Bjarne Stroustrup describes some rules that he uses for the design of C++. Knowing the rules helps to understand why C++ is the way it is. The following is a summary of the rules. Much more detail can be found in

The Design and Evolution of C++.
- C++ is designed to be a statically typed, general-purpose language that is as efficient and portable as C.
- C++ is designed to directly and comprehensively support multiple programming styles (procedural programming, data abstraction, object-oriented programming, and generic programming).
- C++ is designed to give the programmer choice, even if this makes it possible for the programmer to choose incorrectly.
- C++ is designed to be as compatible with C as possible, therefore providing a smooth transition from C.
- C++ avoids features that are platform specific or not general purpose.
- C++ does not incur overhead for features that are not used.
- C++ is designed to function without a sophisticated programming environment.

Please refer to the in depth book on C++ Internals by Stanley B. Lippman (he worked on implementing/maintaining C-front the original C++ implementation at Bell Labs). "Inside the C++ Object Model" documents how the C++ compiler converts your program statements into an in-memory layout.

10.3.6 History of C++

Stroustrup began work on C with Classes in 1979. The idea of creating a new language originated from Stroustrup's experience programming for his Ph.D. thesis. Stroustrup found that Simula had features that were very helpful for large software development but was too slow for practical uses, while BCPL was fast but too low level and unsuitable for large software development. When Stroustrup started working in Bell Labs, he had the problem of analyzing the UNIX kernel with respect to distributed computing. Remembering his Ph.D. experience, Stroustrup set out to enhance the C language with Simula-like features. C was chosen because it is general-purpose, fast, and portable. At first, class (with data encapsulation), derived class, strong type checking, inlining, and default argument were features added to C. The first commercial release occurred in October 1985.

In 1983, the name of the language was changed from C with Classes to C++. New features that were added to the language included virtual functions, function name and operator overloading, references, constants, user-controlled free-store memory control, improved type checking, and new comment style (//). In 1985, the first edition of The C++ Programming Language was released, providing an important reference to the language, as there was not yet an official standard. In 1989, Release 2.0 of C++ was released. New features included multiple inheritance, abstract classes, static member functions, const member functions, and protected members. In 1990, The Annotated C++ Reference Manual was released and provided the basis for the future standard. Late addition of features included templates, exceptions, namespaces, new casts, and a Boolean type.

As the C++ language evolved, a standard library also evolved with it. The first addition to the C++ standard library was the stream I/O library which provided facilities to replace the traditional C functions such as printf and scanf. Later, among the most significant additions to the standard library, was the Standard Template Library.

After years of work, a joint ANSI-ISO committee standardized C++ in 1998 (ISO/IEC 14882:1998). For some years after the official release of the standard in 1998, the committee processed defect reports, and published a corrected version of the C++ standard in 2003.

No one owns the C++ language; it is royalty-free. The standard document itself is, however, not available for free.

1. Future Development

C++ continues to evolve to meet future requirements. One group in particular works to make the most of C++ in its current form and advise the C++ standards committee which features work well and which need improving: Boost.org. Current work indicates that C++ will capitalize on its multi-paradigm nature more and more. The work at Boost.org, for example, is greatly expanding C++'s functional and metaprogramming capabilities. The C++ standard does not cover implementation of name decoration, exception handling, and other implementation-specific features, making object code produced by different compilers incompatible; there are, however, 3rd-party standards for particular machines or OSs which attempt to standardise compilers on those platforms, for example C++ compilers still struggle to support the entire C++ standard, especially in the area of templates-a part of the language that was more-or-less entirely conceived by the standards committee. One particular point of contention is the export keyword, intended to allow template definitions to be separated from their declarations. The first compiler to implement export was Comeau C++, in early 2003 (5 years after the release of the standard); in 2004, beta compiler of Borland C++ Builder X was also released with export. Both of these compilers are based on the EDG C++ frontend. It should also be noted that many C++ books provide example code for implementing the keyword export (Ivor Horton's Beginning ANSI C++, pg. 827) which will not compile, but there is no reference to the problem with the keyword export mentioned. Other compilers such as Microsoft Visual C++ and GCC do not support it at all. Herb Sutter, secretary of the C++ standards committee, has recommended that export be removed from future versions of the C++ standard \, but finally the decision was made to leave it in the C++ standard.

Other template issues include constructions such as partial template specialisation, which was poorly supported for several years after the C++ standard was released.

2. History of the Name "C++"

This name is credited to Rick Mascitti (mid-1983) and was first used in December 1983. Earlier, during the research period, the developing language had been referred to as "C with Classes". The final name stems from C's "++" operator (which increments the value of a variable) and a common naming convention of using "+" to indicate an enhanced computer program, for example: "Wikipedia+". According to Stroustrup: "the name signifies the evolutionary nature of the changes from C". C+ was the name of an earlier, unrelated programming language.

Some C programmers have noted that if the statements x=3; and y=x++; are executed, then x==4 and y==3; x is incremented after its value is assigned to y. However, if the second statement is y=++x;, then y=4 and x=4. Following such reasoning, a more proper name for C++ might actually be ++C. However, c++ and ++c both increment c, and, on its own line, the form c++ is more common than ++c. However, the introduction of C++ did not change the C language itself, so an even more accurate name might be "C+1".

10.3.7 C++ is not a Superset of C

While most C code is also valid C++, C++ does not form a superset of C: there exists valid C code that is not valid C++. This is in contrast to Objective-C, another extension of C to support object-oriented programming, which is a superset of C.

Furthermore, code that is both valid C and valid C++ may produce different results depending upon whether it is compiled as C or C++. For example, the following program prints "C" when compiled with a C compiler but prints "C++" when compiled with a C++ compiler. This is because in C the type of a character literal (such as 'a') is int, whereas in C++ it is char.

```
#include <stdio.h>
int main()
{
  printf("%s\n", (sizeof('a') ==sizeof(char)) ? "C++" : "C");
  return 0;
}
```

There are other differences as well. For example, C++ forbids calling the "main" function from within the program, whereas this is legal in C. In addition, C++ is much stricter about various features; for example, it lacks implicit type conversion between unrelated pointer types and does not allow a function to be used that has not yet been declared.

A common portability issue from C to C++ are the numerous additional keywords that C++ introduced. This makes C code that uses them as identifiers illegal in C++. For example:

```
struct template {
    int new;
    struct class    class;
};
```
is legal C code, but is rejected by a C++ compiler, since the keywords "template", "new" and "class" are not appropriate in the corresponding places.

Chapter 11
Artificial Intelligence

Artificial intelligence, also known as machine intelligence, is defined as intelligence exhibited by any manufactured (ie. artificial) system. The term is often applied to general purpose computers, and also in the field of scientific investigation into the theory and practical application of AI.

11.1 Overview

The concept of "artificial intelligence" is something which can be considered in two parts: "what is the nature of artifice" and "what is intelligence"? The first question is relatively easy to answer, although it also necessarily leads to an examination of what it is possible to manufacture. For example, the limitations of certain types of systems, such as classical computational systems, or of available manufacturing processes, or of human intellect, may all place constraints on what can be manufactured.

The second question raises fundamental ontological issues of consciousness and self and mind (including the unconscious mind). It also raises questions about the nature of intelligence as displayed by humans, as intelligent behavior in humans is complex and often difficult to study or understand. Study of animals and artificial systems which are not simply models of what already exists are also considered highly relevant.

Several distinct types of artificial intelligence are discussed below, along with divisions, history, proponents, opponents, and applied research in the field. Lastly, references to fictional and non-fictional descriptions of AI are provided.

11.2 Strong AI and Weak AI

One popular and early definition of artificial intelligence research, put forth by John McCarthy at the Dartmouth Conference in 1956, is "making a machine behave in ways that would be called intelligent if a human were so behaving.", repeating the claim put forth by Alan Turing in "Computing machinery and intelligence" (Mind, October 1950). However this definition seems to ignore the possibility of strong AI (see below). Another definition of artificial intelligence is intelligence arising from an artificial device. Most definitions could be categorized as concerning either systems that think like humans, systems that act like humans, systems that think rationally or systems that act rationally.

11.2.1 Strong Artificial Intelligence

Strong artificial intelligence research deals with the creation of some form of computer-based artificial intelligence that can truly reason and solve problems; a strong form of AI is said to be sentient, or self-aware. In theory, there are two types of strong AI:

Human-like AI, in which the computer program thinks and reasons much like a human mind.

Non-human-like AI, in which the computer program develops a totally non-human sentience, and a non-human way of thinking and reasoning.

11.2.2 Weak Artificial Intelligence

Weak artificial intelligence research deals with the creation of some form of computer-based artificial intelligence that can reason and solve problems only in a limited domain; such a machine would, in some ways, act as if it were intelligent, but it would not possess true intelligence or sentience. The classical test for such abilities is the Turing test.

There are several fields of weak AI, one of which is natural language. Many weak AI fields have specialized software or programming languages created for them. For example, the "most-human" natural language chatterbot A.L.I.C.E. uses a programming language AIML that is specific to its program, and the various clones, named Alicebots. Jabberwacky is a little closer to strong AI, since it learns how to converse from the ground up based solely on user interactions.

To date, much of the work in this field has been done with computer simulations of intelligence based on predefined sets of rules. Very little progress has been made in strong AI. Depending on how one defines one's goals, a moderate amount of progress has been made in weak AI.

When viewed with a moderate dose of cynicism, weak artificial intelligence can be viewed as "the set of computer science problems without good solutions at this point." Once a sub-discipline results in useful work, it is carved out of artificial intelligence and given its own name. Examples of this are pattern recognition, image processing, neural networks, natural language processing, robotics and game theory. While the roots of each of these disciplines is firmly established as having been part of artificial intelligence, they are now thought of as somewhat separate.

11.2.3 Philosophical Criticism and Support of Strong AI

The term "Strong AI" was originally coined by John Searle and was applied to digital computers and other information processing machines. Searle defined strong AI:

"According to strong AI, the computer is not merely a tool in the study of the mind; rather, the appropriately programmed computer really is a mind" (J. Searle in Minds Brains and Programs. The Behavioral and Brain Sciences, vol. 3, 1980).

Searle and most others involved in this debate are addressing the problem of whether a machine that works solely through the transformation of encoded data could be a mind, not the wider issue of Monism versus Dualism (ie: whether a machine of any type, including biological machines, could contain a mind).

Searle states in his Chinese Room argument that information processors carry encoded data which describe other things. The encoded data itself is meaningless without a cross reference to the things it describes. This leads Searle to point out that there is no meaning or understanding in an information processor itself. As a result Searle claims to demonstrate that even a machine that passed the Turing test would not necessarily be conscious in the human sense.

Some philosophers hold that if Weak AI is accepted as possible then Strong AI must also be possible. Daniel C. Dennett argues in Consciousness Explained that if there is no magic spark or soul, then Man is just a machine, and he asks why the Man-machine should have a privileged position over all other possible machines when it comes to intelligence or "mind". Simon Blackburn in his introduction to philosophy, Think, points out that you might appear intelligent but there is no way of telling if that intelligence is real (ie: a "mind"). However, if the discussion is limited to strong AI rather than artificial consciousness it may be possible to identify features of human minds that do not occur in information processing computers.

Strong AI seems to involve the following assumptions about the mind and brain:

the mind is software, a finite state machine so the Church-Turing thesis applies to it.

the brain is purely hardware (i.e. only follows the rules of a classical computer).

The first assumption is particularly problematic because of the old adage that any computer is just a glorified abacus. It is indeed possible to construct any type of information processor out of balls and wood, although such a device would be very slow and prone to failure, it would be able to do anything that a modern computer can do. This means that the proposition that information processors can be minds is equivalent to proposing that minds can exist as devices made of rolling balls in wooden channels.

Some (including Roger Penrose) attack the applicability of the Church-Turing thesis directly by drawing attention to the halting problem in which certain types of computation cannot be performed by information systems yet seem to be performed by human minds.

Ultimately the truth of Strong AI depends upon whether information processing machines can include all the properties of minds such as Consciousness. However, Weak AI is independent of the Strong AI problem and there can be no doubt that many of the features of modern computers such as multiplication or database searching might have been considered "intelligent" only a century ago.

11.3 History Development of AI Theory

Much of the (original) focus of artificial intelligence research draws from an experimental approach to psychology, and emphasizes what may be called linguistic intelligence (best exemplified in the Turing test).

Approaches to artificial intelligence that do not focus on linguistic intelligence include robotics and collective intelligence approaches, which focus on active manipulation of an environment, or consensus decision making, and draw from biology and political science when seeking models of how "intelligent" behavior is organized.

Artificial intelligence theory also draws from animal studies, in particular with insects, which are easier to emulate as robots, as well as animals with more complex cognition, including apes, who resemble humans in many ways but have less developed capacities for planning and cognition. AI researchers argue that animals, which are simpler than humans, ought to be considerably easier to mimic. But satisfactory computational models for animal intelligence are not available.

Seminal papers advancing the concept of machine intelligence include A Logical Calculus of the Ideas Immanent in Nervous Activity(1943), by Warren McCulloch and Walter Pitts, and On Computing Machinery and Intelligence(1950), by Alan Turing, and Man-Computer Symbiosis by J.C.R. Licklider.

There were also early papers which denied the possibility of machine intelligence on logical or philosophical grounds such as Minds, Machines and Gdel(1961) by John Lucas.

With the development of practical techniques based on AI research, advocates of AI have argued that opponents of AI have repeatedly changed their position on tasks such as computer chess or speech recognition that were previously regarded as "intelligent" in order to deny the accomplishments of AI. They point out that this moving of the goalposts effectively defines "intelligence" as "whatever humans can do that machines cannot".

John von Neumann anticipated this in 1948 by saying, in response to a comment at a lecture that it was impossible for a machine to think: "You insist that there is something a machine cannot do. If you will tell me precisely what it is that a machine cannot do, then I can always make a machine which will do just that!" . Von Neumann was presumably alluding to the Church-Turing thesis which states that any effective procedure can be simulated by a (generalized) computer.

In 1969 McCarthy and Hayes started the discussion about the frame problem with their essay, "Some Philosophical Problems from the Standpoint of Artificial Intelligence".

11.4 Experimental AI Research

Artificial intelligence began as an experimental field in the 1950s with such pioneers as Allen Newell and Herbert Simon, who founded the first artificial intelligence laboratory at Carnegie-Mellon University, and McCarthy and Marvin Minsky, who founded the MIT AI Lab in 1959. They all attended the aforementioned Dartmouth College summer AI conference in 1956, which was organized by McCarthy, Minsky, Nathan Rochester of IBM and Claude Shannon.

Historically, there are two broad styles of AI research-the "neats" and "scruffies". "Neat", classical or symbolic AI research, in general, involves symbolic manipulation of abstract concepts, and is the methodology used in most expert systems. Parallel to this are the "scruffy", or "connectionist", approaches, of which neural networks are the best-known example, which try to "evolve" intelligence through building systems and then improving them through some automatic process rather than systematically designing something to complete the task. Both approaches appeared very early in AI history. Throughout the 1960s and 1970s scruffy approaches were pushed to the background, but interest was regained in the 1980s when the limitations of the "neat" approaches of the time became clearer. However, it has become clear that contemporary methods using both broad approaches have severe limitations.

Artificial intelligence research was very heavily funded in the 1980s by the Defense Advanced Research Projects Agency in the United States and by the fifth generation computer systems project in Japan. The failure of the work funded at the time to produce immediate results, despite the grandiose promises of some AI practitioners, led to correspondingly large cutbacks in funding by government agencies in the late 1980s, leading to a general downturn in activity in the field known as AI winter. Over the following decade, many AI researchers moved into related areas with more modest goals such as machine learning, robotics, and computer vision, though research in pure AI continued at reduced levels.

Practical applications of AI techniques

Whilst progress towards the ultimate goal of human-like intelligence has been slow, many spinoffs have come in the process. Notable examples include the languages LISP and Prolog, which were invented for AI research but are now used for non-AI tasks. Hacker culture first sprang from AI laboratories, in particular the MIT AI Lab, home at various times to such luminaries as McCarthy, Minsky, Seymour Papert (who developed Logo there), Terry Winograd (who abandoned AI after developing SHRDLU).

Many other useful systems have been built using technologies that at least once were active areas of AI research. Some examples include:

(1) Deep Blue, a chess-playing computer, beat Garry Kasparov in a famous match in

1997.

(2) InfoTame, a text analysis search engine developed by the KGB for automatically sorting millions of pages of communications intercepts.

(3) Fuzzy logic, a technique for reasoning under uncertainty, has been widely used in industrial control systems.

(4) Expert systems are being used to some extent industrially.

(5) Machine translation systems such as SYSTRAN are widely used, although results are not yet comparable with human translators.

(6) Neural networks have been used for a wide variety of tasks, from intrusion detection systems to computer games.

(7) Optical character recognition systems can translate arbitrary typewritten European script into text.

(8) Handwriting recognition is used in millions of personal digital assistants.

(9) Speech recognition is commercially available and is widely deployed.

(10) Computer algebra systems, such as Mathematica and Macsyma, are commonplace.

(11) Machine vision systems are used in many industrial applications ranging from hardware verification to security systems.

(12) AI Planning methods were used to automatically plan the deployment of US forces during Gulf War I. This task would have cost months of time and millions of dollars to perform manually, and DARPA stated that the money saved on this single application was more than their total expenditure on AI research over the last 50 years.

The vision of artificial intelligence replacing human professional judgment has arisen many times in the history of the field, in science fiction and today in some specialized areas where "expert systems" are used to augment or to replace professional judgment in some areas of engineering and of medicine.

Keywords

Artificial intelligence	人工智能
term	术语
investigation	调研
examination	研究
limitations	局限性
ontological	本体论
consciousness	意识
proponent	支持者
opponent	反对者
fictional	虚拟的
rationally	理性地

sentient	有知觉的
possess	具备
specialized	专有的
chatterbot	聊天机器人
interactions	互动
simulations	仿真
predefined	预先定义的
cynicism	玩世不恭
sub-discipline	分支科学
pattern recognition	模式识别
image processing	图形处理
neural networks	神经网络
natural language processing	自然语言处理
robotics	机器人理论
game theory	博弈论
biological	生物的
demonstrate	表明
philosophers	哲学家
privileged	拥有特权的
finite	有限
adage	谚语
glorified abacus	名字好听的算盘
proposition	主张
laboratory	实验室
manipulation	操作
abstract	抽象
scruffy	不简洁的
defense	防御
grandiose	雄伟
cutbacks	削减
ultimate	最终的

11.5 Exercise 11

Multiple or single choices.

1. To answer the question about "what is the nature of artifice" will leads to an examination of what it is possible to manufacture such as _____ which may constraints on what can be

manufactured.
   A. the limitations of certain types of systems
   B. the limitations of available manufacturing processes
   C. the limitations of manufacturing material
   D. the limitations of human intellect

2. Most definitions could be categorized as concerning either ____.
   A. systems that think like humans
   B. systems that act like humans
   C. systems that think rationally
   D. systems that act rationally

3. About strong artificial intelligence, we are aware of ____.
   A. a strong form of AI is said to be sentient, or self-aware
   B. there are only one types of strong AI in theory
   C. the computer program thinks and reasons much like a human mind
   D. the computer program thinks and reasons less like a human mind

4. About weak artificial intelligence, we are aware of ____.
   A. it act as if it were intelligent, but it would not possess true intelligence or sentience
   B. it act and think like a human being
   C. the classical test for such abilities is the Turing test
   D. one of the research field is natural language

5. Which of the following sub-discipline are carved out of weak artificial intelligence ____.
   A. pattern recognition
   B. image processing
   C. neural networks
   D. natural language processing

True or False.

1. The classical test for a machine would, in some ways, act as if it were intelligent, but it would not possess true intelligence or sentience, is the Turing test.

2. Searle address that the problem of whether a machine that works solely through the transformation of encoded data could be a mind.

3. The truth of Strong AI depends upon whether information processing machines can include all the properties of minds such as Consciousness.

4. Deep Blue, a chess-playing computer, beat Garry Kasparov in a famous match in 1997 is one of the famous event in the history of the AI development.

5. Strong AI seems to assume that the mind is purely hardware.

11.6 Further Reading: Alan Turing

Alan Turing (see Figure 11.1) is often considered the father of modern computer science.

Alan Mathison Turing(June 23, 1912—June 7, 1954, see Figure 11.1) was a British mathematician, logician, cryptographer, and is often considered a father of modern computer science. With the Turing Test, he made a significant and characteristically provocative contribution to the debate regarding synthetic consciousness: whether it will ever be possible to say that a machine is conscious and can think. He provided an influential formalisation of the concept of algorithm and computation with the Turing machine, formulating the now widely accepted "Turing" version of the Church-Turing thesis, namely that any practical computing model has either the equivalent or a subset of the capabilities of a Turing machine.

Figure 11.1 Alan Turing

Turing's contributions during World War II were never publicly acknowledged during his lifetime because his work was classified. At Bletchley Park he was a pivotal player in breaking German cyphers, becoming the head of Hut 8, the group tasked with breaking Naval Enigma.

After the war, he designed one of the earliest electronic programmable digital computers at the National Physical Laboratory and actually built another early machine at the University of Manchester. The Turing Award was created in his honour.

Turing was later tried and convicted for "gross indecency and sexual perversion" because of his homosexuality. He began a government-mandated hormonal treatment programme and died shortly thereafter in what is generally considered to have been a suicide.

11.6.1 Childhood and Youth

Turing was conceived in 1911 in Chatrapur, India. His father, Julius Mathison Turing, was a member of the Indian civil service. Julius and wife Ethel (née Stoney) wanted Alan to be brought up in Britain, so they returned to Paddington, London. His father's civil service commission was still active, and during Turing's childhood years his parents travelled between

Guildford, England and India, leaving their two sons to stay with friends in England, rather than risk their health in the British colony. Very early in life, Turing showed signs of the genius he was to display more prominently later. He is said to have taught himself to read in three weeks, and to have shown an early affinity for numbers and puzzles.

His parents enrolled him at St. Michael's, a day school, at six years of age. The headmistress recognized his genius early on, as did many of his subsequent educators. In 1926, at the age of 14, he went on to the Sherborne boarding school in Dorset. His first day of term coincided with a general strike in England, and so determined was he to attend his first day that he rode his bike unaccompanied over sixty miles from Southampton to school, stopping overnight at an inn—a feat reported in the local press.

Turing's natural inclination toward mathematics and science did not earn him respect with the teachers at Sherborne, a famous and expensive public school (a British private school with charitable status), whose definition of education placed more emphasis on the classics. His headmaster wrote to his parents: "I hope he will not fall between two schools. If he is to stay at Public School, he must aim at becoming educated. If he is to be solely a Scientific Specialist, he is wasting his time at a Public School," (Alan Turing: The Enigma by Andrew Hodges, Walker Publishing Company edition (2000), p. 26).

But despite this, Turing continued to show remarkable ability in the studies he loved, solving advanced problems in 1927 without having even studied elementary calculus. In 1928, aged sixteen, Turing encountered Albert Einstein's work; not only did he grasp it, but he extrapolated Einstein's questioning of Newton's laws of motion from a text in which this was never made explicit.

Turing's hopes and ambitions at school were raised by his strong feelings for his friend Christopher Morcom, with whom he fell in love, though the feeling was not reciprocated. Morcom died only a few weeks into their last term at Sherborne, from complications of bovine tuberculosis, contracted after drinking infected cow's milk as a boy. Turing was heart-broken.

The computer room at King's is now named after Turing, who became a student there in 1931 and a Fellow in 1935.

11.6.2 College and his Work on Computability

Due to his unwillingness to work as hard on his classical studies as on science and mathematics, Turing failed to win a scholarship to Trinity College, Cambridge, and went on to the college of his second choice, King's College, Cambridge. He was an undergraduate from 1931 to 1934, graduating with a distinguished degree, and, in 1935 he was elected a Fellow at King's on the strength of a dissertation on the Gaussian error function.

Alan Turing, on the steps of the bus, with members of the Walton Athletic Club, 1946.

In his momentous paper "On Computable Numbers, with an Application to the Entscheidungsproblem" (submitted on May 28, 1936), Turing reformulated Kurt Gdel's 1931 results on the limits of proof and computation, substituting Gdel's universal arithmetics-based formal language by what are now called Turing machines, formal and simple devices. He proved that such a machine would be capable of performing any conceivable mathematical problem if it were representable as an algorithm, even if no actual Turing machine would be likely to have practical applications, being much slower than alternatives. Turing machines are to this day the central object of study in theory of computation. He went on to prove that there was no solution to the Entscheidungsproblem by first showing that the halting problem for Turing machines is uncomputable: it is not possible to algorithmically decide whether a given Turing machine will ever halt. While his proof was published subsequent to Alonzo Church's equivalent proof in respect to his lambda calculus, Turing's work is considerably more accessible and intuitive. It was also novel in its notion of a "Universal (Turing) Machine," the idea that such a machine could perform the tasks of any other machine. The paper also introduces the notion of definable numbers.

Most of 1937 and 1938 he spent at Princeton University, studying under Alonzo Church. In 1938 he obtained his Ph. D. from Princeton; his dissertation introduced the notion of hypercomputation where Turing machines are augmented with so-called oracles, allowing a study of problems that cannot be solved algorithmically.

Back in Cambridge in 1939, he attended lectures by Ludwig Wittgenstein about the foundations of mathematics. The two argued and disagreed vehemently, with Turing defending formalism and Wittgenstein arguing that mathematics is overvalued and does not discover any absolute truths.

11.6.3 Cryptanalysis (Code Breaking)

During World War II he was a major participant in the efforts at Bletchley Park to break German ciphers. Turing's codebreaking work was kept secret until the 1970s; not even his close friends knew about it. He contributed several mathematical insights into breaking both the Enigma machine and the Lorenz SZ 40/42 (a teletype cipher attachment codenamed "Tunny" by the British).

Two cottages in the stable yard at Bletchley Park. Turing worked here from 1939—1940 until he moved to Hut 8.

Turing realized that it was not necessary to test all the possible combinations to crack the Enigma machine. He proved that it was possible to test for the correct settings of the rotors (approximately one million combinations) without having to consider the settings of the plugboard (approximately 157 million combinations). Whilst still a formidable task, one million combinations was achievable using an electromechanical machine—the bombe, named after the Polish-designed bomba—which could be used to eliminate large numbers of candidate Enigma settings. For each possible setting, a chain of logical deductions was implemented electrically, and it was possible to detect when a contradiction had occurred and rule out that setting. Turing's bombe, with an enhancement suggested by mathematician Gordon Welchman, was the primary tool used by British and American codebreakers to read Enigma traffic, with over 200 bombes in operation by the end of the war. The design and production of the machine itself was undertaken by Harold Keen of the British Tabulating Machine company. For a time, Turing was head of Hut 8, the section responsible for cryptanalysing German Naval signals. Turing also invented the technique of Banburismus to assist in breaking Enigma.

Turing devised some methods for attacking Tunny, termed Turingismus or Turingery, although other methods were also used. To assist in the codebreaking, the first digital programmable electronic computer was developed, Colossus. Turing, however, was not directly involved—Colossus was designed and built at the Post Office Research Station at Dollis Hill by a team led by Thomas Flowers in 1943.

In the later part of the war, Turing undertook (with engineer Donald Bayley) the design of a portable machine codenamed Delilah to allow secure voice communications, teaching himself electronic theory at the same time. Intended for different applications, Delilah lacked the ability to be used over long-distance radio transmissions. Delilah was completed too late to be used in the war. While Turing demonstrated it to officials by encoding/decoding a recording of a Winston Churchill speech, it was not adopted for use.

11.6.4　Work on Early Computers and the Turing Test

Turing achieved world-class Marathon standards. His best time of 2 hours, 46 minutes, 3 seconds, was only 11 minutes slower than the winner in the 1948 Olympic Games.

From 1945 to 1947 he was at the National Physical Laboratory, where he worked on the design of ACE (Automatic Computing Engine). He presented a paper on February 19, 1946, which was the first complete design of a stored-program computer. Although he succeeded in designing the ACE, there were delays in starting the project and he became disillusioned. In late 1947 he returned to Cambridge for a "sabbatical" year. While he was at Cambridge work on building the ACE stopped before it was ever begun. In 1949 he became deputy director of the computing laboratory at the University of Manchester, and worked on software for one of the earliest true computers—the Manchester Mark I. During this time he continued to do more abstract work, and in "Computing machinery and intelligence" (Mind, October 1950), Turing addressed the problem of artificial intelligence, and proposed an experiment now known as the Turing test, an attempt to define a standard for a machine to be called "sentient". In 1948, Turing, working with his former undergraduate colleague, D.G. Champernowne, began writing a chess program for a computer that did not yet exist. In 1952, lacking a computer powerful enough to execute the program, Turing played a game in which he simulated the computer, taking about half an hour per move. The game was recorded; the program lost to a colleague of Turing, however, it is said that the programme won a game against Champernowne's wife.

11.6.5 Work on Pattern Formation and Mathematical Biology

Turing worked from 1952 until his death in 1954 on mathematical biology, specifically morphogenesis. He published one paper on the subject called "The Chemical Basis of Morphogenesis" in 1952. His central interest in the field was understanding Fibonacci phyllotaxis, the existence of Fibonacci numbers in plant structures. He used reaction-diffusion equations which are now central to the field of pattern formation. Later papers went unpublished until 1992 when Collected Works of A.M. Turing was published.

11.6.6 Prosecution for Homosexuality and Turing's Death

In the book, Zeroes and Ones, author Sadie Plant speculates that the rainbow Apple logo with a bite taken out of it was a homage to Turing. This seems to be an urban legend as the Apple logo was designed in 1976, two years before Gilbert Baker's rainbow pride flag.

Prosecution of Turing for his homosexuality crippled his career. In 1952, his male lover helped an accomplice to break into Turing's house and commit larceny. Turing went to the police to report the crime. As a result of the police investigation, Turing was said to have had a sexual relationship with a 19-year-old man, and Turing was charged with "gross indecency and sexual perversion." He unapologetically offered no defence, and was convicted. Following the well-publicised trial, he was given a choice between incarceration and libido-reducing hormonal treatment. He chose the oestrogen hormone injections, which lasted for a year, with side effects including the development of breasts.

In 1954, he died of cyanide poisoning, apparently from a cyanide-laced apple he left half-eaten. Most believe that his death was intentional, and the death was ruled a suicide. His mother, however, strenuously argued that the ingestion was accidental due to his careless storage of laboratory chemicals. Friends of his have said that Turing may have killed himself in this ambiguous way quite deliberately, to give his mother some plausible deniability. The possibility of assassination has also been suggested, owing to Turing's involvement in the secret service and the perception of Turing as a security risk due to his homosexuality.

Chapter 12
Machine Learning

Machine learning is a sub-field of computer science that evolved from the study of pattern recognition and computational learning theory in artificial intelligence. In 1959, Arthur Samuel defined machine learning as a "Field of study that gives computers the ability to learn without being explicitly programmed". Machine learning explores the study and construction of algorithms that can learn from and make predictions on data. Such algorithms operate by building a model from example inputs in order to make data-driven predictions or decisions, rather than following strictly static program instructions.

Machine learning is closely related to and often overlaps with computational statistics; a discipline which also focuses in prediction-making through the use of computers. It has strong ties to mathematical optimization, which delivers methods, theory and application domains to the field. Machine learning is employed in a range of computing tasks where designing and programming explicit algorithms is infeasible. Example applications include spam filtering, optical character recognition (OCR), search engines and computer vision. Machine learning is sometimes conflated with data mining, where the latter sub-field focuses more on exploratory data analysis and is known as unsupervised learning.

Within the field of data analytics, machine learning is a method used to devise complex models and algorithms that lend themselves to prediction. These analytical models allow researchers, data scientists, engineers, and analysts to "produce reliable, repeatable decisions and results" and uncover "hidden insights" through learning from historical relationships and trends in the data.

Machine learning can be thought of as a set of tools and methods that attempt to infer patterns and extract insight from observations made of the physical world. For example, if you wanted to predict the price of a house based on the number of rooms, number of bathrooms, square footage, and lot size, you can use a simple machine learning algorithm (e.g. linear regression) to learn from an existing real estate sales data set where the price of each house is known, and then based on what you've learned, you can predict the price of other houses where the price is unknown.

A support vector machine is a classifier that divides its input space into two regions, separated by a linear boundary. Here, it has learned to distinguish black and white circles.

An example of SVM is shown in Figure 12.1.

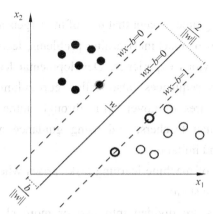

Figure 12.1   an example of SVM

12.1   Overview

Tom M. Mitchell provided a widely quoted, more formal definition: "A computer program is said to learn from experience E with respect to some class of tasks T and performance measure P if its performance at tasks in T, as measured by P, improves with experience E". This definition is notable for its defining machine learning in fundamentally operational rather than cognitive terms, thus following Alan Turing's proposal in his paper "Computing Machinery and Intelligence" that the question "Can machines think?" be replaced with the question "Can machines do what we (as thinking entities) can do?"

12.1.1   Types of problems and tasks

Machine learning tasks are typically classified into three broad categories, depending on the nature of the learning "signal" or "feedback" available to a learning system. These are:
- Supervised learning: The computer is presented with example inputs and their desired outputs, given by a "teacher", and the goal is to learn a general rule that maps inputs to outputs.
- Unsupervised learning: No labels are given to the learning algorithm, leaving it on its own to find structure in its input. Unsupervised learning can be a goal in itself (discovering hidden patterns in data) or a means towards an end (feature learning).
- Reinforcement learning: A computer program interacts with a dynamic environment in which it must perform a certain goal (such as driving a vehicle), without a teacher explicitly telling it whether it has come close to its goal. Another example is learning to play a game by playing against an opponent.

Between supervised and unsupervised learning is semi-supervised learning, where the teacher gives an incomplete training signal: a training set with some (often many) of the target outputs missing. Transduction is a special case of this principle where the entire set of problem instances is known at learning time, except that part of the targets are missing.

Among other categories of machine learning problems, learning to learn learns its own inductive bias based on previous experience. Developmental learning, elaborated for robot learning, generates its own sequences (also called curriculum) of learning situations to cumulatively acquire repertoires of novel skills through autonomous self-exploration and social interaction with human teachers, and using guidance mechanisms such as active learning, motor synergies, and imitation.

Another categorization of machine learning tasks arises when one considers the desired output of a machine-learned system:

In classification, inputs are divided into two or more classes, and the learner must produce a model that assigns unseen inputs to one (or multi-label classification) or more of these classes. This is typically tackled in a supervised way. Spam filtering is an example of classification, where the inputs are e-mail (or other) messages and the classes are "spam" and "not spam".

In regression, also a supervised problem, the outputs are continuous rather than discrete.

In clustering, a set of inputs is to be divided into groups. Unlike in classification, the groups are not known beforehand, making this typically an unsupervised task.

Density estimation finds the distribution of inputs in some space.

Dimensionality reduction simplifies inputs by mapping them into a lower-dimensional space. Topic modeling is a related problem, where a program is given a list of human language documents and is tasked to find out which documents cover similar topics.

12.1.2 History and relationships to other fields

As a scientific endeavor, machine learning grew out of the quest for artificial intelligence. Already in the early days of AI as an academic discipline, some researchers were interested in having machines learn from data. They attempted to approach the problem with various symbolic methods, as well as what were then termed "neural networks"; these were mostly perceptions and other models that were later found to be reinventions of the generalized linear models of statistics. Probabilistic reasoning was also employed, especially in automated medical diagnosis.

However, an increasing emphasis on the logical, knowledge-based approach caused a rift between AI and machine learning. Probabilistic systems were plagued by theoretical and practical problems of data acquisition and representation. By 1980, expert systems had come to dominate AI, and statistics was out of favor. Work on symbolic/knowledge-based learning did continue within AI, leading to inductive logic programming, but the more statistical line of research was now outside the field of AI proper, in pattern recognition and information retrieval. Neural networks research had been abandoned by AI and computer science around the same time. This line, too, was continued outside the AI/CS field, as "connectionism", by researchers from other disciplines including Hopfield, Rumelhart and Hinton. Their main success came in the mid-1980s with the reinvention of back propagation.

Machine learning, reorganized as a separate field, started to flourish in the 1990s. The field changed its goal from achieving artificial intelligence to tackling solvable problems of a practical nature. It shifted focus away from the symbolic approaches it had inherited from AI, and toward methods and models borrowed from statistics and probability theory. It also benefited from the increasing availability of digitized information, and the possibility to distribute that via the Internet.

Machine learning and data mining often employ the same methods and overlap significantly. They can be roughly distinguished as follows:

Machine learning focuses on prediction, based on known properties learned from the training data.

Data mining focuses on the discovery of (previously) unknown properties in the data. This is the analysis step of Knowledge Discovery in Databases.

The two areas overlap in many ways: data mining uses many machine learning methods, but often with a slightly different goal in mind. On the other hand, machine learning also employs data mining methods as "unsupervised learning" or as a preprocessing step to improve learner accuracy. Much of the confusion between these two research communities (which do often have separate conferences and separate journals, ECML PKDD being a major exception) comes from the basic assumptions they work with: in machine learning, performance is usually evaluated with respect to the ability to reproduce known knowledge, while in Knowledge Discovery and Data Mining (KDD) the key task is the discovery of previously unknown knowledge. Evaluated with respect to known knowledge, an uninformed (unsupervised) method will easily be outperformed by supervised methods, while in a typical KDD task, supervised methods cannot be used due to the unavailability of training data.

Machine learning also has intimate ties to optimization: many learning problems are formulated as minimization of some loss function on a training set of examples. Loss functions express the discrepancy between the predictions of the model being trained and the actual problem instances (for example, in classification, one wants to assign a label to instances, and models are trained to correctly predict the pre-assigned labels of a set examples). The difference between the two fields arises from the goal of generalization: while optimization algorithms can minimize the loss on a training set, machine learning is concerned with minimizing the loss on unseen samples.

12.1.3 Theory

A core objective of a learner is to generalize from its experience. Generalization in this context is the ability of a learning machine to perform accurately on new, unseen examples/tasks after having experienced a learning data set. The training examples come from some generally unknown probability distribution (considered representative of the space of occurrences) and the learner has to build a general model about this space that enables it to produce sufficiently accurate predictions in new cases.

The computational analysis of machine learning algorithms and their performance is a branch of theoretical computer science known as computational learning theory. Because training sets are finite and the future is uncertain, learning theory usually does not yield guarantees of the performance of algorithms. Instead, probabilistic bounds on the performance are quite common. The bias-variance decomposition is one way to quantify generalization error.

How well a model trained with existing examples and predicts the output properly for unknown instances called generalization. For best generalization, complexity of the hypothesis should match complexity of the function underlying the data. If hypothesis is less complex than the function, we've underfitted. Then, we increase the complexity, the training error decreases. But if we've too complex hypothesis, we've overfitted. After then, we should find the hypothesis that has the minimum training error.

In addition to performance bounds, computational learning theorists study the time complexity and feasibility of learning. In computational learning theory, a computation is considered feasible if it can be done in polynomial time. There are two kinds of time complexity results. Positive results show that a certain class of functions can be learned in polynomial time. Negative results show that certain classes cannot be learned in polynomial time.

There are many similarities between machine learning theory and statistical inference, although they use different terms.

12.1.4 Approaches

1. Decision tree learning

Decision tree learning uses a decision tree as a predictive model, which maps observations about an item to conclusions about the item's target value.

2. Association rule learning

Association rule learning is a method for discovering interesting relations between variables in large databases.

3. Artificial neural networks

An artificial neural network (ANN) learning algorithm, usually called "neural network" (NN), is a learning algorithm that is inspired by the structure and functional aspects of biological neural networks. Computations are structured in terms of an interconnected group of artificial neurons, processing information using a connectionist approach to computation. Modern neural networks are non-linear statistical data modeling tools. They are usually used to model complex relationships between inputs and outputs, to find patterns in data, or to capture the statistical structure in an unknown joint probability distribution between observed variables.

4. Inductive logic programming

Inductive logic programming (ILP) is an approach to rule learning using, logic programming as a uniform representation for input examples, background knowledge, and hypotheses. Given an encoding of the known background knowledge and a set of examples represented as a logical database of facts, an ILP system will derive a hypothesized logic program that entails all positive and no negative examples. Inductive programming is a related field that considers any kind of programming languages for representing hypotheses (and not only logic programming), such as functional programs.

5. Support vector machines

Support vector machines (SVMs) are a set of related supervised learning methods used for classification and regression. Given a set of training examples, each marked as belonging to one of two categories, an SVM training algorithm builds a model that predicts whether a new example falls into one category or the other.

6. Clustering

Cluster analysis is the assignment of a set of observations into subsets (called clusters) so that observations within the same cluster are similar according to some predestinated criterion or criteria, while observations drawn from different clusters are dissimilar. Different clustering techniques make different assumptions on the structure of the data, often defined by some similarity metric and evaluated for example by internal compactness (similarity between members of the same cluster) and separation between different clusters. Other methods are based on estimated density and graph connectivity. Clustering is a method of unsupervised learning, and a common technique for statistical data analysis.

7. Bayesian networks

A Bayesian network, belief network or directed acyclic graphical model is a probabilistic graphical model that represents a set of random variables and their conditional independencies via a directed acyclic graph (DAG). For example, a Bayesian network could represent the probabilistic relationships between diseases and symptoms. Given symptoms, the network can be used to compute the probabilities of the presence of various diseases. Efficient algorithms exist that perform inference and learning.

8. Reinforcement learning

Reinforcement learning is concerned with how an agent ought to take actions in an environment so as to maximize some notion of long-term reward. Reinforcement learning algorithms attempt to find a policy that maps states of the world to the actions the agent ought to take in those states. Reinforcement learning differs from the supervised learning problem in that correct input/output pairs are never presented, nor sub-optimal actions explicitly corrected.

9. Representation learning

Several learning algorithms, mostly unsupervised learning algorithms, aim at discovering better representations of the inputs provided during training. Classical examples include principal components analysis and cluster analysis. Representation learning algorithms often attempt to preserve the information in their input but transform it in a way that makes it useful, often as a pre-processing step before performing classification or predictions, allowing to reconstruct the inputs coming from the unknown data generating distribution, while not being necessarily faithful for configurations that are implausible under that distribution.

Manifold learning algorithms attempt to do so under the constraint that the learned representation is low-dimensional. Sparse coding algorithms attempt to do so under the constraint that the learned representation is sparse (has many zeros). Multi-linear subspace learning algorithms aim to learn low-dimensional representations directly from tensor representations for multidimensional data, without reshaping them into (high-dimensional) vectors. Deep learning algorithms discover multiple levels of representation, or a hierarchy of features, with higher-level, more abstract features defined in terms of (or generating) lower-level features. It has been argued that an intelligent machine is one that learns a representation that disentangles the underlying factors of variation that explain the observed data.

10. Similarity and metric learning

In this problem, the learning machine is given pairs of examples that are considered similar and pairs of less similar objects. It then needs to learn a similarity function (or a distance metric function) that can predict if new objects are similar. It is sometimes used in Recommendation systems.

11. Sparse dictionary learning

In this method, a datum is represented as a linear combination of basis functions, and the coefficients are assumed to be sparse. Let x be a d-dimensional datum, D be a d by n matrix, where each column of D represents a basis function. r is the coefficient to represent x using D. Mathematically, sparse dictionary learning means solving x≈Dr where r is sparse. Generally speaking, n is assumed to be larger than d to allow the freedom for a sparse representation.

Learning a dictionary along with sparse representations is strongly NP-hard and also difficult to solve approximately. A popular heuristic method for sparse dictionary learning is K-SVD.

Sparse dictionary learning has been applied in several contexts. In classification, the problem is to determine which classes a previously unseen datum belongs to. Suppose a dictionary for each class has already been built. Then a new datum is associated with the class such that it's best sparsely represented by the corresponding dictionary. Sparse dictionary learning has also been applied in image de-noising. The key idea is that a clean image patch can be sparsely represented by an image dictionary, but the noise cannot.

12. Genetic algorithms

A genetic algorithm (GA) is a search heuristic that mimics the process of natural selection, and uses methods such as mutation and crossover to generate new genotype in the hope of finding good solutions to a given problem. In machine learning, genetic algorithms found some uses in the 1980s and 1990s. Vice versa, machine learning techniques have been used to improve the performance of genetic and evolutionary algorithms.

Keywords

Machine learning	机器学习
sub-field	子领域
predictions	预测
overlaps	交叉
ties	联系
mathematical optimization	数学优化
explicit	明确的
infeasible	无法处理的
spam filtering	垃圾邮件过滤
optical character recognition (OCR)	光学文字识别
search engines	搜索引擎
computer vision	计算机视觉
conflate	合并
extract	提取
square footage	占地面积
support vector machine (SVM)	支持向量机
classifier	分类器
cognitive terms	经验性词汇
signal	信号
feedback	反馈
Supervised learning	监督学习
Unsupervised learning	非监督学习
Reinforcement learning	强化学习
semi-supervised learning	半监督学习
dynamic	动态的
inductive bias	学习倾向
mechanisms	机制
maturation	成熟
categorization	分类
multi-label	多标记

tackled	解决
clustering	聚类
regression	回归
beforehand	预先
Density estimation	密度估计
Dimensionality reduction	降维
map	映射
lower-dimensional space	低维空间
scientific endeavor	科学事业
reinventions	再创造
Probabilistic reasoning	概率推理
generalized linear models	广义线性模型
automated medical diagnosis	自动医疗诊断
emphasis	重点
rift	分歧
plagued	困扰
data acquisition	数据采集
underfitted	欠拟合
overfitted	过度拟合
hypothesis	假设
polynomial	多项式
Decision tree learning	决策树学习
Association rule learning	关联学习
Artificial neural networks	人工神经网络
Inductive logic programming	归纳逻辑编程
Bayesian networks	贝叶斯网络
Representation learning	表征学习
Similarity and metric learning	相似性和度量学习
Sparse dictionary learning	稀疏字典学习
Genetic algorithms	遗传算法
genotype	基因型

12.2 Exercise 12

Multiple or single choices.

1. Example applications using the method of machine learning include ____.
    A. spam filtering

B. game theory

C. search engines

D. computer vision

2. Machine learning tasks are typically classified into ____, depending on the nature of the learning "signal" or "feedback" available to a learning system .

   A. Supervised learning

   B. Unsupervised learning

   C. Similarity and metric learning

   D. Reinforcement learning

3. In the process of spam filtering, we may know that ____.

   A. it is typically tackled in a supervised way

   B. it is typically tackled in a unsupervised way

   C. inputs are divided into two or more classes

   D. the classes are "spam" and "not spam"

4. Which of the following options belong to unsupervised learning ____.

   A. regression

   B. classification

   C. clustering

   D. searching

5. The overlap between machine learning and data mining include ____.

   A. data mining uses many machine learning methods

   B. have the same goal in mind

   C. machine learning also employs "unsupervised learning" to improve learner accuracy

   D. they are focus on prediction

True or False.

1. Machine learning evolved from the study of pattern recognition and computational learning theory in artificial intelligence.

2. Machine learning explores the study and construction of algorithms that can learn from and make predictions on data.

3. Machine learning is equivalent to computational statistics.

4. Machine learning tasks are typically classified into two broad categories which are supervised learning and unsupervised learning.

5. Machine learning focuses on what data mining attach importance to.

12.3 Further Reading: Applications for machine learning

12.3.1 Adaptive websites

An adaptive website adjusts the structure, content, or presentation of information in response to measured user interaction with the site, with the objective of optimizing future user interactions. Adaptive websites "are web sites that automatically improve their organization and presentation by learning from their user access patterns." User interaction patterns may be collected directly on the website or may be mined from Web server logs. A model or models are created of user interaction using artificial intelligence and statistical methods. The models are used as the basis for tailoring the website for known and specific patterns of user interaction.

12.3.2 Affective computing

Affective computing is the study and development of systems and devices that can recognize, interpret, process, and simulate human affects. It is an interdisciplinary field spanning computer science, psychology, and cognitive science. While the origins of the field may be traced as far back as to early philosophical enquiries into emotion, the more modern branch of computer science originated with Rosalind Picard's 1995 paper on affective computing. A motivation for the research is the ability to simulate empathy. The machine should interpret the emotional state of humans and adapt its behavior to them, giving an appropriate response for those emotions.

12.3.3 Bioinformatics

Bioinformatics is an interdisciplinary field that develops methods and software tools for understanding biological data. As an interdisciplinary field of science, bioinformatics combines computer science,statistics, mathematics, and engineering to analyze and interpret biological data. Bioinformatics has been used for in silico analyses of biological queries using mathematical and statistical techniques.

Bioinformatics is both an umbrella term for the body of biological studies that use computer programming as part of their methodology, as well as a reference to specific analysis "pipelines" that are repeatedly used, particularly in the field of genomics. Common uses of bioinformatics include the identification of candidate genes and nucleotides (SNPs). Often, such identification is made with the aim of better understanding the genetic basis of disease, unique adaptations, desirable properties (esp. in agricultural species), or differences between populations. In a less formal way, bioinformatics also tries to understand the organisational principles within nucleic acid and protein sequences.

12.3.4   Brain–machine interfaces

A brain—computer interface (BCI), sometimes called a mind-machine interface (MMI), direct neural interface (DNI), or brain—machine interface (BMI), is a direct communication pathway between an enhanced or wired brain and an external device. BCIs are often directed at researching, mapping, assisting, augmenting, or repairing human cognitive or sensory-motor functions.

Research on BCIs began in the 1970s at the University of California, Los Angeles (UCLA) under a grant from the National Science Foundation, followed by a contract from DARPA. The papers published after this research also mark the first appearance of the expression brain–computer interface in scientific literature.

The field of BCI research and development has since focused primarily on neuroprosthetics applications that aim at restoring damaged hearing, sight and movement. Thanks to the remarkable cortical plasticity of the brain, signals from implanted prostheses can, after adaptation, be handled by the brain like natural sensor or effector channels. Following years of animal experimentation, the first neuroprosthetic devices implanted in humans appeared in the mid-1990s.

12.3.5   Cheminformatics

Cheminformatics (also known as chemoinformatics, chemioinformatics and chemical informatics) is the use of computer and informational techniques applied to a range of problems in the field of chemistry. These in silico techniques are used in, for example, pharmaceutical companies in the process of drug discovery. These methods can also be used in chemical and allied industries in various other forms.

12.3.6   Classifying DNA sequences

A nucleic acid sequence is a succession of letters that indicate the order of nucleotides within a DNA (using GACT) or RNA(GACU) molecule. By convention, sequences are usually presented from the 5' end to the 3' end. For DNA, the sense strand is used. Because nucleic acids are normally linear (unbranched) polymers, specifying the sequence is equivalent to defining the covalent structure of the entire molecule. For this reason, the nucleic acid sequence is also termed the primary structure.

The sequence has capacity to represent information. Biological deoxyribonucleic acid represents the information which directs the functions of a living thing. In that context, the term genetic sequence is often used. Sequences can be read from the biological raw material through DNA sequencing methods.

Nucleic acids also have a secondary structure and tertiary structure. Primary structure is sometimes mistakenly referred to as primary sequence. Conversely, there is no parallel concept of secondary or tertiary sequence.

12.3.7 Computational anatomy

Computational anatomy is a discipline within medical imaging focusing on the study of anatomical shape and form at the morphome scale of morphology. It involves the development and application of computational, mathematical and data-analytical methods for modeling and simulation of biological structures. The field is broadly defined and includes foundations in anatomy, applied mathematics and pure mathematics, machine learning, computational mechanics, computational science, medical imaging, neuroscience, physics, probability, and statistics; it also has strong connections with fluid mechanics and geometric mechanics. Additionally, it complements newer, interdisciplinary fields like bioinformatics and neuroinformatics in the sense that its interpretation uses metadata derived from the original sensor imaging modalities (of which Magnetic Resonance Imaging is one example). It focuses on the anatomical structures being imaged, rather than the medical imaging devices. It is similar in spirit to the history of Computational linguistics, a discipline that focuses on the linguistic structures rather than the detectors acting as the transmission and communication medium(s).

In Computational anatomy, the diffeomorphism group for coordinate transformations is generated via the Lagrangian and Eulerian velocities of flow in $\mathbb{R}^3$. The flows between coordinates in Computational anatomy are constrained to be geodesic flows satisfying the principle of least action for the Kinetic energy of the flow defined via a Sobolev smoothness norm with more than two finite square-integrable derivatives for each component of the velocity of flow. This, in turn, guarantees that the flows in $\mathbb{R}^3$ are diffeomorphisms; it also implies that the diffeomorphic shape momentum in Computational anatomy, which satisfies the Euler-Lagrange equation for geodesics, is determined by its velocity and spatial derivatives. This separates the discipline from the case of incompressible fluids for which momentum is a pointwise function of velocity. Computational anatomy intersects the study of Riemannian manifolds and nonlinear global analysis, where groups of diffeomorphisms are the central focus. Emerging high-dimensional theories of shape are central to many studies in Computational anatomy, as are questions emerging from the fledgling field of shape statistics. The metric structures in Computational anatomy are related in spirit tomorphometrics, with the distinction that Computational anatomy focuses on an infinite-dimensional space of coordinate systems transformed by a diffeomorphism, hence the central use of the terminology diffeomorphometry, the metric space study of coordinate systems via diffeomorphisms.

12.3.8 Computational finance

Computational finance is a branch of applied computer science that deals with problems of practical interest in finance. Some slightly different definitions are the study of data and algorithms currently used in finance and the mathematics of computer programs that realize financial models or systems.

Computational finance emphasizes practical numerical methods rather than mathematical proofs and focuses on techniques that apply directly to economic analyses. It is an interdisciplinary field between mathematical finance and numerical methods. Two major areas are efficient and accurate computation of fair values of financial securities and the modeling of stochastic price series.

12.3.9 Computer vision, including object recognition

Computer vision is a field that includes methods for acquiring, processing, analyzing, and understanding images and, in general, high-dimensional data from the real world in order to produce numerical or symbolic information, e.g., in the forms of decisions. A theme in the development of this field has been to duplicate the abilities of human vision by electronically perceiving and understanding an image. Understanding in this context means the transformation of visual images (the input of retina) into descriptions of world that can interface with other thought processes and elicit appropriate action. This image understanding can be seen as the disentangling of symbolic information from image data using models constructed with the aid of geometry, physics, statistics, and learning theory. Computer vision has also been described as the enterprise of automating and integrating a wide range of processes and representations for vision perception.

As a scientific discipline, computer vision is concerned with the theory behind artificial systems that extract information from images. The image data can take many forms, such as video sequences, views from multiple cameras, or multi-dimensional data from a medical scanner. As a technological discipline, computer vision seeks to apply its theories and models to the construction of computer vision systems.

Sub-domains of computer vision include scene reconstruction, event detection, video tracking, object recognition, object pose estimation, learning, indexing, motion, and image restoration.

12.3.10 Detecting credit card fraud

Credit card fraud is a wide-ranging term for theft and fraud committed using or involving a payment card, such as a credit card or debit card, as a fraudulent source of funds in a transaction. The purpose may be to obtain goods without paying, or to obtain unauthorized funds from an account. Credit card fraud is also an adjunct to identity theft. According to the United States Federal Trade Commission, while identity theft had been holding steady for the last few years, it saw a 21 percent increase in 2008. However, credit card fraud, that crime which most people associate with ID theft, decreased as a percentage of all ID theft complaints for the sixth year in a row.

Although incidence of credit card fraud is limited to about 0.1% of all card transactions, this has resulted in huge financial losses as the fraudulent transactions have been large value transactions. In 1999, out of 12 billion transactions made annually, approximately 10 million—or one out of every 1200 transactions—turned out to be fraudulent. Also, 0.04% (4 out of every 10,000) of all monthly active accounts were fraudulent. Even with tremendous volume and value increase in credit card transactions since then, these proportions have stayed the same or have decreased due to sophisticated fraud detection and prevention systems. Today's fraud detection systems are designed to prevent one twelfth of one percent of all transactions processed which still translates into billions of dollars in losses.

In the decade to 2008, general credit card losses have been 7 basis points or lower (i.e. losses of $0.07 or less per $100 of transactions). In 2007, fraud in the United Kingdom was estimated at £535 million.

12.3.11 Software

Open-source software suites containing a variety of machine learning algorithms include the following:

- dlib
- ELKI
- Encog
- GNU Octave
- H2O
- Mahout
- Mallet (software project)
- mlpy
- MLPACK
- MOA (Massive Online Analysis)
- ND4J with Deeplearning4j
- NuPIC
- OpenCV
- OpenNN
- Orange

- R
- scikit-learn
- scikit-image
- Shogun
- TensorFlow
- Torch (machine learning)
- Spark
- Yooreeka
- Weka
- VOMBAT
- Commercial software with open-source editions
- KNIME
- RapidMiner
- Commercial software
- Angoss KnowledgeSTUDIO
- Databricks
- Google Prediction API
- IBM SPSS Modeler
- KXEN Modeler
- LIONsolver
- Mathematica
- MATLAB
- Microsoft Azure Machine Learning
- Neural Designer
- NeuroSolutions
- Oracle Data Mining
- RCASE
- SAS Enterprise Miner
- STATISTICA Data Miner

Chapter 13

How DSL Works

13.1 Overview

When you connect to the Internet, you might connect through a regular modem, through a local-area network connection in your office, through a cable modem or through a digital subscriber line (DSL) connection. DSL (as shown in Figure 13.1) is a very high-speed connection that uses the same wires as a regular telephone line.

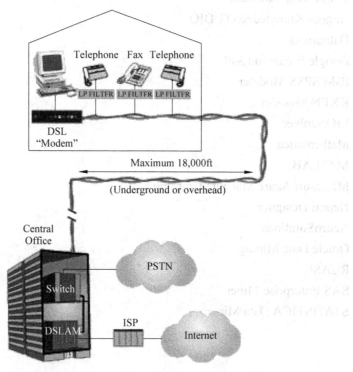

Figure 13.1   DSL

Here are some advantages of DSL:
- You can leave your Internet connection open and still use the phone line for voice calls.
- The speed is much higher than a regular modem (1.5 Mbps v s. 56 kbps).
- DSL doesn't necessarily require new wiring; it can use the phone line you already

have.
- The company that offers DSL will usually provide the modem as part of the installation. But there are disadvantages:
- A DSL connection works better when you are closer to the provider's central office.
- The connection is faster for receiving data than it is for sending data over the Internet.
- The service is not available everywhere.

In this article, we explain how a DSL connection manages to squeeze more information through a standard phone line—and lets you make regular telephone calls even when you're online!

13.2 Telephone Lines

A standard telephone installation in the United States consists of a pair of copper wires that the phone company installs in your home. The copper wires have lots of room for carrying more than your phone conversations—they are capable of handling a much greater bandwidth, or range of frequencies, than that demanded for voice. DSL exploits this "extra capacity" to carry information on the wire without disturbing the line's ability to carry conversations. The entire plan is based on matching particular frequencies to specific tasks.

To understand DSL, you first need to know a couple of things about a normal telephone line—the kind that telephone professionals call POTS, for Plain Old Telephone Service. One of the ways that POTS makes the most of the telephone company's wires and equipment is by limiting the frequencies that the switches, telephones and other equipment will carry. Human voices, speaking in normal conversational tones, can be carried in a frequency range of 0 to 3400 Hertz. This range of frequencies is tiny. For example, compare this to the range of most stereo speakers, which cover from roughly 20 Hertz to 20,000 Hertz. And the wires themselves have the potential to handle frequencies up to several million Hertz in most cases. The use of such a small portion of the wire's total bandwidth is historical—remember that the telephone system has been in place, using a pair of copper wires to each home, for about a century. By limiting the frequencies carried over the lines, the telephone system can pack lots of wires into a very small space without worrying about interference between lines. Modern equipment that sends digital rather than analog data can safely use much more of the telephone line's capacity.

13.3 Asymmetrical DSL

Most homes and small business users are connected to an asymmetric DSL(ADSL) line. ADSL divides up the available frequencies in a line on the assumption that most Internet users look at, or download, much more information than they send, or upload. Under this assumption, if the connection speed from the Internet to the user is three to four times faster than the connection from the user back to the Internet, then the user will see the most benefit (most of the time).

Other types of DSL include:
- Very high bit-rate DSL (VDSL)—This is a fast connection, but works only over a short distance.
- Symmetric DSL (SDSL)—This connection, used mainly by small businesses, doesn't allow you to use the phone at the same time, but the speed of receiving and sending data is the same.
- Rate-adaptive DSL (RADSL)—This is a variation of ADSL, but the modem can adjust the speed of the connection depending on the length and quality of the line.

13.4 Distance Limitations

Precisely how much benefit you see will greatly depend on how far you are from the central office of the company providing the ADSL service. ADSL is a distance-sensitive technology: As the connection's length increases, the signal quality decreases and the connection speed goes down. The limit for ADSL service is 18,000 feet (5460 meters), though for speed and quality of service reasons many ADSL providers place a lower limit on the distances for the service. At the extremes of the distance limits, ADSL customers may see speeds far below the promised maximums, while customers nearer the central office have faster connections and may see extremely high speeds in the future. ADSL technology can provide maximum downstream (Internet to customer) speeds of up to 8 megabits per second (Mbps) at a distance of about 6000 feet (1820 meters), and upstream speeds of up to 640 kilobits per second (kbps). In practice, the best speeds widely offered today are 1.5 Mbps downstream, with upstream speeds varying between 64 and 640 kbps.

You might wonder, if distance is a limitation for DSL, why it's not also a limitation for voice telephone calls. The answer lies in small amplifiers called loading coils that the telephone company uses to boost voice signals. Unfortunately, these loading coils are incompatible with ADSL signals, so a voice coil in the loop between your telephone and the telephone company's central office will disqualify you from receiving ADSL. Other factors that might disqualify you from receiving ADSL include:
- Bridge taps—These are extensions, between you and the central office, that extend service to other customers. While you wouldn't notice these bridge taps in normal phone service, they may take the total length of the circuit beyond the distance limits of the service provider.
- Fiber-optic cables—ADSL signals can't pass through the conversion from analog to digital and back to analog that occurs if a portion of your telephone circuit comes through fiber-optic cables.

- Distance—Even if you know where your central office is (don't be surprised if you don't—the telephone companies don't advertise their locations), looking at a map is no indication of the distance a signal must travel between your house and the office.

13.5 Splitting the Signal: CAP

There are two competing and incompatible standards for ADSL. The official ANSI standard for ADSL is a system called discrete multi tone, or DMT. According to equipment manufacturers, most of the ADSL equipment installed today uses DMT. An earlier and more easily implemented standard was the carrierless amplitude/phase(CAP) (as shown in Figure 13.2) system, which was used on many of the early installations of ADSL.

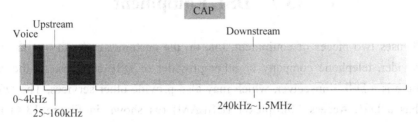

Figure 13.2 CAP

CAP operates by dividing the signals on the telephone line into three distinct bands: Voice conversations are carried in the 0 to 4 kHz (kilohertz) band, as they are in all POTS circuits. The upstream channel (from the user back to the server) is carried in a band between 25 and 160 kHz. The downstream channel (from the server to the user) begins at 240 kHz and goes up to a point that varies depending on a number of conditions (line length, line noise, number of users in a particular telephone company switch) but has a maximum of about 1.5 MHz (megahertz). This system, with the three channels widely separated, minimizes the possibility of interference between the channels on one line, or between the signals on different lines.

13.6 Splitting the Signal: DMT

DMT (as shown in Figure 13.3) also divides signals into separate channels, but doesn't use two fairly broad channels for upstream and downstream data. Instead, DMT divides the data into 247 separate channels, each 4 kHz wide.

Figure 13.3 DMT

One way to think about it is to imagine that the phone company divides your copper line into 247 different 4-kHz lines and then attaches a modem to each one. You get the equivalent of 247 modems connected to your computer at once! Each channel is monitored and, if the quality is too impaired, the signal is shifted to another channel. This system constantly shifts signals between different channels, searching for the best channels for transmission and reception. In addition, some of the lower channels (those starting at about 8 kHz), are used as bidirectional channels, for upstream and downstream information. Monitoring and sorting out the information on the bidirectional channels, and keeping up with the quality of all 247 channels, makes DMT more complex to implement than CAP, but gives it more flexibility on lines of differing quality.

13.7 DSL Equipment

ADSL uses two pieces of equipment, one on the customer end and one at the Internet service provider, telephone company or other provider of DSL services. At the customer's location there is a DSL transceiver, which may also provide other services. The DSL service provider has a DSL Access Multiplexer (DSLAM) (as shown in Figure 13.4) to receive customer connections.

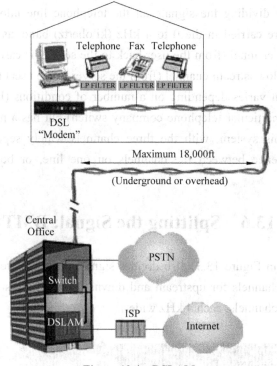

Figure 13.4 DSLAM

In the next couple of sections, we'll take a look at these two pieces of equipment.

13.7.1 DSL Equipment: Transceiver

Most residential customers call their DSL transceiver a "DSL modem." as shown in Figure 13.5. The engineers at the telephone company or ISP call it an ATU-R.

Regardless of what it's called, it's the point where data from the user's computer or network is connected to the DSL line.

The transceiver can connect to a customer's equipment in several ways, though most residential installation uses USB or 10 base-T Ethernet connections. While most of the ADSL transceivers sold by ISPs and telephone companies are simply transceivers, the devices used by businesses may combine network routers, network switches or other networking equipment in the same platform.

Figure 13.5   DSL modem

13.7.2 DSL Equipment: DSLAM

The DSLAM at the access provider is the equipment that really allows DSL to happen. A DSLAM takes connections from many customers and aggregates them onto a single, high-capacity connection to the Internet. DSLAMs are generally flexible and able to support multiple types of DSL in a single central office, and different varieties of protocol and modulation—both CAP and DMT, for example—in the same type of DSL. In addition, the DSLAM may provide additional functions including routing or dynamic IP address assignment for the customers.

The DSLAM provides one of the main differences between user service through ADSL and through cable modems. Because cable-modem users generally share a network loop that runs through a neighborhood, adding users means lowering performance in many instances. ADSL provides a dedicated connection from each user back to the DSLAM, meaning that users won't see a performance decrease as new users are added-until the total number of users begins to saturate the single, high-speed connection to the Internet. At that point, an upgrade by the service provider can provide additional performance for all the users connected to the DSLAM.

Keywords

digital subscriber line (DSL)	数字用户线路
Internet	因特网
cable	电缆
squeeze	压缩
copper	铜

bandwidth	带宽
Plain Old Telephone Service（POTS）	普通老式电话业务
switches	开关
tiny	小的
stereo	立体声
Hertz	赫兹
equipment	设备
asymmetric DSL(ADSL)	非对称数字用户线路
send	发送
upload	上传
Very high bit-rate DSL	高速数字用户线路
Symmetric DSL	对称数字用户线路
Rate-adaptive DSL	可调节速率数字用户线路
distance-sensitive	距离敏感型
downstream	下行速率
upstream	上行速率
bridge taps	桥接抽头
fiber-optic cables	光纤电缆
discrete multi tone (DMT)	离散多音频
carrierless	无载波
amplitude	振幅
transceiver	收发器
multiplexer	复用器
residential	住宅
Ethernet	以太网络
routers	路由器
saturate	饱和

13.8   Exercise 13

Multiple or single choices.

1. The character of DSL include: ____.
    A. use the internet and phone in the same time
    B. require new wiring
    C. the connection is faster for receiving data than it is for sending data
    D. the service is not available everywhere

2. One of the ways that POTS makes the most of the telephone company's wires and equipment is by _____.
   A. limiting the frequencies of switches
   B. limiting the frequencies of telephones
   C. limiting the frequencies of other equipment
   D. limiting the frequencies of radios

3. The character of asymmetric DSL(ADSL) line is _____.
   A. the upload speed is more faster than download speed
   B. the download speed is more faster than upload speed
   C. the upload speed is the same speed compared with download speed
   D. the upload speed and download speed will be different according to subscriber

4. Which of the following describe about other DSL is right _____.
   A. VDSL—a fast connection, but works only over a short distance
   B. SDSL—used mainly by small businesses and allow you to use the phone at the same time
   C. ADSL—Internet users look at, or download, much more information than they send, or upload
   D. RADSL—the modem can adjust the speed of the connection depending on the length and quality of the line

5. Which of the following statements may be true about ADSL _____.
   A. as the connection's length increases, the signal quality decreases
   B. as the connection's length increases, the connection speed goes down
   C. the limit for ADSL service is 10,000 meters
   D. customers nearer the central office have faster connections and may see extremely high speeds in the future

True or False.

1. About DSL, the connection is faster for receiving data than it is for sending data over the Internet.

2. Human voices, speaking in normal conversational tones, can be carried in a frequency range of 0 to 20,000 Hertz.

3. Symmetric DSL—the speed of receiving and sending data is the same.

4. ADSL is a distance-sensitive technology.

5. DSLAMs are generally flexible and able to support multiple types of DSL in a single central office.

13.9 Further Reading: How Telephones Work

Although most of us take it completely for granted, the telephone you have in your house is one of the most amazing devices ever created. If you want to talk to someone, all you have to do is pick up the phone and dial a few digits. You are instantly connected to that person, and you can have a two-way conversation. The telephone network extends worldwide, so you can reach nearly anyone on the planet. When you compare that to the state of the world just 100 years ago, when it might have taken several weeks to get a one-way written message to someone, you realize just how amazing the telephone is!

Figure 13.6 shows the entire telephone network, including a home connection, cell phone towers, long distance exchanges and transcontinental connections.

Figure 13.6   Telephone Network

In this article, we will look at the telephone device that you have in your house as well as the telephone network it connects to so you can make and receive calls.

13.9.1   A Simple Telephone

Surprisingly, a telephone is one of the simplest devices you have in your house. It is so simple because the telephone connection to your house has not changed in nearly a century. If you have an antique phone from the 1920s, you could connect it to the wall jack in your house and it would work fine!

As shown in Figure 13.7, the very simplest working telephone would look like this inside:

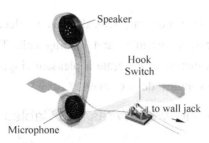

Figure 13.7 Simple Telephone

As you can see, it only contains three parts and they are all simple:

- A switch to connect and disconnect the phone from the network—This switch is generally called the hook switch. It connects when you lift the handset.
- A speaker—This is generally a little 50-cent, 8-ohm speaker of some sort.
- A microphone—In the past, telephone microphones have been as simple as carbon granules compressed between two thin metal plates. Sound waves from your voice compress and decompress the granules, changing the resistance of the granules and modulating the current flowing through the microphone.

That's it! You can dial this simple phone by rapidly tapping the hook switch-all telephone switches still recognize "pulse dialing". If you pick the phone up and rapidly tap the switch hook four times, the phone company's switch will understand that you have dialed a "4."

13.9.2 A Real Telephone

The only problem with the phone shown on the previous page is that when you talk, you will hear your voice through the speaker. Most people find that annoying, so any "real" phone contains a device called a duplex coil or something functionally equivalent to block the sound of your own voice from reaching your ear. A modern telephone also includes a bell so it can ring and a touch-tone keypad and frequency generator. A "real" phone looks like in Figure 13.8.

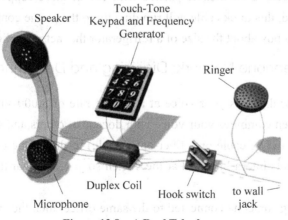

Figure 13.8 A Real Telephone

Still, it's pretty simple. In a modern phone there is an electronic microphone, amplifier and circuit to replace the carbon granules and loading coil. The mechanical bell is often replaced by a speaker and a circuit to generate a pleasant ringing tone. But a regular $6.95 telephone remains one of the simplest devices ever.

13.9.3 The Telephone Network: Wires and Cables

The telephone network starts in your house. A pair of copper wires runs from a box at the road to a box (often called an entrance bridge) at your house.

As shown in Figure 13.9, A typical phone company box that you see by the side of the road.

From there, the pair of wires is connected to each phone jack in your house (usually using red and green wires). If your house has two phone lines, then two separate pairs of copper wires run from the road to your house. The second pair is usually colored yellow and black inside your house.

Figure 13.9 A Typical phone company Box

Along the road runs a thick cable packed with 100 or more copper pairs. Depending on where you are located, this thick cable will run directly to the phone company's switch in your area or it will run to a box about the size of a refrigerator that acts as a digital concentrator.

13.9.4 The Telephone Network: Digitizing and Delivering

The concentrator digitizes your voice at a sample rate of 8,000 samples per second and 8-bit resolution. It then combines your voice with dozens of others and sends them all down a single wire (usually a coax cable or a fiber-optic cable) to the phone company office. Either way, your line connects into a line card at the switch so you can hear the dial tone when you pick up your phone.

If you are calling someone connected to the same office, then the switch simply creates a loop between your phone and the phone of the person you called. If it's a long-distance call, then your voice is digitized and combined with millions of other voices on the long-distance network. Your voice normally travels over a fiber-optic line to the office of the receiving party, but it may also be transmitted by satellite or by microwave towers.

13.9.5 Creating Your Own Telephone Network

Not only is a telephone a simple device, but the connection between you and the phone company is even simpler. In fact, you can easily create your own intercom system using two telephones, a 9-volt battery (or some other simple power supply) and a 300-ohm resistor that you can get for a dollar at Radio Shack. You can wire it up as shown in Figure 13.10.

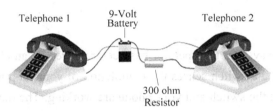

Figure 13.10 Wire it up

Your connection to the phone company consists of two copper wires. Usually they are red and green. The green wire is common, and the red wire supplies your phone with 6 to 12 volts DC at about 30 milliamps. If you think about a simple carbon granule microphone, all it is doing is modulating that current (letting more or less current through depending on how the sound waves compress and relax the granules), and the speaker at the other end "plays" that modulated signal. That's all there is to it!

Hand Generated! You know the hand crank on those old-fashioned telephones? It was used to generate the ring-signal AC wave and sound the bell at the other end!

The easiest way to wire up a private intercom like this is to go to a hardware or discount store and buy a 100-foot phone cord. Cut it, strip the wires and hook in the battery and resistor as shown. (Most cheap phone cords contain only two wires, but if the one you buy happens to have four, then use the center two.) When two people pick up the phones together, they can talk to each other just fine. This sort of arrangement will work at distances of up to several miles apart.

The only thing your little intercom cannot do is ring the phone to tell the person at the other end to pick up. The "ring" signal is a 90-volt AC wave at 20 hertz (Hz).

13.9.6 Calling Someone

If you go back to the days of the manual switchboard, it is easy to understand how the larger phone system works. In the days of the manual switchboard, there was a pair of copper wires running from every house to a central office in the middle of town. The switchboard operator sat in front of a board with one jack for every pair of wires entering the office.

Above each jack was a small light. A large battery supplied current through a resistor to each wire pair (in the same way you saw in the previous section). When someone picked up the handset on his or her telephone, the hook switch would complete the circuit and let current flow through wires between the house and the office. This would light the light bulb above that person's jack on the switchboard. The operator would connect his/her headset into that jack and ask who the person would like to talk to. The operator would then send a ring signal to the receiving party and wait for the party to pick up the phone. Once the receiving party picked up, the operator would connect the two people together in exactly the same way the simple intercom on the previous page was connected! It is that simple!

13.9.7 Tones

In a modern phone system, the operator has been replaced by an electronic switch. When you pick up the phone, the switch senses the completion of your loop and it plays a dial tone sound so you know that the switch and your phone are working. The dial tone sound is simply a combination of 350-hertz tone and a 440-hertz tone, and it sounds like this:

You then dial the number using a touch-tone keypad. The different dialing sounds are made of pairs of tones, as shown in Table 13.1.

Table 13.1　Different dialing sounds

Pairs of tones	1209Hz	1336Hz	1477Hz
697Hz	1	2	3
770Hz	4	5	6
852Hz	7	8	9
941Hz	*	0	#

Telephone Bandwidth

In order to allow more long-distance calls to be transmitted, the frequencies transmitted are limited to a bandwidth of about 3000 hertz. All of the frequencies in your voice below 400 hertz and above 3400 hertz are eliminated. That's why someone's voice on a phone has a distinctive sound.

You can prove that this sort of filtering actually happens by using the following sound files:

- 1000-hertz tone.
- 2000-hertz tone.
- 3000-hertz tone.
- 4000-hertz tone.
- 5000-hertz tone.
- 6000-hertz tone.

Call up someone you know and play the 1000-hertz sound file on your computer. The person will be able to hear the tone clearly. The person will also be able to hear the 2000-and 3000-hertz tones. However, the person will have trouble hearing the 4000-hertz tone, and will not hear the 5000-or 6000-hertz tones at all! That's because the phone company clips them off completely.

Chapter 14

Internet Infrastructure

One of the greatest things about the Internet is that nobody really owns it. It is a global collection of networks, both big and small. These networks connect together in many different ways to form the single entity that we know as the Internet. In fact, the very name comes from this idea of interconnected networks.

Since its beginning in 1969, the Internet has grown from four host computer systems to tens of millions. However, just because nobody owns the Internet, it doesn't mean it is not monitored and maintained in different ways. The Internet Society, a non-profit group established in 1992, oversees the formation of the policies and protocols that define how we use and interact with the Internet.

Every computer that is connected to the Internet is part of a network, even the one in your home. For example, you may use a modem and dial a local number to connect to an Internet Service Provider (ISP). At work, you may be part of a local area network (LAN), but you most likely still connect to the Internet using an ISP that your company has contracted with. When you connect to your ISP, you become part of their network. The ISP may then connect to a larger network and become part of their network. The Internet is simply a network of networks.

Most large communications companies have their own dedicated backbones connecting various regions. In each region, the company has a Point of Presence (POP). The POP is a place for local users to access the company's network, often through a local phone number or dedicated line. The amazing thing here is that there is no overall controlling network. Instead, there are several high-level networks connecting to each other through Network Access Points or NAPs.

When you connect to the Internet, your computer becomes part of a network.

14.1 A Network Example

Here's an example as shown in Figure 14.1. Imagine that Company A is a large ISP. In each major city, Company A has a POP. The POP in each city is a rack full of modems that the ISP's customers dial into. Company A leases fiber optic lines from the phone company to connect the POPs together.

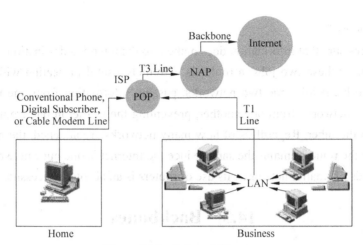

Figure 14.1 Internet Diagram

Imagine that Company B is a corporate ISP. Company B builds large buildings in major cities and corporations locate their Internet server machines in these buildings. Company B is such a large company that it runs its own fiber optic lines between its buildings so that they are all interconnected.

In this arrangement, all of Company A's customers can talk to each other, and all of Company B's customers can talk to each other, but there is no way for Company A's customers and Company B's customers to intercommunicate. Therefore, Company A and Company B both agree to connect to NAPs in various cities, and traffic between the two companies flows between the networks at the NAPs.

In the real Internet, dozens of large Internet providers interconnect at NAPs in various cities, and trillions of bytes of data flow between the individual networks at these points. The Internet is a collection of huge corporate networks that agree to all intercommunicate with each other at the NAPs. In this way, every computer on the Internet connects to every other.

14.2 Bridging The Divide

All of these networks rely on NAPs, backbones and routers to talk to each other. What is incredible about this process is that a message can leave one computer and travel halfway across the world through several different networks and arrive at another computer in a fraction of a second!

The routers determine where to send information from one computer to another. Routers are specialized computers that send your messages and those of every other Internet user speeding to their destinations along thousands of pathways. A router has two separate, but related, jobs:

- It ensures that information doesn't go where it's not needed. This is crucial for keeping large volumes of data from clogging the connections of "innocent

bystanders."
- It makes sure that information does make it to the intended destination.

In performing these two jobs, a router is extremely useful in dealing with two separate computer networks. It joins the two networks, passing information from one to the other. It also protects the networks from one another, preventing the traffic on one from unnecessarily spilling over to the other. Regardless of how many networks are attached, the basic operation and function of the router remains the same. Since the Internet is one huge network made up of tens of thousands of smaller networks, its use of routers is an absolute necessity.

14.3 Backbones

The National Science Foundation (NSF) created the first high-speed backbone in 1987. Called NSFNET, it was a T1 line that connected 170 smaller networks together and operated at 1.544Mbps (million bits per second). IBM, MCI and Merit worked with NSF to create the backbone and developed a T3 (45Mbps) backbone the following year.

Backbones are typically fiber optic trunk lines. The trunk line has multiple fiber optic cables combined together to increase the capacity. Fiber optic cables are designated OC for optical carrier, such as OC-3, OC-12 or OC-48. An OC-3 line is capable of transmitting 155Mbps while an OC-48 can transmit 2488Mbps (2.488Gbps). Compare that to a typical 56kbps modem transmitting 56000 bps and you see just how fast a modern backbone is.

Today there are many companies that operate their own high-capacity backbones, and all of them interconnect at various NAPs around the world. In this way, everyone on the Internet, no matter where they are and what company they use, is able to talk to everyone else on the planet. The entire Internet is a gigantic, sprawling agreement between companies to intercommunicate freely.

14.4 Internet Protocol: IP Addresses & Domain Name System

Every machine on the Internet has a unique identifying number, called an IP Address. The IP stands for Internet Protocol, which is the language that computers use to communicate over the Internet. A protocol is the pre-defined way that someone who wants to use a service talks with that service. The "someone" could be a person, but more often it is a computer program like a Web browser.

A typical IP address looks like this:

216.27.61.137

To make it easier for us humans to remember, IP addresses are normally expressed in decimal format as a dotted decimal number like the one above. But computers communicate in binary form. Look at the same IP address in binary:

11011000.00011011.00111101.10001001

The four numbers in an IP address are called octets, because they each have eight positions when viewed in binary form. If you add all the positions together, you get 32, which is why IP addresses are considered 32-bit numbers. Since each of the eight positions can have two different states (1 or zero), the total number of possible combinations per octet is 2^8 or 256. So each octet can contain any value between zero and 255. Combine the four octets and you get 2^{32} or a possible 4,294,967,296 unique values!

Out of the almost 4.3 billion possible combinations, certain values are restricted from use as typical IP addresses. For example, the IP address 0.0.0.0 is reserved for the default network and the address 255.255.255.255 is used for broadcasts.

The octets serve a purpose other than simply separating the numbers. They are used to create classes of IP addresses that can be assigned to a particular business, government or other entity based on size and need. The octets are split into two sections: Net and Host. The Net section always contains the first octet. It is used to identify the network that a computer belongs to. Host (sometimes referred to as Node) identifies the actual computer on the network. The Host section always contains the last octet. There are five IP classes plus certain special addresses.

When the Internet was in its infancy, it consisted of a small number of computers hooked together with modems and telephone lines. You could only make connections by providing the IP address of the computer you wanted to establish a link with. For example, a typical IP address might be 216.27.22.162. This was fine when there were only a few hosts out there, but it became unwieldy as more and more systems came online.

The first solution to the problem was a simple text file maintained by the Network Information Center that mapped names to IP addresses. Soon this text file became so large it was too cumbersome to manage. In 1983, the University of Wisconsin created the Domain Name System (DNS), which maps text names to IP addresses automatically.

14.5 Uniform Resource Locators

When you use the Web or send an e-mail message, you use a domain name to do it. For example, the Uniform Resource Locator (URL) "http://www.Yiren.com" contains the domain name Yiren.com. So does this e-mail address: example@Yiren.com. Every time you use a domain name, you use the Internet's DNS servers to translate the human-readable domain name into the machine-readable IP address.

Top-level domain names, also called first-level domain names, include .COM, .ORG, .NET, .EDU and .GOV. Within every top-level domain there is a huge list of second-level domains. For example, in the .COM first-level domain there is:

- Yiren.
- Yahoo.
- Microsoft.

Every name in the .COM top-level domain must be unique. The left-most word, like www, is the host name. It specifies the name of a specific machine (with a specific IP address) in a domain. A given domain can, potentially, contain millions of host names as long as they are all unique within that domain.

DNS servers accept requests from programs and other name servers to convert domain names into IP addresses. When a request comes in, the DNS server can do one of four things with it:

- It can answer the request with an IP address because it already knows the IP address for the requested domain.
- It can contact another DNS server and try to find the IP address for the name requested. It may have to do this multiple times.
- It can say, "I don't know the IP address for the domain you requested, but here's the IP address for a DNS server that knows more than I do."
- It can return an error message because the requested domain name is invalid or does not exist.

14.6   Clients, Servers and Ports

Internet servers make the Internet possible. All of the machines on the Internet are either servers or clients. The machines that provide services to other machines are servers. And the machines that are used to connect to those services are clients. There are Web servers, e-mail servers, FTP servers and so on serving the needs of Internet users all over the world.

When you connect to www.Yiren.com to read a page, you are a user sitting at a client's machine. You are accessing the Yiren Web server. The server machine finds the page you requested and sends it to you. Clients that come to a server machine do so with a specific intent, so clients direct their requests to a specific software server running on the server machine. For example, if you are running a Web browser on your machine, it will want to talk to the Web server on the server machine, not the e-mail server.

A server has a static IP address that does not change very often. A home machine that is dialing up through a modem, on the other hand, typically has an IP address assigned by the ISP every time you dial in. That IP address is unique for your session—it may be different the next time you dial in. This way, an ISP only needs one IP address for each modem it supports, rather than one for each customer.

Any server machine makes its services available using numbered ports—one for each service that is available on the server. For example, if a server machine is running a Web server and a file transfer protocol (FTP) server, the Web server would typically be available on port 80, and the FTP server would be available on port 21. Clients connect to a service at a specific IP address and on a specific port number.

Once a client has connected to a service on a particular port, it accesses the service using a specific protocol. Protocols are often text and simply describe how the client and server will have their conversation. Every Web server on the Internet conforms to the hypertext transfer protocol (HTTP).

Networks, routers, NAPs, ISPs, DNS and powerful servers all make the Internet possible. It is truly amazing when you realize that all this information is sent around the world in a matter of milliseconds! The components are extremely important in modern life—without them, there would be no Internet. And without the Internet, life would be very different indeed for many of us.

Keywords

maintain	保持
non-profit group	非营利组织
dial	拨号
Internet Service Provider (ISP)	互联网服务提供商
local area network	局域网
dedicated	专用的
backbones	骨干
Point of Presence (POP)	存在点
Network Access Points	网络接入点
lease	租用
arrangement	布局
intercommunicate	互通
interconnect	互联
a fraction of	几分之一
ensure	确保
crucial	关键
gigantic	巨大的
sprawl	蔓延
Protocol	协议
octets	八位字节
infancy	初期
hooked together	连接起来

14.7　Exercise 14

Multiple or single choices.

1. About a router has two separate assignment, we can know that ＿＿＿.
   A. the router ensures that information can go anywhere it's needed
   B. the router ensures that information doesn't go where it's not needed
   C. the router ensure that information does make it to the intended destination.
   D. the basic operation and function of the router will change with the increase of the networks

2. Which of the following IP address may be a user's IP address? ＿＿＿.
   A. A.0.0.0.0
   B. 192.168.1.1
   C. 192.168.1.2
   D. 255.255.255.255

3. The first-level domain names you know may include ＿＿＿ and so on.
   A. .com
   B. .org
   C. .cx
   D. .info

4. When a request comes in, the DNS server can ＿＿＿.
   A. answer the request with an IP address
   B. contact another DNS server and try to find the IP address for the name requested
   C. return an error message
   D. introduce other DNS server for a unknown IP address request

True or False.

1. Backbones are fiber optic trunk lines which has multiple fiber optic cables combined together to increase the capacity.

2. Every machine on the Internet has a unique identifying number, called an MAC Address.

3. IP addresses are considered 32-bit numbers.

4. The address 0.0.0.0 is usually used for broadcasts.

5. The Web server would typically be available on port 21.

14.8 Further Reading: Modem

Modem or Modulator/demodulator is an electronic device for converting between serial data (typically EIA-232) from a computer and an audio signal suitable for transmission over a telephone line connected to another modem. In one scheme the audio signal is composed of silence (no data) or one of two frequencies representing zero and one.

Modems are distinguished primarily by the maximum data rate they support. Data rates can range from 75 bits per second up to 56,000 and beyond. Data from the user (i.e. flowing from the local terminal or computer via the modem to the telephone line) is sometimes at a lower rate than the other direction, on the assumption that the user cannot type more than a few characters per second.

Various data compression and error correction algorithms are required to support the highest speeds. Other optional features are auto-dial (auto-call) and auto-answer which allow the computer to initiate and accept calls without human intervention. Most modern modems support a number of different protocols, and two modems, when first connected, will automatically negotiate to find a common protocol (this process may be audible through the modem or computer's loudspeakers). Some modem protocols allow the two modems to renegotiate ("retrain") if the initial choice of data rate is too high and gives too many transmission errors.

A modem may either be internal (connected to the computer's bus) or external ("stand-alone", connected to one of the computer's serial ports). The actual speed of transmission in characters per second depends not just the modem-to-modem data rate, but also on the speed with which the processor can transfer data to and from the modem, the kind of compression used and whether the data is compressed by the processor or the modem, the amount of noise on the telephone line (which causes retransmissions), the serial character format (typically 8N1: one start bit, eight data bits, no parity, one stop bit).

Fiber-optic transmission of light depends on preventing light from escaping from the fiber. When a beam of light encounters a boundary between two transparent substances, some of the light is normally reflected, while the rest passes into the new substance. How much of the beam is reflected, and how much enters the second substance, depends on the angle at which the light strikes the boundary. When the Sun shines down on the ocean from directly overhead, for example, much of its light penetrates the water. When the Sun is setting, however, its light strikes the surface of the water at a shallow angle, and most of it is reflected. Fiber optics makes use of certain special conditions, under which all of the light encountering the surface between two materials is reflected, to reduce loss.

The Internet and the Web are each a series of interconnected computer networks. Personal computers or workstations are connected to a Local Area Network (LAN) either by a dial-up connection through a modem and standard phone line, or by being directly wired into the LAN. Other modes of data transmission that allow for connection to a network include T1 connections and dedicated lines. Bridges and hubs link multiple networks to each other. Routers transmit data through networks and determine the best path of transmission.

Networks are connections between groups of computers and associated devices that allow users to transfer information electronically. The local area network shown on the left is representative of the setup used in many offices and companies. Individual computers, called work stations (W.S.), communicate to each other via cable or telephone line linking to servers. Servers are computers exactly like the W.S., except that they have an administrative function and are devoted entirely to monitoring and controlling W.S. access to part or all of the network and to any shared resources (such as printers). The red line represents the larger network connection between servers, called the backbone; the blue line shows local connections. A modem (modulator/demodulator) allows computers to transfer information across standard telephone lines. Modems convert digital signals into analog signals and back again, making it possible for computers to communicate, or network, across thousands of miles.

DNS

DNS is a general-purpose distributed, replicated, data query service chiefly used on Internet for translating hostnames into Internet addresses. Also, the style of hostname used on the Internet, though such a name is properly called a fully qualified domain name. DNS can be configured to use a sequence of name servers, based on the domains in the name being looked for, until a match is found.

The name resolution client (e.g. UNIX's gethostbyname() library function) can be configured to search for host information in the following order: first in the local /etc/hosts file, second in NIS and third in DNS. This sequencing of Naming Services is sometimes called "name service switching". Under Solaris is configured in the file /etc/nsswitch.conf.

DNS can be queried interactively using the command nslookup. It is defined in STD 13, RFC 1034, RFC 1035, RFC 1591.

BIND is a common DNS server.

Uniform Resource Locator

URL, previously "Universal Resource Locator", is a standard way of specifying the location of an object, typically a web page, on the Internet. Other types of object are described below. URLs are the form of address used on the World-Wide Web. They are used in HTML documents to specify the target of a hyperlink which is often another HTML document (possibly stored on another computer).

Here are some example URLs:

http://www.w3.org/default.html
http://www.acme.co.uk:8080/images/map.gif
http://www.foldoc.org/?Uniform+Resource+Locator
http://www.w3.org/default.html#Introduction
ftp://wuarchive.wustl.edu/mirrors/msdos/graphics/gifkit.zip
ftp://spy:secret@ftp.acme.com/pub/topsecret/weapon.tgz
mailto:fred@doc.ic.ac.uk
news:alt.hypertext
telnet://dra.com

The part before the first colon specifies the access scheme or protocol. Commonly implemented schemes include: ftp, http (World-Wide Web), gopher or WAIS. The "file" scheme should only be used to refer to a file on the same host. Other less commonly used schemes include news, telnet or mailto (e-mail).

The part after the colon is interpreted according to the access scheme. In general, two slashes after the colon introduce a hostname (host:port is also valid, or for FTP user:passwd@host or user@host). The port number is usually omitted and defaults to the standard port for the scheme, e.g. port 80 for HTTP.

For an HTTP or FTP URL the next part is a pathname which is usually related to the pathname of a file on the server. The file can contain any type of data but only certain types are interpreted directly by most browsers. These include HTML and images in gif or jpeg format. The file's type is given by a MIME type in the HTTP headers returned by the server, e.g. "text/html", "image/gif", and is usually also indicated by its filename extension. A file whose type is not recognized directly by the browser may be passed to an external "viewer" application, e.g. a sound player.

The last (optional) part of the URL may be a query string preceded by "?" or a "fragment identifier" preceded by "#". The later indicates a particular position within the specified document.

Only alphanumerics, reserved characters (:/?#"<>%+) used for their reserved purposes and "$", "-", "_", ".", "&", "+" are safe and may be transmitted unencoded. Other characters are encoded as a "%" followed by two hexadecimal digits. Space may also be encoded as "+". Standard SGML "&<name>;" character entity encodings (e.g. "é") are also accepted when URLs are embedded in HTML. The terminating semicolon may be omitted if &<name> is followed by a non-letter character.

Client-Server(C/S)

C/S is a common form of distributed system in which software is split between server tasks and client tasks. A client sends requests to a server, according to some protocol, asking for information or action, and the server responds.

This is analogous to a customer (client) who sends an order (request) on an order form to a supplier (server) who dispatches the goods and an invoice (response). The order form and invoice are part of the "protocol" used to communicate in this case.

There may be either one centralized server or several distributed ones. This model allows clients and servers to be placed independently on nodes in a network, possibly on different hardware and operating systems appropriate to their function, e.g. fast server/cheap client.

Examples are the name-server/name-resolver relationship in DNS, the file-server/file-client relationship in NFS and the screen server/client application split in the X Window System.

Chapter 15
How Internet Search Engines Work

The good news about the Internet and its most visible component, the World Wide Web, is that there are hundreds of millions of pages available, waiting to present information on an amazing variety of topics. The bad news about the Internet is that there are hundreds of millions of pages available, most of them titled according to the whim of their author, almost all of them sitting on servers with cryptic names. When you need to know about a particular subject, how do you know which pages to read? If you're like most people, you visit an Internet search engine.

Internet search engines are special sites on the Web that are designed to help people find information stored on other sites. There are differences in the ways various search engines work, but they all perform three basic tasks:

- They search the Internet—or select pieces of the Internet-based on important words.
- They keep an index of the words they find, and where they find them.
- They allow users to look for words or combinations of words found in that index.

Early search engines held an index of a few hundred thousand pages and documents, and received maybe one or two thousand inquiries each day. Today, a top search engine will index hundreds of millions of pages, and respond to tens of millions of queries per day. In this article, we'll tell you how these major tasks are performed, and how Internet search engines put the pieces together in order to let you find the information you need on the Web.

15.1   Looking at the Web

When most people talk about Internet search engines, they really mean World Wide Web search engines. Before the Web became the most visible part of the Internet, there were already search engines in place to help people find information on the Net. Programs with names like "gopher" and "Archie" kept indexes of files stored on servers connected to the Internet, and dramatically reduced the amount of time required to find programs and documents. In the late 1980s, getting serious value from the Internet meant knowing how to use gopher, Archie, Veronica and the rest.

Today, most Internet users limit their searches to the Web, so we'll limit this article to search engines that focus on the contents of Web pages.

1. An Itsy-Bitsy Beginning

Before a search engine can tell you where a file or document is, it must be found. To find information on the hundreds of millions of Web pages that exist, a search engine employs special software robots, called spiders, to build lists of the words found on Web sites. When a spider is building its lists, the process is called Web crawling. In order to build and maintain a useful list of words, a search engine's spiders have to look at a lot of pages.

How does any spider start its travels over the Web? The usual starting points are lists of heavily used servers and very popular pages. The spider will begin with a popular site, indexing the words on its pages and following every link found within the site. In this way, the spidering system quickly begins to travel, spreading out across the most widely used portions of the Web.

Google.com began as an academic search engine. In the paper that describes how the system was built, Sergey Brin and Lawrence Page give an example of how quickly their spiders can work. They built their initial system to use multiple spiders, usually three at one time. Each spider could keep about 300 connections to Web pages open at a time. At its peak performance, using four spiders, their system could crawl over 100 pages per second, generating around 600 kilobytes of data each second.

Keeping everything running quickly meant building a system to feed necessary information to the spiders. The early Google system had a server dedicated to providing URLs to the spiders. Rather than depending on an Internet service provider for the domain name server (DNS) that translates a server's name into an address, Google had its own DNS, in order to keep delays to a minimum.

When the Google spider looked at an HTML page, it took note of two things:
- The words within the page.
- Where the words were found.

As shown in Figure 15.1, "Spiders" take a Web page's content and create key search words that enable online users to find pages they're looking for.

Words occurring in the title, subtitles, meta tags and other positions of relative importance were noted for special consideration during a subsequent user search. The Google spider was built to index every significant word on a page, leaving out the articles "a," "an" and "the." Other spiders take different approaches.

These different approaches usually attempt to make the spider operate faster, allow users to search more efficiently, or both. For example, some spiders will keep track of the words in the title, sub-headings and links, along with the 100 most frequently used words on the page and each word in the first 20 lines of text. Lycos is said to use this approach to spidering the Web.

Other systems, such as AltaVista, go in the other direction, indexing every single word on a page, including "a," "an," "the" and other "insignificant" words. The push to completeness

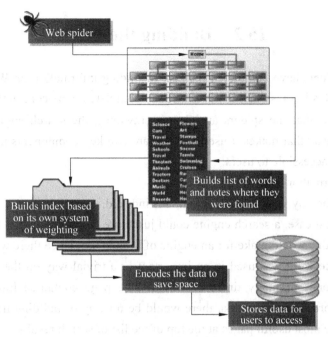

Figure 15.1 Web spider

in this approach is matched by other systems in the attention given to the unseen portion of the Web page, the meta tags.

2. Meta Tags

Meta tags allow the owner of a page to specify key words and concepts under which the page will be indexed. This can be helpful, especially in cases in which the words on the page might have double or triple meanings—the meta tags can guide the search engine in choosing which of the several possible meanings for these words is correct. There is, however, a danger in over-reliance on meta tags, because a careless or unscrupulous page owner might add meta tags that fit very popular topics but have nothing to do with the actual contents of the page. To protect against this, spiders will correlate meta tags with page content, rejecting the meta tags that don't match the words on the page.

All of this assumes that the owner of a page actually wants it to be included in the results of a search engine's activities. Many times, the page's owner doesn't want it showing up on a major search engine, or doesn't want the activity of a spider accessing the page. Consider, for example, a game that builds new, active pages each time sections of the page are displayed or new links are followed. If a Web spider accesses one of these pages, and begins following all of the links for new pages, the game could mistake the activity for a high-speed human player and spin out of control. To avoid situations like this, the robot exclusion protocol was developed. This protocol, implemented in the meta-tag section at the beginning of a Web page, tells a spider to leave the page alone—to neither index the words on the page nor try to follow its links.

15.2 Building the Index

Once the spiders have completed the task of finding information on Web pages (and we should note that this is a task that is never actually completed—the constantly changing nature of the Web means that the spiders are always crawling), the search engine must store the information in a way that makes it useful. There are two key components involved in making the gathered data accessible to users:

- The information stored with the data.
- The method by which the information is indexed.

In the simplest case, a search engine could just store the word and the URL where it was found. In reality, this would make for an engine of limited use, since there would be no way of telling whether the word was used in an important or a trivial way on the page, whether the word was used once or many times or whether the page contained links to other pages containing the word. In other words, there would be no way of building the ranking list that tries to present the most useful pages at the top of the list of search results.

To make for more useful results, most search engines store more than just the word and URL. An engine might store the number of times that the word appears on a page. The engine might assign a weight to each entry, with increasing values assigned to words as they appear near the top of the document, in sub-headings, in links, in the meta tags or in the title of the page. Each commercial search engine has a different formula for assigning weight to the words in its index. This is one of the reasons that a search for the same word on different search engines will produce different lists, with the pages presented in different orders.

Regardless of the precise combination of additional pieces of information stored by a search engine, the data will be encoded to save storage space. For example, the original Google paper describes using 2 bytes, of 8 bits each, to store information on weighting—whether the word was capitalized, its font size, position, and other information to help in ranking the hit. Each factor might take up 2 or 3 bits within the 2-byte grouping (8 bits=1 byte). As a result, a great deal of information can be stored in a very compact form. After the information is compacted, it's ready for indexing.

An index has a single purpose: It allows information to be found as quickly as possible. There are quite a few ways for an index to be built, but one of the most effective ways is to build a hash table. In hashing, a formula is applied to attach a numerical value to each word. The formula is designed to evenly distribute the entries across a predetermined number of divisions. This numerical distribution is different from the distribution of words across the alphabet, and that is the key to a hash table's effectiveness.

In English, there are some letters that begin many words, while others begin fewer. You'll find, for example, that the "M" section of the dictionary is much thicker than the "X" section.

This inequity means that finding a word beginning with a very "popular" letter could take much longer than finding a word that begins with a less popular one. Hashing evens out the difference, and reduces the average time it takes to find an entry. It also separates the index from the actual entry. The hash table contains the hashed number along with a pointer to the actual data, which can be sorted in whichever way allows it to be stored most efficiently. The combination of efficient indexing and effective storage makes it possible to get results quickly, even when the user creates a complicated search.

15.3 Building a Search

Searching through an index involves a user building a query and submitting it through the search engine. The query can be quite simple, a single word at minimum. Building a more complex query requires the use of Boolean operators that allow you to refine and extend the terms of the search.

The Boolean operators most often seen are:
- AND—All the terms joined by "AND" must appear in the pages or documents. Some search engines substitute the operator "+" for the word AND.
- OR—At least one of the terms joined by "OR" must appear in the pages or documents.
- NOT—The term or terms following "NOT" must not appear in the pages or documents. Some search engines substitute the operator "-" for the word NOT.
- FOLLOWED BY—One of the terms must be directly followed by the other.
- NEAR—One of the terms must be within a specified number of words of the other.
- Quotation Marks—The words between the quotation marks are treated as a phrase, and that phrase must be found within the document or file.

15.4 Future Search

The searches defined by Boolean operators are literal searches—the engine looks for the words or phrases exactly as they are entered. This can be a problem when the entered words have multiple meanings. "Bed," for example, can be a place to sleep, a place where flowers are planted, the storage space of a truck or a place where fish lay their eggs. If you're interested in only one of these meanings, you might not want to see pages featuring all of the others. You can build a literal search that tries to eliminate unwanted meanings, but it's nice if the search engine itself can help out.

One of the areas of search engine research is concept-based searching. Some of this research involves using statistical analysis on pages containing the words or phrases you search for, in order to find other pages you might be interested in. Obviously, the information stored about each page is greater for a concept-based search engine, and far more processing is required for each search. Still, many groups are working to improve both results and performance of this type of search engine. Others have moved on to another area of research, called natural-language queries.

The idea behind natural-language queries is that you can type a question in the same way you would ask it to a human sitting beside you—no need to keep track of Boolean operators or complex query structures. The popular natural language query site AskJeeves.com parses the query for keywords that it then applies to the index of sites it has built. It only works with simple queries; but competition is heavy to develop a natural-language query engine that can accept a query of great complexity.

Keywords

component	组成
whim	怪念头
cryptic	神秘
index	索引
inquiries	查询
dramatically	显著地
crawling	爬行
spider	蜘蛛
academic	学术的
subtitles	字幕
subsequent	随后
sub-headings	副标题
meta	元
over-reliance	过度依赖
unscrupulous	不道德的
match	匹配
exclusion	排除
Hash table	哈希表
formula	公式
distribution	分配
alphabet	字母
inequity	不平衡
submit	提交
quotation	引文
phrase	短语
ambidextrous	怀有二心的

literal 字面的
statistical analysis 统计分析

15.5　Exercise 15

Multiple or single choices.

1. Various search engines perform three basic tasks that they can ____.
   A. select pieces of the Internet based on important words
   B. keep an index of the words they find
   C. download the information relevant with your input words
   D. allow users to look for words or combinations of words found in that index

2. Google.com began as ____.
   A. an document search engine
   B. an image search engine
   C. an academic search engine
   D. an music search engine

3. When the Google spider looked at an HTML page, it mark ____.
   A. the word within the page
   B. account the appearance of the word within the page
   C. where the words were found
   D. the time it take when search the word at an HTML

4. When Google spider search for a key word, some "insignificant" words may be ignored for saving searching time such as ____.
   A. a
   B. an
   C. the
   D. as

5. The two key components involved in making the gathered data accessible to users are ____.
   A. store the URL
   B. information stored with the data
   C. method by which the information is indexed
   D. store the number of times that the word appears on a page

True or False.

1. The usual starting points are lists of heavily used servers and very popular pages.

2. Google.com began as an academic search engine.

3. Using the Meta tags may search some unscrupulous page which have nothing deal with the actual contents.

4. The most effective ways of building an index is to build a hash table.

5. The "X" section of the dictionary is much thicker than the "M" section.

15.6 Further Reading: Web crawler

A Web crawler is an Internet bot which systematically browses the World Wide Web, typically for the purpose of Web indexing (web spidering).

Web search engines and some other sites use Web crawling or spidering software to update their web content or indices of others sites' web content. Web crawlers can copy all the pages they visit for later processing by a search engine which indexes the downloaded pages so the users can search much more efficiently.

Crawlers consume resources on the systems they visit and often visit sites without tacit approval. Issues of schedule, load, and "politeness" come into play when large collections of pages are accessed. Mechanisms exist for public sites not wishing to be crawled to make this known to the crawling agent. For instance, including a robots.txt file can request bots to index only parts of a website, or nothing at all.

As the number of pages on the internet is extremely large, even the largest crawlers fall short of making a complete index. For that reason search engines were bad at giving relevant search results in the early years of the World Wide Web, before the year 2000. This is improved greatly by modern search engines, nowadays very good results are given instantly.

Crawlers can validate hyperlinks and HTML code. They can also be used for web scraping (see also data-driven programming).

15.6.1 Nomenclature

A Web crawler may also be called a Web spider, an ant, an automatic indexer, or (in the FOAF software context) a Web scutter.

15.6.2 Overview

A Web crawler starts with a list of URLs to visit, called the seeds. As the crawler visits these URLs, it identifies all the hyperlinks in the page and adds them to the list of URLs to visit, called the crawl frontier. URLs from the frontier are recursively visited according to a set of policies. If the crawler is performing archiving of websites it copies and saves the information as it goes. The archives are usually stored in such a way they can be viewed, read and navigated as they were on the live web, but are preserved as "snapshots".

The large volume implies the crawler can only download a limited number of the Web pages within a given time, so it needs to prioritize its downloads. The high rate of change can imply the pages might have already been updated or even deleted.

The number of possible URLs crawled being generated by server-side software has also made it difficult for web crawlersto avoid retrieving duplicate content. Endless combinations of HTTP GET (URL-based) parameters exist, of which only a small selection will actually return unique content. For example, a simple online photo gallery may offer three options to users, as specified through HTTP GET parameters in the URL. If there exist four ways to sort images, three choices of thumbnail size, two file formats, and an option to disable user-provided content, then the same set of content can be accessed with 48 different URLs, all of which may be linked on the site. This mathematical combination creates a problem for crawlers, as they must sort through endless combinations of relatively minor scripted changes in order to retrieve unique content.

As Edwards et al. noted, "Given that the bandwidth for conducting crawls is neither infinite nor free, it is becoming essential to crawl the Web in not only a scalable, but efficient way, if some reasonable measure of quality or freshness is to be maintained." A crawler must carefully choose at each step which pages to visit next.

15.6.3 Crawling policy

The behavior of a Web crawler is the outcome of a combination of policies:
- a selection policy which states the pages to download.
- a re-visit policy which states when to check for changes to the pages.
- a politeness policy that states how to avoid overloading Web sites.
- a parallelization policy that states how to coordinate distributed web crawlers.

1. Selection policy

Given the current size of the Web, even large search engines cover only a portion of the publicly available part. A 2009 study showed even large-scale search engines index no more than 40%~70% of the indexable Web; a previous study by Steve Lawrence and Lee Giles showed that no search engine indexed more than 16% of the Web in 1999. As a crawler always downloads just a fraction of the Web pages, it is highly desirable for the downloaded fraction to contain the most relevant pages and not just a random sample of the Web.

This requires a metric of importance for prioritizing Web pages. The importance of a page is a function of its intrinsic quality, its popularity in terms of links or visits, and even of its URL (the latter is the case of vertical search engines restricted to a single top-level domain, or search engines restricted to a fixed Web site). Designing a good selection policy has an added difficulty: it must work with partial information, as the complete set of Web pages is not known during crawling.

Cho *et al.* made the first study on policies for crawling scheduling. Their data set was a 180,000-pages crawl from the stanford.edu domain, in which a crawling simulation was done with different strategies.The ordering metrics tested were breadth-first, backlink count and partial Pagerank calculations. One of the conclusions was that if the crawler wants to download pages with high Pagerank early during the crawling process, then the partial Pagerank strategy is the better, followed by breadth-first and backlink-count. However, these results are for just a single domain. Cho also wrote his Ph.D. dissertation at Stanford on web crawling.

Najork and Wiener performed an actual crawl on 328 million pages, using breadth-first ordering. They found that a breadth-first crawl captures pages with high Pagerank early in the crawl (but they did not compare this strategy against other strategies). The explanation given by the authors for this result is that "the most important pages have many links to them from numerous hosts, and those links will be found early, regardless of on which host or page the crawl originates."

Abiteboul designed a crawling strategy based on an algorithm called OPIC (On-line Page Importance Computation). In OPIC, each page is given an initial sum of "cash" that is distributed equally among the pages it points to. It is similar to a Pagerank computation, but it is faster and is only done in one step. An OPIC-driven crawler downloads first the pages in the crawling frontier with higher amounts of "cash". Experiments were carried in a 100,000-pages synthetic graph with a power-law distribution of in-links. However, there was no comparison with other strategies nor experiments in the real Web.

Boldi used simulation on subsets of the Web of 40 million pages from the .it domain and 100 million pages from the WebBase crawl, testing breadth-first against depth-first, random ordering and an omniscient strategy. The comparison was based on how well PageRank computed on a partial crawl approximates the true PageRank value. Surprisingly, some visits that accumulate PageRank very quickly (most notably, breadth-first and the omniscient visit) provide very poor progressive approximations.

Baeza-Yates *et al.* used simulation on two subsets of the Web of 3 million pages from the .gr and .cl domain, testing several crawling strategies. They showed that both the OPIC strategy and a strategy that uses the length of the per-site queues are better than breadth-first crawling, and that it is also very effective to use a previous crawl, when it is available, to guide the current one.

Daneshpajouh *et al.* designed a community based algorithm for discovering good seeds. Their method crawls web pages with high PageRank from different communities in less iteration in comparison with crawl starting from random seeds. One can extract good seed from a previously-crawled-Web graph using this new method. Using these seeds a new crawl can be very effective.

2. Re-visit policy

The Web has a very dynamic nature, and crawling a fraction of the Web can take weeks or months. By the time a Web crawler has finished its crawl, many events could have happened, including creations, updates, and deletions.

From the search engine's point of view, there is a cost associated with not detecting an event, and thus having an outdated copy of a resource. The most-used cost functions are freshness and age.

3. Politeness policy

Crawlers can retrieve data much quicker and in greater depth than human searchers, so they can have a crippling impact on the performance of a site. Needless to say, if a single crawler is performing multiple requests per second and/or downloading large files, a server would have a hard time keeping up with requests from multiple crawlers.

4. Parallelization policy

A parallel crawler is a crawler that runs multiple processes in parallel. The goal is to maximize the download rate while minimizing the overhead from parallelization and to avoid repeated downloads of the same page. To avoid downloading the same page more than once, the crawling system requires a policy for assigning the new URLs discovered during the crawling process, as the same URL can be found by two different crawling processes.

15.6.4 Architectures

A crawler must not only have a good crawling strategy, as noted in the previous sections, but it should also have a highly optimized architecture.

While it is fairly easy to build a slow crawler that downloads a few pages per second for a short period of time, building a high-performance system that can download hundreds of millions of pages over several weeks presents a number of challenges in system design, I/O and network efficiency, and robustness and manageability.

Web crawlers are a central part of search engines, and details on their algorithms and architecture are kept as business secrets. When crawler designs are published, there is often an important lack of detail that prevents others from reproducing the work. There are also emerging concerns about "search engine spamming", which prevent major search engines from publishing their ranking algorithms.

15.6.5　Security

While most of the website owners are keen to have their pages indexed as broadly as possible to have strong presence in search engines, web crawling can also have unintended consequences and lead to a compromise or data breach if search engine indexes resources that shouldn't be publicly available or pages revealing potentially vulnerable versions of software. In a study from 2013, majority of websites that were victims of opportunistic hacking (mostly website defacements) were well indexed by search engines, which was the main factor that allowed attackers to find potential victims using specific search engine queries.

Apart from standard web application security recommendations website owners can reduce their exposure to opportunistic hacking by only allowing (with robots.txt) search engines to index the public parts of their websites and explicitly blocking indexing of transactional parts (login pages, private pages etc.).

15.6.6　Crawler identification

Web crawlers typically identify themselves to a Web server by using the User-agent field of an HTTP request. Web site administrators typically examine their Web servers' log and use the user agent field to determine which crawlers have visited the web server and how often. The user agent field may include a URL where the Web site administrator may find out more information about the crawler. Examining Web server log is tedious task, and therefore some administrators use tools to identify, track and verify Web crawlers. Spambots and other malicious Web crawlers are unlikely to place identifying information in the user agent field, or they may mask their identity as a browser or other well-known crawler.

It is important for Web crawlers to identify themselves so that Web site administrators can contact the owner if needed. In some cases, crawlers may be accidentally trapped in a crawler trap or they may be overloading a Web server with requests, and the owner needs to stop the crawler. Identification is also useful for administrators that are interested in knowing when they may expect their Web pages to be indexed by a particular search engine.

15.6.7　Crawling the deep web

A vast amount of web pages lie in the deep or invisible web.These pages are typically only accessible by submitting queries to a database, and regular crawlers are unable to find these pages if there are no links that point to them. Google's Sitemaps protocol and mod are intended to allow discovery of these deep-Web resources.

Deep web crawling also multiplies the number of web links to be crawled. Some crawlers only take some of the URLs in <a href="URL"> form. In some cases, such as the Googlebot, Web crawling is done on all text contained inside the hypertext content, tags, or text.

Strategic approaches may be taken to target deep Web content. With a technique called screen scraping, specialized software may be customized to automatically and repeatedly query a given Web form with the intention of aggregating the resulting data. Such software can be used to span multiple Web forms across multiple Websites. Data extracted from the results of one Web form submission can be taken and applied as input to another Web form thus establishing continuity across the Deep Web in a way not possible with traditional web crawlers.

Pages built on AJAX are among those causing problems to web crawlers. Google has proposed a format of AJAX calls that their bot can recognize and index.

15.6.8 Visual vs programmatic crawlers

There are a number of "visual web scraper/crawler" products available on the web which will crawl pages and structure data into columns and rows based on the users requirements. One of the main difference between a classic and a visual crawler is the level of programming ability required to set up a crawler. The latest generation of "visual scrapers" like Diffbot, outwithub, and import.io remove the majority of the programming skill needed to be able to program and start a crawl to scrape web data.

The visual scraping/crawling methodology relies on the user "teaching" a piece of crawler technology, which then follows patterns in semi-structured data sources. The dominant method for teaching a visual crawler is by highlighting data in a browser and training columns and rows. While the technology is not new, for example it was the basis of Needlebase which has been bought by Google (as part of a larger acquisition of ITA Labs), there is continued growth and investment in this area by investors and end-users.

Chapter 16

Encryption

The incredible growth of the Internet has excited businesses and consumers alike with its promise of changing the way we live and work. But a major concern has been just how secure the Internet is, especially when you're sending sensitive information through it.

Information security is provided on computers and over the Internet by a variety of methods. A simple but straightforward security method is to only keep sensitive information on removable storage media like floppy disks. But the most popular forms of security all rely on encryption, the process of encoding information in such a way that only the person (or computer) with the key can decode it.

In this article, you will learn about encryption and authentication. You will also learn about public-key and symmetric-key systems, as well as hash algorithms.

16.1 In the Key of...

Computer encryption is based on the science of cryptography, which has been used throughout history. Before the digital age, the biggest users of cryptography were governments, particularly for military purposes. The existence of coded messages has been verified as far back as the Roman Empire. But most forms of cryptography in use these days rely on computers, simply because a human-based code is too easy for a computer to crack.

Most computer encryption systems belong in one of two categories.

1. Symmetric Key

In symmetric-key encryption, each computer has a secret key (code) that it can use to encrypt a packet of information before it is sent over the network to another computer. Symmetric-key requires that you know which computers will be talking to each other so you can install the key on each one. Symmetric-key encryption is essentially the same as a secret code that each of the two computers must know in order to decode the information. The code provides the key to decoding the message.

Think of it like this: You create a coded message to send to a friend in which each letter is substituted with the letter that is two down from it in the alphabet. So "A" becomes "C," and "B" becomes "D". You have already told a trusted friend that the code is "Shift by 2". Your friend gets the message and decodes it. Anyone else who sees the message will see only nonsense.

2. Public Key

Public-key encryption uses a combination of a private key and a public key. The private key is known only to your computer, while the public key is given by your computer to any computer that wants to communicate securely with it. To decode an encrypted message, a computer must use the public key, provided by the originating computer, and its own private key. A very popular public-key encryption utility is called Pretty Good Privacy (PGP), which allows you to encrypt almost anything. You can find out more about PGP at the PGP site.

public-key encryption on a large scale, such as a secure Web server might need, requires a different approach. This is where digital certificates come in. A digital certificate is basically a bit of information that says that the Web server is trusted by an independent source known as a certificate authority. The certificate authority acts as a middleman that both computers trust. It confirms that each computer is in fact who it says it is, and then provides the public keys of each computer to the other.

As shown in Figure 16.1, a popular implementation of public-key encryption is the Secure Sockets Layer (SSL). Originally developed by Netscape, SSL is an Internet security protocol used by Internet browsers and Web servers to transmit sensitive information. SSL became part of an overall security protocol known as Transport Layer Security (TLS).

Figure 16.1   Look for the "s" after "http" in the address whenever you are about to enter sensitive information, such as a credit-card number, into a form on a Web site

As shown in Figure 16.2, The padlock symbol lets you know that you are using encryption.

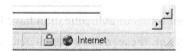

Figure 16.2   Padlock symbol

16.2   Hash This

The key in public-key encryption is based on a hash value. This is a value that is computed from a base input number using a hashing algorithm. Essentially, the hash value is a summary of the original value. The important thing about a hash value is that it is nearly impossible to derive the original input number without knowing the data used to create the hash value. Here's a simple example:

Input number	Hashing algorithm	Hash value
10,667	Input # x 143	1,525,381

16.3 Are You Authentic

As stated earlier, encryption is the process of taking all of the data that one computer is sending to another and encoding it into a form that only the other computer will be able to decode. Another process, authentication, is used to verify that the information comes from a trusted source. Basically, if information is "authentic," you know who created it and you know that it has not been altered in any way since that person created it. These two processes, encryption and authentication, work hand-in-hand to create a secure environment.

There are several ways to authenticate a person or information on a computer:

- Password—The use of a user name and password provides the most common form of authentication. You enter your name and password when prompted by the computer. It checks the pair against a secure file to confirm. If either the name or the password does not match, then you are not allowed further access.
- Digital signatures—A digital signature is basically a way to ensure that an electronic document (e-mail, spreadsheet, text file) is authentic. The Digital Signature Standard (DSS) is based on a type of public-key encryption method that uses the Digital Signature Algorithm (DSA). DSS is the format for digital signatures that has been endorsed by the U.S. government. The DSA algorithm consists of a private key, known only by the originator of the document (the signer), and a public key. If anything at all is changed in the document after the digital signature is attached to it, it changes the value that the digital signature compares to, rendering the signature invalid.

Recently, more sophisticated forms of authentication have begun to show up on home and office computer systems. Most of these new systems use some form of biometrics for authentication. Biometrics uses biological information to verify identity.

To make sure no error appear in the progress of transmission or encryption, two common ways is:

- Checksum—Probably one of the oldest methods of ensuring that data is correct, checksums also provide a form of authentication because an invalid checksum suggests that the data has been compromised in some fashion. A checksum is determined in one of two ways. Let's say the checksum of a packet is 1 byte long. A byte is made up of 8 bits, and each bit can be in one of two states, leading to a total of 256 (2^8) possible combinations. Since the first combination equals zero, a byte can have a maximum value of 255.

Byte 1	Byte 2	Byte 3	Byte 4	Byte 5	Byte 6	Byte 7	Byte 8	Total	Checksum
212	232	54	135	244	15	179	80	1151	127

$$1\,151/256=4.496\ (\text{round to } 4)$$
$$4\times 256=1\,024$$
$$1\,151-1\,024=127$$

- Cyclic Redundancy Check (CRC)—CRCs are similar in concept to checksums, but they use polynomial division to determine the value of the CRC, which is usually 16 or 32 bits in length. The good thing about CRC is that it is very accurate. If a single bit is incorrect, the CRC value will not match up. Both checksum and CRC are good for preventing random errors in transmission but provide little protection from an intentional attack on your data. Symmetric-and public-key encryption techniques are much more secure.

Keywords

encryption	加密
sensitive	敏感的
straightforward	直接的
removable	抽取式的
throughout history	贯穿历史
digital age	数字时代
military	军事的
verify	验证
crack	破解
symmetric-key	对称密钥
public-key	公共密钥
nonsense	胡言乱语
Pretty Good Privacy (PGP)	良好隐私
digital certificates	数字证书
middleman	中间人
Secure Sockets Layer (SSL)	加密套接字协议层
authentication	认证
prompt	提示
Digital signatures	数字签名
endorse	认可
biometrics	生物识别技术
checksum	校验

Cyclic Redundancy Check (CRC)   循环冗余校验

16.4 Exercise 16

Multiple or single choices.

1. Which of the following simple but straightforward security method do you think is more safe? ____.
    A. Keep sensitive information on U disk
    B. Keep sensitive information on floppy disk
    C. Keep sensitive information on company computer
    D. Keep sensitive information on mobile hard disk

2. You send a message which is "HSL" and you told him that the code is "Shift by 3". Your friend gets the message and he will know what you really want to say is ____.
    A. EDL
    B. DKS
    C. SIO
    D. KVO

3. There are several ways to authenticate a person or information on a computer include ____.
    A. password
    B. digital signature
    C. biometrics
    D. hashing algorithm

4. Biometric authentication methods include ____.
    A. speech recognition
    B. fingerprint recognition
    C. retina recognition
    D. face scanning

True or False.

1. Before the digital age, the biggest users of cryptography were governments, particularly for military purposes.

2. Public-key encryption uses a combination of a private key and a public key.

3. A digital certificate is basically a bit of information that says that the Web server is trusted by an independent source known as a certificate authority.

16.5   Further Reading: Identity Theft

In this article, we'll look into the dark world of identity theft to which we can all fall victim. We'll find out how others can get access to your personal identification information, how you can protect yourself, and what to do if you become a victim.

16.5.1   Types of Identity Theft

Identity theft can enter into many areas of our lives. It involves any instance where a person uses someone else's identification documents or other identifiers in order to impersonate that person for whatever reason. According to a September 2003 survey conducted by the Federal Trade Commission, an estimated 10 million people in the United States found out they were victims of identity theft in the previous year. More appropriately titled identity fraud, your identity might be stolen in order for someone to commit:

- Financial fraud—This type of identity theft includes bank fraud, credit card fraud, computer and telecommunications fraud, social program fraud, tax refund fraud, mail fraud, and several more. In fact, a total of 25 types of financial identity fraud are investigated by the United States Secret Service. While financial identity theft is the most prevalent (of the approximate 10,000 financial crime arrests that Secret Service agents made in 1997, 94 percent involved identity theft), it certainly isn't the only type. Other types of identity theft, however, usually involve a financial element as well-typically to fund some sort of criminal enterprise.
- Criminal activities—This type of identity fraud involves taking on someone else's identity in order to commit a crime, enter a country, get special permits, hide one's own identity, or commit acts of terrorism. These criminal activities can include:

    a. Computer and cyber crimes.

    b. Organized crime.

    c. Drug trafficking.

    d. Alien smuggling.

    e. Money laundering.

16.5.2   Stealing Your Identity

Have you ever eaten at a restaurant, paid with a credit card, and forgotten to get your copy of the credit card receipt? Did you know that many of these receipts have your credit card number printed right there for anyone to see (and use)? And, if you've signed them, your signature (as example as shown in Figure 16.3) is also right there for someone to carefully copy. This can lead to the most simple form of identity theft. With this bit of information, some unscrupulous person can be well on his way to making purchases either by phone or on the Internet using your credit card number. You won't know about it until you get your statement (a good reason why you should always study the charges on your credit card statements!). All they have to have, in most cases, is your mailing address, which can be looked up in a phone book or easily found on the Internet.

Figure 16.3   Signature

Credit card fraud is identity theft in its most simple and common for. It can be accomplished either through a scenario like the one we just mentioned, or it can happen when your pre-approved credit card offers fall into the wrong hands. All a person has to do is get these out of your mailbox (or trash can) and mail them in with a change of address request and start spending. Someone can even apply for a credit card in your name if they have the right information. You won't know a thing about it until the credit card company tracks you down and demands payment for the purchases "you" have a racked up.

16.5.3   Accessing Your Personal Information

Your personal information can be found in many places. It can be:

Dug out of trash cans and dumpsters, known as "dumpster diving."

Basically, anywhere you've provided that information can be a target. Often, employees who have access to the information are bribed or offered a cut of the profits in exchange for personal information about other employees. The more sophisticated the perpetrator, the more money is stolen and the more people scammed. Clerks can even put skimmers on the credit card machines that will record credit card information for later use. Temporary employees seem to be more frequently involved in identity theft scandals than permanent employees, simply because fewer background checks are done on them.

16.5.4 Public Information

What about all of the publicly available information someone can access about you? Sources for this information include:

While some information about your life is pretty well protected, such as medical, financial and academic records, your other identifying information (social security number, home address, etc) is not so protected. One scary statistic: According to the Federal Trade Commission (FTC), in 2000, 19 percent (as opposed to 13 percent in 2001) of all victims of identity theft who completed that section of the FTC complaint form had a personal relationship with the thief; 10 percent of those thieves were family members.

16.5.5 How To Protect Yourself

Protecting yourself from identity theft takes proactive effort. You can't simply assume it's not going to happen to you and go on about your life—it can happen to anyone. It even happens to celebrities. Oprah Winfrey, Tiger Woods, Robert De Niro and Martha Stewart have all had their identities stolen. While you can't ever totally protect yourself from these thieves, you can at least make yourself less attractive as a victim by doing what you can to make it more difficult for them to access your information. Here are some things you can do to protect yourself:

DON'T give out your Social Security number unless it is absolutely necessary. Many companies collect more information than they really need. Make sure that it's something they have to have and make sure they'll protect your privacy.

REVIEW your monthly credit card statement each month to make sure there aren't any charges showing up that aren't yours. Also, make sure you get a monthly statement. If the statement is late, contact the credit card company. You never know when someone may have turned in a change-of-address form so they could make a few more weeks of purchases on your credit card without you noticing.

Identity Theft Insurance?

Some insurance companies offer identity theft insurance. While these policies don't cover everything, they certainly help out by covering a portion of lost wages for time spent dealing with the theft, mailing and other costs associated with filing paperwork to correct the problem, loan re-application fees, phone charges and even some attorney fees.

These steps can help lessen your chances of becoming a victim of identity fraud, but nothing is a sure thing. The thing to remember is that documents you throw away often have all the information a thief needs to steel your identity and wreak havoc on your life.

16.5.6 Internet Transactions

The ease of shopping and comparing products and prices online has made it an attractive option for many shoppers. How can you make sure your transactions are safe and your credit card information going only where you intend it to? There are several ways to help ensure safe transactions on the Internet, and more are becoming possible all the time. Some of these include:

- Use the latest Internet browser. The program that you use to surf the Internet is called a browser. This software has built-in encryption capabilities that scramble the information you send to a server. Using the most recent browser ensures that the data is protected using the latest encryption technology. This technology also uses a Secure Sockets Layer (SSL), which is an Internet security protocol used by Internet browsers and Web servers to transmit sensitive information. The server receiving the data uses special "keys" to decode it. You can make sure you are on an SSL by checking the URL—the http at the beginning of the address should have changed to https. Also, you should notice a small lock icon in the status bar at the bottom of your browser window.
- Look for digital certificates that authenticate the entity you are dealing with. Independent services like VeriSign will authenticate the identity of the Web site you are visiting. Web sites that use this service (usually those that sell items or services online) will have the VeriSign logo. By clicking on the logo, you can be assured that the site is legitimate, rather than a clone of the legitimate company set up to collect your personal and financial information.
- Read the privacy policy. The information you enter on the Web site should be kept confidential. Make sure you read the company's privacy policy to ensure that your personal information won't be sold to others. Services like Trust-E review a company's privacy policy (for a fee) and then allow the company to post the Trust-E logo if its privacy policy follows certain industry standards for consumer protection.
- Only use one credit card for all of your online purchases.
- Never give out passwords or user ID information online unless you know who you are dealing with and why they need it. Don't give it out to your Internet service provider if you get an e-mail requesting it. This is a relatively recent scam used to access your account and get your credit card number, along with whatever other personal information is there.
- Keep records of all of your Internet transactions. Watch your credit card statement for the charges and make sure they're accurate.
- After you've made purchases online, check your e-mail. Merchants often send confirmation e-mails or other communications about your order.

16.5.7　If It Happens To You

What if you find out through a phone call from a creditor, a review of your credit report, or even a visit from the police, that your identity has been stolen. The first thing to do is report the crime to the police and get a copy of your police report or case number. Most credit card companies, banks, and others may ask you for it in order to make sure a crime has actually occurred.

You should then immediately contact your credit card issuers, close your existing accounts and get replacement cards with new account numbers. Make sure you request that the old account reflect that it was "closed at consumer's request" for credit report purposes. It is also smart to follow up your telephone conversation with letters to the credit card companies that summarize your request in writing.

Close any accounts the thief has opened in your name. If you open new accounts yourself, make sure you request that passwords be put on those accounts. As with any password, make sure you use something that is not obvious to others. Don't use your mother's maiden name, the last four digits of your social security number, or anything else that would be obvious.

Next, call the fraud units of the three credit reporting bureaus and report the theft of your credit cards and/or numbers. Ask that your accounts be flagged with a "fraud alert." This usually means that someone can't set up a new account in your name without the creditor calling you at a phone number you specify. Verify with the credit bureau representative you speak with that this will happen, and provide them with the number at which you want to be reached. The down side of this is that you won't be able to get "instant credit" at department stores. This flag, also known as a "victim's statement," is the best way to prevent unauthorized accounts.

The Credit Bureaus
Equifax Credit Information Services-Consumer Fraud Div.
P.O. Box 105496
Atlanta, Georgia 30348-5496
Tel: (800) 997-2493
www.equifax.com

Experian
P.O. Box 2104
Allen, Texas 75013-2104
Tel: (888) EXPERIAN (397-3742)
www.experian.com

TransUnion Fraud Victim Assistance Dept.
P.O. Box 390
Springfield, PA 19064-0390

Tel: (800) 680-7289
www.transunion.com

Reporting to the FTC

Consumer Response Center
Federal Trade Commission
600 Pennsylvania Ave, NW
Washington, DC 20580
Toll-free 877-FTC-HELP (382-4357)
On the Web: www.ftc.gov/ftc/complaint.htm
For consumer information: www.ftc.gov/ftc/consumer.htm

16.5.8　What Congress Is Doing About It

Congress declared identity theft a federal crime in 1998 when it passed the Identity Theft and Assumption Deterrence Act. This offense, in most circumstances, carries a maximum term of 15 years imprisonment, a fine, and criminal forfeiture of any personal property used or intended to be used to commit the offense.

16.5.9　What the Future Holds

Future efforts for preventing identity theft will most likely come through technological advancements that incorporate some physical aspect of a person's body in order to verify identity. Known as biometrics, this type of authentication uses individually unique physical attributes such as fingerprints, iris/retina, facial structure, speech, facial thermograms, hand geometry and written signature. It can be used to authenticate both your identity and the party you are dealing with.

Chapter 17
Taking a Closer Look at the DCE

Citibank is relying on an integrated set of operating system and net-independent services called the Distributed Computing Environment (DCE) to achieve transparent connections between the systems. By employing the DCE, Citibank developers can concentrate on the application itself rather than devising application enablers.

The DCE, which the Open Software Foundation, Inc. (OSF) announced in May 1990, is a technology many users will likely end up licensing from systems and network vendors. Most of the major players have committed to using the DCE as the basis for future systems and network software.

Currently, the DCE consists of a set of services organized into two categories.

The first grouping is fundamental distributed services, which provide programmers with the tools necessary to create distributed computing applications. These include threads, remote procedure calls (RPC), directory, time and security services.

The other category, data-sharing services, provides users with capabilities built on top of the basis for efficient information use across a network. They include a distributed file system and personal computer integration consisting of MS-DOS file and printer support services.

In the future, OSF plans to address spooling, transaction processing and distributed, object-oriented environments.

17.1 Common Threads

The DCE Threads Service provides portable facilities that allow a programmer to build an application that performs many actions simultaneously. For instance while one thread executes an RPC, another can process user input. By contrast, applications traditionally have dealt with a single thread of control. Threads Service is used by other DCE components, such as RPC, Security, Naming, Time, and the Distributed File System.

Threads Service includes operations to create and control multiple threads of execution within a single process and synchronize across to global data within an application.

This threading capability becomes particularly important within the context of an RPC, for instance. RPC, by nature, is synchronous operations. A client makes a call for a remote function and then waits until the request is fulfilled. With threads, however, one thread can make the request while another begins to process data from a different request. Threading can greatly improve the performance of a distributed application.

Threads Service puts less demand on the skill of a programmer than on other alternatives such as explicit asynchronous operations or shared memory. Asynchronous interfaces, although they have existed in some environments for some time, can be a major cost drain, the less retraining a new technology requires, the better.

17.2 Remote Calls

RPC are one of the tried-and-true modes of implementing distributed processing. Their function is to make procedures in an application run on a computer any where in the network.

RPC handle the nuts and bolts of distribution, such as the semantics of the call, binding the client to the serve or communications failures. In theory, the programmer doesn't have to become a communications expert to write a distributed network application. Programmers use an interface specification language to detail remote operations, Compiling those routines to produces code for both the client and the server.

The benefit of such an RPC approach is that it provides simplicity to the programmer. RPC adhere to the local procedure model as closely as possible while providing the distributed aspects of applications in a straightforward manner. In other words, it foists less of a conceptual change on developers, thus reducing retraining time. This is especially important for inhouse corporate development teams.

Regardless of the transport protocol used, RPC provide identical behavior within applications and keep the management of connections invisible. This means developers don't have to rewrite applications to support different transport services. The PRC interface supports a variety of transports simultaneously and allows the introduction of new transports and protocols without affecting the coding of the application.

17.3 Directory Services

Locating object—users, servers, data and applications—in a distributed network is the task of the DCE's Distributed Naming Service (DNS). The service enables programmers and end users to identify resources such as server, files, disks or print queues by name, without knowing their location in a network.

OSF specified a two-tiered architecture for DNS to address both intracell and worldwide communications. The cell is a fundamental organizational unit for systems in the DCE. They can map to social, political or organizational boundaries and consist of computers that must frequently communicate with one another—such as in work groups, departments of divisions of companies. Generally, computers in a cell are geographically close and each cell ranges in size from two to thousands of computers.

There are four elements in the DCE's DNS: the Cell Directory Service (CDS), Global Directory Agents (GDA), the Global Directory Service (GDS) and the X/Open Directory Service (XDS).

The DCE's CDS handles directory queries from clients in a cell. It looks at the first part of a file name, for instance, to determine if the data resides in the cell. If it does, it supplies the data. If not, it passes the request to a GDA, which does the look up in the GDS and feeds it back to the client through the CDS. The requesting client can then issue a direct call to the CDS with the file location data. CDSs typically reside on multiple servers in cells across the network.

17.4 Distributed Security Service

The DCE's Distributed Security Service (DSS) offers several levels of security. The DCE uses the Kerberos authentication system, which was developed by the Massachusetts Institute of Technology under its Project Athena.

Kerberos uses private key encryption to provide three levels of protection. The lowest requires only that user authenticity be established at the initiation of a connection, assuming that subsequent network messages flow from the authenticated principal. The next level requires the authentication of each network message. On the level beyond these, safe messages are private messages, where each is encrypted as well as authenticated.

End users should be minimally affected by the net-based service. In other words, you shouldn't have to memorize dozens of passwords or codes. A great deal of the security benefits stems from this network service managing a user's access, or authorization.

DSS also comes with an authorization service. The DCE supports authorization checks based on Portable Operating System Interface-conformant access control lists (ACL) and an authentication interface to RPCs.

17.5 Distributed File System

The DCE's Distributed File System (DFS) is intended to provide transparent access to any file sitting on any node on the net, provided clearance is obtained.

DFS is based on Transarc Corp.'s. Andrew File System (AFS) Version 4.0, which OSF opted for over Sun's Network File System (NFS). DFS software resides on each net node. DFS integrates each node's file system with DCE's directory services, ensuring a uniform naming convention for all files stored in DFS.

OSF chose AFS because it allows users to address files with the same path name from anywhere in the networks, regardless of the user's computer.

It uses DCE's Security System with ACLs, which control access to individual files. An RPC streaming function allows DFS to move large amounts of data through a wide-area net work in one burst rather than dribbling it across in smaller packers. This capability is important because of the latencies inherent in WANs.

To maximize file access performance, DFS caches frequently accessed files on a workstation's local disk. When a user accesses date on the file server, a copy of the data is cached locally. When the user is finished working with the data, a local cache manager writes the data back to the server. To prevent problems from arising when multiple users on different computers access and modify the same data, DFS uses a token management scheme to coordinate file modification.

File tokens allocated to client by a serve when the client caches data locally. If a client wishes to modify a file, it must request a write token from the server, which allows it to make changes. This setup ensures that clients holding read-only tokens will be notified their files are no longer valid in the event of a file change.

17.6 Distributed Time Service

Distributed network systems need a consistent time service to synchronize operations on computers across the network.

In the DCE, a time server is used to provide time to other systems for the purpose of synchronization. Any non-time server system is called a clerk. Distributed Time Service (DTS) uses three types of servers to coordinate network time. A Local Server synchronizes with other local servers of the same local-area networks. A Global Server is available across an extended LAN or WAN. A Courier is a designated local server that regularly coordinates with global servers.

At periodic intervals, servers synchronize with every other local server on the LAN via the DTS protocol.

17.7 Extending and Using the DCE

This is not to say that the DCE is perfect. Plenty of opportunities exist for enhancement, some of which OSF will handle, and others that vendors and users will tackle.

With all the discussion about the DCE over the past few years, it is easy to forget the software is just now making its impact on the market. Tool kits are appearing, and pioneering users—such as Citibank—are beginning pilot developments.

With users beginning to explore the technology and vendors bringing out new products that support the DCE, it appears the industry is taking its first step down the road to truly transparent interoperability.

Keywords

Distributed Computing Environment	分布式计算环境
vendors	供应商
remote procedure calls (RPC)	远程过程调用
spooling	后台
portable	可携带
simultaneously	同时
synchronize	同步
asynchronous	异步
fulfill	落实
cost drain	开支
nuts and bolts	螺母和螺栓
adhere	采取
inhouse	室内
Distributed Naming Service	分布式命名服务
intracell	小区内
Cell Directory Service	单元目录服务
Massachusetts Institute of Technology	麻省理工学院
principal	主要
latencies	潜伏期

17.8   Exercise 17

Multiple or single choices.

1. The fundamental distributed services of DCE, which provide programmers with the tools necessary to create distributed computing applications include ____.
   A. threads
   B. remote procedure calls (RPC)
   C. directory
   D. security services.

2. Which of the following answer is the elements in the DCE's DNS? ____.
   A. CDS
   B. ADS
   C. GDS
   D. XDS

3. Servers can obtain the official Universal Coordinated Time from ____.
   A. standards organizations via short-wave radio
   B. dial-up lines
   C. current computer time
   D. Satellite

True or False.

1. DCE Threads Service provides portable facilities that allow a programmer to build an application that performs many actions simultaneously.

2. RPC is asynchronous operations.

3. RPC makes procedures in an application run on a computer any where in the network.

4. Distributed network systems need a consistent time service to synchronize operations.

5. Kerberos authentication system is developed by the Massachusetts Institute of Technology under its Project Athena.

17.9 Further Reading: How to Kerberize Your Site

17.9.1 Introduction

There's a lot of talk on the Internet about security and the lack of it on UNIX systems. This is, in part, a by-product of the world of the net in which we choose to do business. We can do some things to help counteract the possibility of attack on our systems.

One way to help is with Kerberos. Kerberos is an authentication and encryption scheme that allows a user to become "known" by an authenticating server and then use that authentication to access systems and services on the net. The services can then transpire in an encrypted fashion to further secure transactions occurring over the net. The philosophy behind the creation of Kerberos, and a short summary of how it works is available, but here we assume that you know what Kerberos is, and wish to implement a Kerberos domain on your network. But, we also assume that you are not a hot-shot UNIX programmer, so we intend to lead you by the hand in a step-by-step fashion through the entire process. In other words, this is our version of "Kerberos for Dummies."

Several commercial integrators provide enterprise Kerberos solutions as well as technical support and maintenance. In particular, perhaps the easiest way to install Kerberos V5 is to use Kerbnet from Cygnus solutions. Kerbnet is free and has clients for Win32 machines, Macintoshes and UNIX hosts, and has KDC software for UNIX and NT as well as host servers for UNIX platforms.

17.9.2 Pick a Kerberos Server Machine (KDC)

The machine that you pick as your Kerberos server is the key to your network security. Therefore, it must be as secure as possible. Here is a non-exhaustive list of things to consider when selecting your server machine:

The server should be physically secure.

The operating system should be up to date with all the latest patches applied.

There should be no user accounts on the machine except for the Kerberos administrator.

There should as few processes as possible running on the server other than the Kerberos key server demons.

You probably need additional machines to serve as alternate servers in case there is a hardware problem or a network outage.

17.9.3 DCE and Kerberos

There is alternate mode of operation which uses the DCE Security Service as the K5 KDC. Several sites (such as ESnet) are doing this. This is a good idea if you have DCE at your site because it allows you to have just one database of users rather than having users registered in two places.

17.9.4 Install the Kerberos Server

Be sure to get Kerberos version 5 patch level 1 (or greater) to fix two serious security holes.

In these instructions, your typing is shown in italics.

Consider obtaining the Kerbnet code from Cygnus Solutions. This code is prebuilt and well-documented.

17.9.5 Obtain the Necessary Code

(1) Create a directory for your kerberos code.
mkdir /kerberos
cd /kerberos
(2) ftp to MIT to obtain the code. It is easiest if you use your Web browser.
ftp://athena-dist.mit.edu/pub/kerberos
An alternate site for all of the software is ftp://prep.ai.mit.edu/pub/kerberos.

(3) In the /pub/kerberos directory, get the README.KRB5 file:

(4) Read the file to find out the name of the directory containing the code (it changes periodically)

For me, this was dist/961209, but it may have changed. You must go to this directory in one step!

ftp://athena-dist.mit.edu/pub/kerberos/dist/961209

(5) Retrieve all the files (in binary!). The .asc files are the digital signatures for the corresponding files so that you can ensure that they have been unchanged.

(6) gunzip all the files and extract them. If you do not have gunzip, you can obtain it from the same place:

ftp://athena-dist.mit.edu/pub/gnu/gzip-1.2.4.tar

17.9.6 Do You Need More Code

Kerberos requires an ANSI-compatible C compiler and you will encounter fewer problems if you use the free gnu C compiler, gcc. So, you may have to make a temporary detour and obtain and compile all of the gnu tools. You can get a complete list at ftp://athena-dist.mit.edu/pub/gnu/.

To properly build the C compiler (gcc), you will first need to get and install bison and the gnu assembler (as). The assembler is found in the binutils package. Without the gnu assembler, I obtained numerous warnings during the compilation procedure, all of which disappeared when the gnu as was used. You might also be more successful if you use the gnu make facility. If you do not have a C library on your machine, you will also need to obtain glibc from the gnu distribution (this should NOT be necessary on a UNIX machine). For debugging, obtain the gnu debugger, gdb.

It will probably take an afternoon to build all of these tools. In each case, installation is fairly straight forward. As root, gunzip and untar each of the above .tar.gz files and switch into the program's root directory, which is always the utility name followed by the release number. To be safe, read the INSTALL file or README file if the former does not exist. You can use gzcat to save disk space by doing both of these steps at once:

gzcat filename.tar.gz | tar-xpf-

17.9.7 Building the Gnu Tools

I suggest making the tools in the order bison, make, binutils. In each case, the process is similar. Switch to the program's main source directory and

(1) Create the Makefile by running

./configure

(2) Make the program by doing

make

(3) Install the utilities in /usr/local/bin by doing

make install

17.9.8 Building the Gnu C Compiler

You are now ready to make the gnu C compiler. This is fairly complicated, so you need to read the INSTALL file and follow the instructions. Be sure to use the gnu assembler, as.

(1) In the gcc_version directory, run

./configure-with-gnu-as

(2) Build the stage 1 compiler. Just type make LANGUAGES=c in the compiler directory. This will take some time.

(3) Copy the initial compiler build into the stage1 subdirectory by typing

make stage1

(4) Copy the gnu assembler into the stage1 directory. For example,

cp /usr/local/bin/as ~/gnu/gcc-2.7.2.1/stage1/as

(5) Compile the compiler with itself to make the stage2 compiler.

make CC="stage1/xgcc-Bstage1/" CFLAGS="-g -O2" LANGUAGES="C C++"

If you do not wish to have a C++ compiler, just use LANGUAGES="C".

(6) Finally, to be safe, test the compiler by compiling it with itself one more time. Install any other necessary GNU tools (such as the gnu assembler or the GNU linker) in the "stage2" subdirectory as you did in the "stage1" subdirectory, then

make stage2

make CC="stage2/xgcc -Bstage2/" CFLAGS="-g -O2"

This is called making the stage3 compiler. Aside from the "-B" option, the compiler options should be the same as when you made the stage2 compiler. But the "LANGUAGES" option need not be the same. The command shown above builds compilers for all the supported languages; if you don't want them all, you can specify the languages to build by typing the argument "LANGUAGES="LIST"", as described above.

(7) You can compare the stage2 and stage3 compilers (they should be identical) by running make compare

However, I have had no luck doing this. The two stages are always quite different on my machine.

(8) Install the compiler driver, the compiler's passes and run-time support with "make install". Use the same value for "CC", "CFLAGS" and "LANGUAGES" that you used when compiling the files that are being installed. One reason this is necessary is that some versions of make have bugs and recompile files gratuitously when you perform this step. If you use the same variable values, those files will be recompiled properly.

For example, if you have built the stage 2 compiler, you can use the following command:

make install CC="stage2/xgcc -Bstage2/" CFLAGS="-g -O" LANGUAGES="C C++" This should copy the files "cc1", "cpp" and "libgcc.a" to files "cc1", "cpp" and "libgcc.a" in the directory "/usr/local/lib/gcc-lib/TARGET/VERSION", which is where the compiler driver program looks for them. Here TARGET is the target machine type specified when you ran 'configure', and VERSION is the version number of GNU CC. This naming scheme permits various versions and/or cross-compilers to coexist. This step also copies the driver program "xgcc" into "/usr/local/bin/gcc", so that it appears in typical execution search paths.

On my system, the make install step hung when it tried to use the makeinfo utility that was not present on my system. So, I had to edit the Makefile. I changed the lines for the action install-normal:

install-normal: install-common $(INSTALL_HEADERS) $(INSTALL_LIBGCC) \\ install-libobjc install-man lang.install-normal install-driver to eliminate install-info from the second line. Then gcc was successfully installed!

17.9.9 Compiling Kerberos

In general, you will have the best results if you use the GNU version of make. If you get weird errors during the make process, try the GNU make and see if they don't go away.

It is also good to get advice from experts. So, obtain the README file from Doug Engert's ftp site at Argonne National Laboratory:

ftp://achilles.ctd.anl.gov/pub/kerberos.v5/README

For example, for my HP-UX 10 system, Doug suggests many options in the configure command:

../src/configure -with-cc=gcc \\
--with-ccopts="-O" --prefix=/krb5\\
--with-cppopts='-DANL_DCE -DAFS524'

The prefix option places the resulting source into the directory /krb5 rather than the default. In general, I had much better success getting configure to work properly if I put the --prefix command near the beginning of the configure argument string rather than at the end. It shouldn't make any difference, but it did.

For Solaris 2.6 I used the Sun c89 compiler and make (no gnu utilities) and had no problems at all. I used the configure command:

../src/configure --with-cc=c89 \\
--enable-shared \\
--with-ccopts="-O"\\
--with-cppopts="-DANL_DCE -DANL_AFS_PAG -DANL_DFS_PAG -DAFS524 -DNO_MOTD"\\
--prefix=/krb5

For AIX 3.2.x, I had the IBM ANSI compiler, but the build only worked if I used

--with-cc=cc as opposed to xlc or c89. On this platform, all components built properly with the configure command:

../src/configure --with-cc=cc \\
--with-ccopts="-O" \\
--with-cppopts='-DANL_DCE -DAFS524'\\
--prefix=/krb5

For AIX 4.1.3, I used the IBM (cc) compiler and the configure command:

../src/configure --with-cc=cc --prefix=/krb5 \\
--with-cppopts='-DANL_DCE -DAFS524 -ULOGIN_CAP_F'

For AIX 4.2 with the IBM C/C++ compiler, configure would not work unless I used the command:

../src/configure --with-cc=cc --prefix=/krb5 \\
--with-cppopts='-DANL_DCE -DAFS524 -ULOGIN_CAP_F'

However, make will fail in the /src/util/pty directory unless you edit the Makefile to remove the two switches -DHAVE_SETUTXENT=1 -DHAVE_UTMPX_H=1

Having both utmp.h and utmpx.h included causes the utmp structure to be multiple defined, and the definition in utmpx is not the one that is needed.

The code for the telnetd will not compile because the include in the file src/appl/telnet/telnetd/termios-tn.c is incorrect. Change #include <termios.h> to #include <sys/termio.h>.

17.9.10 For All Platforms

In the /krb5-1.0/src directory, run the above configure command.
(1) Make the code by doing
make
in the src directory.
If you are using the GNU make, and have renamed it to gmake, use the command
gmake MAKE=gmake
(2) Install Kerberos using
make install
(or gmake MAKE=gmake install)
which puts the code into /krb5, as specified by the prefix option in configure.

17.9.11 Configuring the Kerberos KDC

The next step is to print out and read the three Kerberos manuals, located in the docs directory of the Kerberos distribution. Unfortunately, the configuration process is not totally obvious.

1. Create the Configuration Files

Create the krb5.conf and kdc.conf files using mine as boiler plate. krb5.conf needs to be in /etc, and the location of kdc.conf can be specified in the krb5.conf file. However, initially, Kerberos cannot find the file unless it is in the var directory inder the root kerberos installation location. Therefore, we put it in /krb5/var/krb5kdc/kdc.conf. Note that the realm names are case sensitive! Use the same case in both files and when making a connection to the realm.

Edit these files to reflect your Kerberos domain instead of mine (dsdoe.ornl.gov).

2. Create the Kerberos Databases

Create the Kerberos database using the kdb5_util command
/krb5/sbin/kdb5_util create -r dsdoe.ornl.gov -s

Initializing database '/krb5/lvar/krb5kdc/principal' for realm 'dsdoe.ornl.gov',
master key name 'K/M@dsdoe.ornl.gov'

You will be prompted for the database Master Password. It is important that you NOT FORGET this password.

Enter KDC database master key:
your_master_key

Re-enter KDC database master key to verify:
your_master_key

Replace our domain name with yours. The -s creates a stash file which is used to authenticate the KDC to itself.

Create an administrator kadm5.acl file following the instructions in the Kerberos manual. Put it in the location specified in the 'acl_file=' section of kdc.conf.

Add your administrator(s) to the KDC database as per the manual
/krb5:738: sbin/kadmin.local

kadmin.local: addprinc admin/admin@dsdoe.ornl.gov

Enter password for principal "admin/admin@dsdoe.ornl.gov": your_password
Re-enter password for principal "admin/admin@dsdoe.ornl.gov": your_password
Principal "admin/admin@dsdoe.ornl.gov" created./krb5/sbin/kadmin.local

Create the keytab file on the server. kadmind uses this to determine what access it should give to administrators. The manual is wrong here. Stay in kadmin.local and give the command:

kadmin.local: ktadd -k /krb5/var/krb5kdc/kadm5.keytab kadmin/admin kadmin/changepw

Entry for principal kadmin/admin with kvno 3, encryption type DES-CBC-CRC added to keytab WRFILE:/krb5/var/krb5kdc/kadm5.keytab.

Entry for principal kadmin/changepw with kvno 3, encryption type DES-CBC-CRC added to keytab WRFILE:/krb5/var/krb5kdc/kadm5.keytab.

kadmin.local: quit

Edit the /etc/services file to include the following kerberized servcies. This list shows all the available servcies. Your key server should only have the uncommented lines on the key server machine. The other services are used for Kerberized hosts.

```
# # Kerberos (Project Athena/MIT) services
#
#kerberos 88/udp kdc # Kerberos 5 kdc
#kerberos 88/tcp kdc # Kerberos 5 kdc
#klogin 543/tcp # Kerberos rlogin -kfall
#kshell 544/tcp krcmd # Kerberos remote shell -kfall
krb5_prop 754/tcp # Kerberos v5 slave propagation
kerberos-adm 749/tcp # Kerberos v5 admin/chpwd
kerberos-adm 749/udp # Kerberos v5 admin/chpwd
#eklogin 2105/tcp # Kerberos encrypted rlogin -kfall
kpasswd 761/tcp kpwd # Kerberos "passwd" -kfall
#ktelnet 545/tcp # Kerberized telnet v4/v5
#kftp-data 546/tcp # Kerberized ftp data V5
#kftp 547/tcp # Kerberized ftp v5
#
```

3. Start the Kerberos Key Servers

Start the master KDC and the Kerberos administration demons:

/krb5/sbin/krb5kdc

/krb5/sbin/kadmind

If you want the two servers to start up automatically when your kdc machine is rebooted, you need to add them to your rc.local, inittab, or init.d or whatever your system uses to start processes at boot time.

4. Getting the Kadmin Demon to Work

kadmin allows you to administer the Kerberos databases remotely (and securely). If you just run kadmin, you will obtain an error message like:

kadmin: Client not found in Kerberos database while initializing kadmin interface

To be able to use the kadmin interface, you need to register yourself as a database administrator.

On the KDC machine, in kadmin.local add an administrator role for yourself:

kadmin.local: addprinc jar/admin@dsdoe.ornl.gov

Enter password for principal "jar/admin@dsdoe.ornl.gov": your_password

Re-enter password for principal "jar/admin@dsdoe.ornl.gov": your_password

Principal "jar/admin@dsdoe.ornl.gov" created.

kadmin.local: quit

Now, on a remote machine (on which you have also installed Kerberos), you can get a ticket as an administrator.

dsrocf:/krb5/bin: ./kinit jar/admin

Password for jar/admin@dsdoe.ornl.gov: your_password
dsrocf:/krb5/bin: ./klist
Now you can check to see that you have the correct ticket
Ticket cache:
/tmp/krb5cc_0
Default principal: jar/admin@dsdoe.ornl.gov
Valid starting Expires Service principal
18 Dec 96 14:13:52 19 Dec 96 00:13:26 krbtgt/dsdoe.ornl.gov@dsdoe.ornl.gov
NOTE: This HP-UX machine has DCE clients installed as part of the operating system. BE SURE TO USE THE PROGRAMS IN THE /krb5 DIRECTORY TREE. THE DCE VERSIONS ARE NOT COMPATIBLE WITH KERBEROS V5.

Now you can access kadmin on the Kerberos server (dsroc3) from dsrocf.
dsrocf:/krb5/sbin:409: ./kadmin
Enter password: your_password
kadmin:

5. Tip

A very easy way to see whats going wrong is to use strace. You can see what the program is trying to do and where it fails. To use strace on kadmin for example: strace kadmin.

17.9.12 Setting Up a Host Server

A host is a machine that offers kerberized services (telnet, ftp) to users. We want to configure dsrocf as a Kerberized host. To allow this, we need to create a /etc/krb5.keytab file. This is easy now that we can run kadmin remotely on the KDC. You need to be root on the host machine in order to be able to write into /etc. In kadmin, add the host as a principal:

kadmin: add_principal host/dsrocf.dsdoe.ornl.gov
Enter password for principal "host/dsrocf.dsdoe.ornl.gov@dsdoe.ornl.gov":
Re-enter password for principal "host/dsrocf.dsdoe.ornl.gov@dsdoe.ornl.gov":
Principal "host/dsrocf.dsdoe.ornl.gov@dsdoe.ornl.gov" created.

Then add its keytab entry in the LOCAL (dsrocf) /etc/krb5.keytab file. This process securely shares a secret key to be used for communication between the Kerberized host and the KDC server.

kadmin: ktadd host/dsrocf.dsdoe.ornl.gov
Entry for principal host/dsrocf.dsdoe.ornl.gov with kvno 4,
encryption type DES-CBC-CRC added to keytab WRFILE:/etc/krb5.keytab.
Repeat this process for every host in your realm.

Finally, you should add the following lines to the end of the /etc/inetd.conf file on each host so that the Kerberos daemons start up automatically when your host is rebooted:

#

```
#Kerberos daemons
#
klogin    stream tcp nowait root /krb5/sbin/klogind klogind-ki
eklogin   stream tcp nowait root /krb5/sbin/klogind klogind-eki
kshell    stream tcp nowait root /krb5/sbin/kshd kshd-ki
ktelnet   stream tcp nowait root /krb5/sbin/telnetd telnetd-a user
kftp      stream tcp nowait root /krb5/sbin/ftpd-a
```

If you change the inetd.conf file, you must inform the inet daemon to put them into effect. It is done differently on different systems. For example, on AIX run the refresh -s inetd or kill -1 InetdPID command. On HP-UX, use the command inetd -c.

17.9.13 Domain Names

Kerberos uses gethostbyname( ) to determine the fully qualified domain name of each Kerberos host. On my system, this returned the short name and caused the Kerberos server to deny authorization. To test this function, run

/krb5/src/tests/resolve/resolve \

for each host. I had to edit the /etc/hosts file to be sure that the long host name came before the short name.

17.9.14 Kerberos Clients

A snapshot of the NT (Intel platform) release is available from MIT. This is a preliminary alpha release, but it seems to work well. To use this version, unzip all the files into a subdirectory e.g., D:\\KRB5). Move the dlls to C:\\WINNT\\SYSTEM32. Put your krb5.ini file into the C:\\WINNT directory. krn5.ini has the same contents as the krb5.conf file that is used on UNIX systems.

17.9.15 Getting a Ticket for Another Realm

Doug Engert (ANL) has extended DCE to support Kerberos tickets and to do cross-realm authentication. Here are the steps that I had to use to log on securely to a machine in Doug's domain. Note that we exchanged passwords over the telephone and later changed them in a face-to-face meeting.

Both of us had to perform these steps on our KDCs using our secret shared password:
/krb5/sbin/kadmin.local-e des:v4
addprinc-kvno 1-pw \ krbtgt/dsdoe.ornl.gov@k5.test.anl.gov
addprinc-kvno 1-pw \ krbtgt/k5.test.anl.gov@dsdoe.ornl.gov
Next, I had to run kinit (as myself) to get a ticket from my KDC.
dsrocf:/home/jar:7: /krb5/bin/kinit jar

Password for jar@dsdoe.ornl.gov:
dsrocf:/home/jar:8:
Then I was able to test to see if I could rlogin to a machine at ANL from ORNL:
/krb5/bin/rlogin caliban.ctd.anl.gov -x -l b17783
caliban.ctd.anl.gov% exit
caliban.ctd.anl.gov% logout
Connection closed.
Local flow control on
It worked! We can examine the ticket cache to see how this worked.
dsrocf:/krb5:11: bin/klist
Ticket cache: /tmp/krb5cc_11192
Default principal: jar@dsdoe.ornl.gov Valid starting Expires Service principal
05 Dec 96 16:13:33 06 Dec 96 02:13:33 krbtgt/dsdoe.ornl.gov@dsdoe.ornl.gov
05 Dec 96 16:14:33 06 Dec 96 02:13:33 krbtgt/k5.test.anl.gov@dsdoe.ornl.gov
05 Dec 96 16:14:40 06 Dec 96 02:13:33 host/caliban.ctd.anl.gov@k5.test.anl.gov
dsrocf:/krb5:12:

17.9.16   Kerberos Security Problems

There has been discovered a security-hole in kerberized rsh, rcp and rlogin.

Everyone who has setuid-bits set on these applications is adviced to disable them. The hole allows any user on the system to gain privilegies of any other user including root.

The hole has been successfully tested on kth-kerberos, but is suspected to exist on any other versions of kerberos.

Artur Grabowski (administrator on stacken.kth.se)

Serious buffer overrun vulnerabilities exist in many implementations of Kerberos 4, including implementations included for backwards compatibility in Kerberos 5 implementations. Other less serious buffer overrun vulnerabilites have also been discovered. ALL KNOWN KERBEROS 4 IMPLEMENTATIONS derived from MIT sources are believed to be vulnerable.

17.9.17   Kerberos Authentication Option in SSL

Netscape's Secure Socket Layer (SSL 3.0) has been modified to support the Kerberos authentication option as described in draft-ietf-tls-kerb-cipher-suites-01.txt of the Internet Engineering Task Force (IETF). A reference implementation is available at:
ftp://prospero.isi.edu/pub/ssl-krb

The draft (presented at the IETF's Transport Layer Security (TLS) working group meeting, Dec. 1996) proposes the addition of new cipher suites to the TLS protocol (SSL 3.0) to support Kerberos-based authentication. Kerberos credentials are used to achieve mutual authentication and to establish a master secret which is subsequently used to secure client-server communication.

Note: The reference implementation uses MIT's Kerberos V5 beta 6.

17.9.18 Available Kerberized Goodies

Wolfgang S. Rupprecht has created a version of the CVS code revision control system that uses Kerberos authentication of users. An excellent online CVS manual (for the unkerberized version) is available at http://www.brandonu.ca/~weingart/misc/cvstrain/cvstrain.html.

17.9.19 CygnusKerbnet for NT, Macs, and UNIX

In September, 1997 Cygnus Solutions released a free implementation of Kerberos V5 including clients for Windows 95, Windows NT, and Macintoshes. Kerbnet comes with source code, precompiled binaries and excellent documentation. For the first time, you can set up a KDC on an NT platform which offers some security advantages over UNIX.

Unfortunately, the CygnusKerbnet version of Kerberos is no longer available. However, Cygnus offers a PC environment called Cygwin that emulates UNIX. The Kerberos source code has been modified by Vern Staats to run on Windows 2000 including ssh.

Chapter 18
What is Wi-Fi and How does it work

Wi-Fi (see Figure 18.1) is increasingly becoming the preferred mode of internet connection all over the world. To access this type of connection, one must have a wireless adapter on their computer. Areas which are enabled with Wi-Fi connectivity are known as Hot Spots. One can use advanced softwares like Wirelessmon to detect and request connection to Hotspots. To start a Wireless connection, it is important that the wireless router is plugged into the internet connection and that all the required settings are properly installed.

Wireless technology has widely spread lately and you can get connected almost anywhere; at home, at work, in libraries, schools, airports, hotels and even in some restaurants.

Wireless networking is known as Wi-Fi (Wireless Fidelity) or 802.11 networking as it covers the IEEE 802.11 (1) technologies. The major advantage of Wi-Fi is that it is compatible with almost every operating system, game device and advanced printer.

Figure 18.1   Wi-Fi

18.1   How does Wi-Fi work

Like mobile phones, a Wi-Fi network makes use of radio waves to transmit information across a network. The computer should include a wireless adapter that will translate data sent into a radio signal. This same signal will be transmitted, via an antenna, to a decoder known as the router. Once decoded, the data will be sent to the Internet through a wired Ethernet connection. As the wireless network will work as a two-way traffic, the data received from the Internet will also pass through the router to be coded into a radio signal that will be receipted by the computer's wireless adapter.

18.2 Uses

To connect to a Wi-Fi LAN, a computer has to be equipped with a wireless network interface controller. The combination of computer and interface controller is called a station. For all stations that share a single radio frequency communication channel, transmissions on this channel are received by all stations within range. The transmission is not guaranteed to be delivered and is therefore a best-effort delivery mechanism. A carrier wave is used to transmit the data. The data is organized in packets, referred to as "Ethernet frames".

1. Internet access

Wi-Fi technology may be used to provide Internet access to devices that are within the range of a wireless network that is connected to the Internet. The coverage of one or more interconnected access points (hotspots) can extend from an area as small as a few rooms to as large as many square kilometers. Coverage in the larger area may require a group of access points with overlapping coverage. For example, public outdoor Wi-Fi technology has been used successfully in wireless mesh networks in London, UK.

Organizations and businesses, such as airports, hotels, and restaurants, often provide free-use hotspots to attract customers. Enthusiasts or authorities who wish to provide services or even to promote business in selected areas sometimes provide free Wi-Fi access.

Routers that incorporate a digital subscriber line modem or a cable modem and a Wi-Fi access point, often set up in homes and other buildings, provide Internet access and internetworking to all devices connected to them, wirelessly or via cable.

Many smart phones have a built-in capability of this sort, including those based on Android, BlackBerry, Bada, iOS (iPhone), Windows Phone and Symbian, though carriers often disable the feature, or charge a separate fee to enable it, especially for customers with unlimited data plans. Some laptops that have a cellular modem card can also act as mobile Internet Wi-Fi access points. Wi-Fi also connects places that normally don't have network access, such as kitchens and garden sheds.

2. City-wide Wi-Fi

In the early 2000s, many cities around the world announced plans to construct city-wide Wi-Fi networks (see Figure 18.2). There are many successful examples; in 2004, Mysore became India's first Wi-Fi enabled city. A company called Wi-Fi Net has set up hotspots in Mysore, covering the complete city and a few nearby villages. In 2005, St. Cloud, Florida and Sunnyvale, California, became the first cities in the United States to offer city-wide free Wi-Fi (from Metro Fi). Minneapolis has generated $1.2 million in profit annually for its provider. In May 2010, London, UK, Mayor Boris Johnson pledged to have London-wide Wi-Fi by 2012. Several boroughs including Westminster and Islington already had extensive outdoor Wi-Fi coverage at that point.

Figure 18.2　Wi-Fi city

3. Campus-wide Wi-Fi

Many traditional university campuses in the developed world provide at least partial Wi-Fi coverage. Carnegie Mellon University built the first campus-wide wireless Internet network, called Wireless Andrew, at its Pittsburgh campus in 1993 before Wi-Fi branding originated. By February 1997 the CMU Wi-Fi zone was fully operational. Many universities collaborate in providing Wi-Fi access to students and staff through the durum international authentication infrastructure.

4. Direct computer-to-computer communications

Wi-Fi also allows communications directly from one computer to another without an access point intermediary. This is called ad hoc Wi-Fi transmission. This wireless ad hoc network mode has proven popular with multiplayer handheld game consoles, such as the Nintendo DS, PlayStation Portable, digital cameras, and other consumer electronics devices. Some devices can also share their Internet connection using ad hoc, becoming hotspots or "virtual routers".

Similarly, the Wi-Fi Alliance promotes the specification Wi-Fi Direct for file transfers and media sharing through a new discovery- and security-methodology.

Another mode of direct communication over Wi-Fi is Tunneled Direct Link Setup (TDLS), which enables two devices on the same Wi-Fi network to communicate directly, instead of via the access point.

18.3　Frequencies

A wireless network will transmit at a frequency level of 2.4 GHz or 5 GHz to adapt to the amount of data that is being sent by the user. The 802.11 networking standards (Figure 18.3) will somewhat vary depending mostly on the user's needs, as explained below:

The 802.11a will transmit data at a frequency level of 5 GHz. The Orthogonal Frequency-Division Multiplexing (OFDM) used enhances reception by dividing the radio signals into smaller signals before reaching the router. You can transmit a maximum of 54 megabits of data per second.

- The 802.11b will transmit data at a frequency level of 2.4 GHz, which is a relatively slow speed. You can transmit a maximum of 11 megabits of data per second.
- The 802.11g will transmit data at 2.4 GHz but can transmit a maximum of 54 megabits of data per second as it also uses an OFDM coding.
- The more advanced 802.11n can transmit a maximum of 600 megabits of data per second and uses a frequency level of 5 GHz.

Standard	Frequency band	Bandwidth	Modulation	Maximum data rate
802.11	2.4 GHz	20MHz	DSSS, FHSS	2 Mbps
802.11b	2.4 GHz	20MHz	DSSS	11 Mbps
802.11a	5 GHz	20MHz	OFDM	54 Mbps
802.11g	2.4 Ghz	20MHz	DSSS, OFDM	54 Mbps
802.11n	2.4 GHz, 5 GHz	20MHz, 40MHz	OFDM	600 Mbps
802.11ac	5GHz	20, 40, 80 80+80, 160 MHz	OFDM	6.93 Gbps
802.11ad	60 GHz	2.16 GHz	SC, OFDM	6.76 Gbps

Figure 18.3 IEEE 802.11 Standards

18.4 Advantages and Challenges

1. Commercial advantage

Wi-Fi allows cheaper deployment of local area networks (LANs). Also spaces where cables cannot be run, such as outdoor areas and historical buildings, can host wireless LANs. Manufacturers are building wireless network adapters into most laptops. The price of chipsets for Wi-Fi continues to drop, making it an economical networking option included in even more devices.

Different competitive brands of access points and client network-interfaces can inter-operate at a basic level of service. Products designated as "Wi-Fi Certified" by the Wi-Fi Alliance are backwards compatible. Unlike mobile phones, any standard Wi-Fi device will work anywhere in the world.

2. Restrictions

Spectrum assignments and operational limitations are not consistent worldwide: Australia and Europe allow for an additional two channels beyond the 11 permitted in the United States for the 2.4 GHz band (1~13), while Japan has three more (1~14).

A Wi-Fi signal occupies five channels in the 2.4 GHz band (Figure 13.4). Any two channel numbers that differ by five or more, such as 2 and 7, do not overlap. Channels 1, 6, and 11 are the only group of three non-overlapping channels in America. In Europe and Japan using Channels 1, 5, 9, and 13.

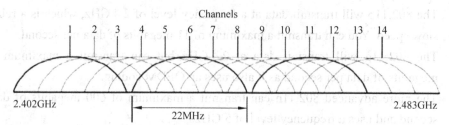

Figure 18.4 Channel distribution in the 2.4 GHz band

3. Transmission distance

The Wi-Fi signal range (Figure 18.5) is limited. An access point compliant with either 802.11b or 802.11g, using the stock antenna might have a range of 100 m (330 ft). IEEE 802.11n, however, can more than double the range. Range also varies with frequency band. Wi-Fi in the 2.4 GHz frequency block has slightly better range than Wi-Fi in the 5 GHz frequency block. On wireless routers with detachable antennas, it is possible to improve range by fitting upgraded antennas which have higher gain in particular directions. Outdoor ranges can be improved to many kilometers through the use of high gain directional antennas. In general, the maximum amount of power that a Wi-Fi device can transmit is limited by local regulations. Equivalent isotropically radiated power (EIRP) in the European Union is limited to 20 dBm (100 mW).

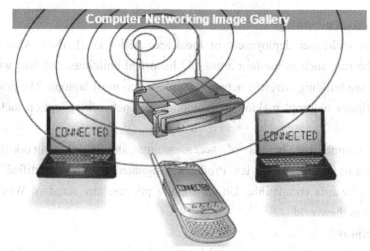

Figure 18.5 Wi-Fi signal range

To reach requirements for wireless LAN applications, Wi-Fi has fairly high power consumption compared to some other standards. Technologies such as Bluetooth (designed to support wireless personal area network (PAN) applications) provide a much shorter propagation range between 1 and 100 m and so in general have a lower power consumption. Other low-power technologies such as ZigBee have fairly long range, but much lower data rate. The high power consumption of Wi-Fi makes battery life in mobile devices a concern.

Researchers have developed a number of "no new wires" technologies to provide alternatives to Wi-Fi for applications in which Wi-Fi's indoor range is not adequate and where installing new wires (such as CAT-6) is not possible or cost-effective. For example, the ITU-T G.hn standard for high speed local area networks uses existing home wiring (coaxial cables, phone lines and power lines). Although G.hn does not provide some of the advantages of Wi-Fi (such as mobility or outdoor use), it is designed for applications (such as IPTV distribution) where indoor range is more important than mobility.

Due to the complex nature of radio propagation at typical Wi-Fi frequencies, particularly the effects of signal reflection off trees and buildings, algorithms can only approximately predict Wi-Fi signal strength for any given area in relation to a transmitter. This effect does not apply equally to long-range Wi-Fi, since longer links typically operate from towers that transmit above the surrounding foliage.

The practical range of Wi-Fi essentially confines mobile use to such applications as inventory-taking machines in warehouses or in retail spaces, barcode-reading devices at check-out stands, or receiving/shipping stations. Mobile use of Wi-Fi over wider range is limited, for instance, the maximum range of a wireless router is as small as 80 sq.m, and the Apple technology AirPort can reach 100~140 sq.m.

18.5 Network Security

The main issue with wireless network security is its simplified access to the network compared to traditional wired networks such as Ethernet. With wired networking, one must either gain access to a building (physically connecting into the internal network), or break through an external firewall. To enable Wi-Fi, one merely needs to be within the range of the Wi-Fi network. Most business networks protect sensitive data and systems by attempting to disallow external access. Enabling wireless connectivity reduces security if the network uses inadequate or no encryption.

An attacker who has gained access to a Wi-Fi network router can initiate a DNS spoofing attack against any other user of the network by forging a response before the queried DNS server has a chance to reply.

1. Data security risks

The most common wireless encryption-standard, Wired Equivalent Privacy (WEP), has been shown to be easily breakable even when correctly configured. Wi-Fi Protected Access (WPA and WPA2) encryption (Figure 18.6), which became available in devices in 2003, aimed to solve this problem. Wi-Fi access points typically default to an encryption-free (open) mode.

Novice users benefit from a zero-configuration device that works out-of-the-box, but this default does not enable any wireless security, providing open wireless access to a LAN. To turn security on requires the user to configure the device, usually via a software graphical user interface (GUI). On unencrypted Wi-Fi networks connecting devices can monitor and record data (including personal information). Such networks can only be secured by using other means of protection, such as a VPN or secure Hypertext Transfer Protocol over Transport Layer Security (HTTPS).

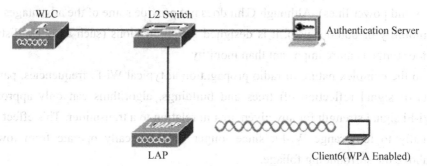

Figure 18.6 Wireless LAN with WPA

Wi-Fi Protected Access encryption (WPA2) is considered secure, provided a strong pass phrase is used. A proposed modification to WPA2 is WPA-OTP or WPA3, which stores an on-chip optically generated onetime pad on all connected devices which is periodically updated via strong encryption then hashed with the data to be sent or received. This would be unbreakable using any (even quantum) computer system as the hashed data is essentially random and no pattern can be detected if it is implemented properly. Main disadvantage is that it would need multi-GB storage chips so would be expensive for the consumers.

2. Securing methods

A common measure to deter unauthorized users involves hiding the access point's name by disabling the SSID broadcast. While effective against the casual user, it is ineffective as a security method because the SSID is broadcast in the clear in response to a client SSID query. Another method is to only allow computers with known MAC addresses to join the network, but determined eavesdroppers may be able to join the network by spoofing an authorized address.

Wired Equivalent Privacy (WEP) encryption was designed to protect against casual snooping but it is no longer considered secure. Tools such as Air Snort or Aircrack-ng can quickly recover WEP encryption keys. Because of WEP's weakness the Wi-Fi Alliance approved Wi-Fi Protected Access (WPA) which uses TKIP. WPA was specifically designed to work with older equipment usually through a firmware upgrade. Though more secure than WEP, WPA has known vulnerabilities.

The more secure WPA2 using Advanced Encryption Standard was introduced in 2004 and is supported by most new Wi-Fi devices. WPA2 is fully compatible with WPA.

A flaw in a feature added to Wi-Fi in 2007, called Wi-Fi Protected Setup, allows WPA and WPA2 security to be bypassed and effectively broken in many situations. The only remedy as of late 2011 is to turn off Wi-Fi Protected Setup, which is not always possible.

Keywords

adapter	适配器
Hotspots	热点
plug	插头
antenna	天线
carrier wave	载波
Enthusiasts	爱好者
carriers	运营商
cellular modem card	无限调节器
pledge	承诺
intermediary	媒介
Orthogonal Frequency-Division Multiplexing (OFDM)	正交频分复用
megabits	兆位
gain-directional	增益定向
Equivalent isotropically radiated power (EIRP)	等效全向辐射功率
Bluetooth	蓝牙
wireless personal area network	无线个人区域网
cost-effective	合算的
propagation	传播
external firewall	外部防火墙
spoofing	欺骗
graphical user interface (GUI)	图形用户界面
Hypertext Transfer Protocol over Transport Layer Security	传输层安全超文本传输协议
deter	阻止

18.6 Exercise 18

Multiple or single choices.

1. The major advantage of Wi-Fi is that it is compatible with almost every ____.
   A. operating system
   B. game device
   C. MP4
   D. advanced printer.

2. Mobile use of Wi-Fi over wider range is limited, and you may only connect to the Wi-Fi in the following distance ____.
   A. 70 sq.m
   B. 80 sq.m
   C. 90 sq.m
   D. 100 sq.m

3. On unencrypted Wi-Fi networks connecting devices can monitor and record data which we can use the following means to protect the networks. ____
   A. VPN
   B. WPA2
   C. HTTPS
   D. WPA-OPT

4. To deter unauthorized users join the networks, we can ____.
   A. hiding the access point's name by disabling the SSID broadcast
   B. only allow computers with known MAC addresses to join the network
   C. WEP encryption
   D. WPA encryption

True or False.

1. A Wi-Fi network makes use of infrared to transmit information across a network.

2. The wireless network will work as a one-way traffic.

3. Wi-Fi allows communications directly from one computer to another without an access point intermediary.

4. An access point compliant with 802.11b, using the stock antenna might have a range of 100 m.

5. Wired Equivalent Privacy (WEP), is a safe encryption method which cannot be breakable easily.

18.7 Further Reading: Wireless Revolution: The History of Wi-Fi

Believe it or not, there was a time when if you wanted to get on the Internet your only option was to have a phone cable or Ethernet cable plugged directly into your computer or laptop. Today, you have the option of sitting in any room in your house cord-free, or even hopping down to your local coffee shop and surfing the Web with relative ease. This seemingly miraculous ability is possible thanks to wireless local area networks, or WLANs. Commonly dubbed Wi-Fi, this advancement allows high-speed data transfers across limited space. But where did it come from?

1. 1980s

In 1985, the technology called 802.11 was made available for use due to a U.S. Federal Communication Commission ruling, which released the three bands of the radio spectrum now used for nearly all wireless communication: 900 MHz, 2.4 GHz, and 5 GHz. Shortly thereafter the IEEE (Institute of Electrical and Electronics Engineers) and the Wi-Fi Alliance (originally called WECA or the Wireless Ethernet Compatibility Alliance) were formed to help develop and regulate wireless technology worldwide.

2. 1990s

When the IEEE was formed in 1990, they chose Vic Hayes, also popularly known as the "Father of Wi-Fi," as its chairman. For the next ten years, Hayes helped direct the development of new wireless protocols as well as market the technology worldwide. His leadership and progressive thinking allowed the Wi-Fi Alliance to spearhead the regulation and widespread use of wireless technology.

The first version of the wireless protocol's legacy is now obsolete and would be considered dreadfully slow by today's standards. It had a maximum data transfer rate of 2 Mbps, or Megabits per second. Most applications created today would not be able to operate efficiently at those speeds.

In 1999, 802.11a and 802.11b were released, and for many years were the standard for Wi-Fi networks. They both operated in the 2.4 GHz range of the radio spectrum, but, unlike 802.11, they were able to transmit data at a much higher rate. The 802.11a protocol could support data transmission up to 54 Mbps, but was designed for much shorter ranges at a much higher cost to produce and maintain. On the other hand, 802.11b had a much lower cost and much longer range than its counterpart, but worked at a much slower speed, maxing out at 11 Mbps. Because both protocols operated in the unregulated 2.4 GHz bandwidth, they were susceptible to interference from other appliances that used the same frequency such as microwave ovens, cordless phones and wireless keyboards.

3. 2000s

In 2003, 802.11g was introduced as the new standard. This new protocol was designed to combine the best of the previous transmission standards—operating at a maximum transfer rate of 54 Mbps while still allowing for the longer range and lower costs. Most devices that incorporate the (g) technology are fully backwards compatible, allowing the use of all three protocols in one device.

The adaptation of 802.11n, sometimes called Wireless-N, saw a huge leap forward in the technology. With the ability to transfer data up to 300 Mbps and the incorporation of multiple wireless signals and antennas (called MIMO technology), people could surf the web even faster and with more stability. The new protocol also allowed data to be transmitted on both the standard 2.4GHz frequency as well as the less populated 5GHz which led to a stronger signal and less interruption.

4. Present Day

The latest technology, 802.11ac, proved to be another huge leap forward. With the advancements in dual-band technology, data can now be transmitted across multiple signals and bandwidths allowing for maximum transmission rates of 1300 Mbps with extended ranges and nearly uninterrupted transmission.

Technology continues to advance at faster and faster rates, making the once amazing seem commonplace. However, it is always good to remember how those technologies started and the innovators that created them. The future of Wi-Fi is predicted to have the potential for even faster speeds and increased stability. Data speeds now make it possible to allow voice and video calls to be transmitted over wireless networks reliably, which can often lead to great savings in phone bills. Service providers like T-Mobile are poised to provide wireless coverage to anyone and everyone in the present and wherever the future takes us.

Chapter 19
Shockwave 3-D Technology

In the past year or so, you may have heard about a new technology that lets you manipulate 3-D images over the Internet. Many Web sites have been using this sort of software for a while, but it has mostly remained a niche market due to a lack of universal 3-D viewer programs.

Macromedia, in conjunction with Intel, NxView and others, hopes to bring this technology to many more Web users with the newest versions of the Shockwave player and the Shockwave authoring program Director.

If you spend much time on the Web, you have probably encountered Shockwave, a graphics format for animation and interactive presentations. Shockwave files are created by a program called Director, which was originally developed for CD-ROM use. The format is very popular with webmasters because it allows them to create elaborate Web content that can be transmitted fairly quickly over the Internet.

In previous editions of Shockwave and Director, Web artists could create only 2-D animation. Two-dimensional animation comes in two forms:

- Frame animation is something like classic cartoons—you see movement as a series of 2-D still images shown in a set sequence. Your viewpoint is set by the movie's creator.
- Vector animation uses 2-D objects (circles, squares, lines) that move with respect to one another. Since it is based on simple geometric equations, vector animation allows artists to create complex movies that have very small file sizes.

The newest edition of Director incorporates Intel Internet 3-D technology developed by Intel Architecture Labs. The program allows Web artists to create interactive 3-D animations and post them on the Web. The newest version of the Shockwave player allows most Internet users, even ones with dial-up connections, to view these intricate animations.

With Shockwave 3-D technology, users can actually download and manipulate 3-D models themselves—they can become the director and move the camera. There are two ways to think about this:

- You can download an object and rotate the object in front of the camera to see it from different perspectives, as shown in Figure 19.1.
- You can download an environment and move the camera through it. This is basically

the same thing you do when you play a first-person video game. The program puts you in a virtual 3-D world, and you control a "camera" in that world by way of your movements. You tell your camera to move left or right, forward or backward, through the environment.

Based on your actions, the computer draws a new frame of the scene from your new, slightly different perspective.

Figure 19.1　The same object viewed from two different perspectives

This is a pretty complex operation: 3-D software must receive input from the user, interpret this input and decide how to redraw the image to create the desired sense of motion. When you're playing a game, your computer or game console can handle this fairly easily, but things get a lot trickier when you're sending this information over the Internet. Additionally, standard Web browsers are not automatically equipped to handle these models, which means that not everybody can access 3-D content. Macromedia's newest Shockwave player is designed to get around both of these problems, allowing most Web users to access 3-D files easily.

19.1　Uses of Shockwave Technology

Adding 3-D to Shockwave enables access to all sorts of new Web content. One of the most obvious applications is Web-based 3-D gaming. First-person adventure games and other games with fully realized 3-D worlds have dominated the PC and game-console market for almost a decade. The new Shockwave capabilities allow this sort of game to be played over the Web.

Web-based 3-D gaming is getting a lot of attention, but it is only one market for the new technology. 3-D capability is perhaps better suited for advancing E-commerce (Figure 19.2). Web merchants can give their customers a much clearer idea of products in their catalog if the customers can see the product as a 3-D image. With 3-D models, online shopping is a little more like in-store shopping-customers can rotate the item around, checking it out from every angle.

Customers can also modify 3-D models for their own particular needs. One of the most useful applications for this is clothes shopping. If an online shopper enters his or her measurements, the 3-D software can generate a model of that person's body, which can be "dressed" with 3-D models of particular clothes. This is a virtual version of the real-world dressing room.

Figure 19.2   A 3-D model used to demonstrate a product on an E-commerce site

This level of user interactivity is also a great addition to educational sites. A 3-D model of an engine that you can turn around and interact with can offer a much clearer illustration of the mechanisms at work than a 2-D model—it's more like actually handling and examining the engine yourself.

For example as shown in Figure 19.3, if you want to understand how a paintball gun works, a 3-D model can be incredibly useful. You can see exactly how the mechanism fits together and fires.

Figure 19.3   This 3-D model of a paintball gun makes it extremely easy to understand how paintball and BB guns work

As shown in Figure 19.4, as a demonstration of the new Shockwave 3-D technology, NxView has created this 3-D model of a four-speed manual transmission.

Figure 19.4   A demonstration of the new shockwave 3-D technology

In all of these applications, the most significant benefit of 3-D is greater user involvement. You can decide what you want to look at instead of just viewing a pre-set movie. The difference is comparable to the difference between watching television and playing a video game.

19.2 Making 3-D Content Accessible

In the last section, we saw that Shockwave's new player is a new format for creating and viewing interactive 3-D content on the Web. The idea of posting this sort of content on the Web is nothing new, but technology companies and Web sites haven't had much luck in bringing 3-D to a lot of viewers. There are two main reasons for this:

- It takes a long time to transmit 3-D "movement" over low-bandwidth connections.
- You often have to download a new plug-in every time you want to view another site's 3-D content.

The new Shockwave player specifically addresses these obstacles, so it could finally make 3-D content a significant component of the Web. The majority of Web users already have the Shockwave player installed and would only need to download the most recent update to add 3-D capabilities. Macromedia has set up partnerships with many Web companies in order to get people using its technology. Previously, Macromedia has had a lot of success with both Shockwave and Flash formats because they work well with all of the main browsers and are easy to install and update. Intel, NxView and other companies partnered with Macromedia because the company has a good track record with disseminating its player technology.

The new format is specially designed to work well with all bandwidth connections, even connection speeds as low as 28.8 kilobytes per second. It does this in a couple of ways.

- When you view 2-D animation on the Web, the Web site sends each successive frame to your computer. In this way, everything in the animation must be transmitted over the Internet individually. In Shockwave 3-D technology, the Web site sends you a complete image only once. Then, when you want to move the image, the site only sends the bare-bones information necessary to make the desired move. It tells your computer how the outer wire frame should be adjusted, and your computer does the rest of the work to fill in the polygons and textures.
- Most personal computers made in the past five years have processors designed to handle the complex 3-D worlds of advanced video games, so they are well-equipped for the job. By relying mostly on the power built into the client machine (your PC), there is much less information that needs to be transmitted from the server machine (the computer storing the Web site). The only hefty download occurs when you bring up the initial image. After that, the site only has to transmit mathematical adjustments, which don't require extensive bandwidth.

But what about this big initial download? Shockwave's new player addresses this problem with something called adaptive 3-D geometry. Adaptive 3-D geometry is a collection of complex algorithms that automatically scales a 3-D model for a particular Internet connection. If you have a slower connection, the Web site transmits an image with simplified textures and fewer polygons. If you have a faster connection, you receive a more complex image.

With these elements, you should be able to access 3-D content no matter kind of Internet connection you use. But how does somebody make Shockwave 3-D content themselves? In the next section, we'll find out what goes into producing a Shockwave 3-D animation and see how webmasters can put 3-D content on their site.

As shown in Figure 19.5, a simpler 3-D model has fewer polygons. This hand is composed of only 862 polygons.

Figure 19.5 A hand with 862 polygons

As shown in Figure 19.6, to create a more detailed model, you have to add more polygons. This hand is composed of 3444 polygons.

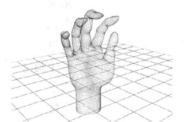

Figure 19.6 A hand with 3444 polygons

19.3 Developing New 3-D Content

We had the opportunity to speak with Miriam Geller, Macromedia's senior product manager for Director and the Shockwave player. To create a 3-D object like the automotive transmission in the example shown above, you use three different tools:

You use a standard 3-D modeling package to create the 3-D object. For example, you might use 3D Studio Max or Maya. With these tools, you create the wireframe image and specify the polygons that cover the wireframe. You export from the 3-D modeling package using a new. W3D file format.

(1) You load the .W3D file into the Macromedia application called Director Shockwave Studio. This application helps you prepare the 3-D object for distribution on the Web. For

example, you can:

- Apply different techniques, such as a multi-resolution mesh or subdivision surfaces, to limit the amount of bandwidth or processing power needed by the 3-D object on the user's machine.
- Add user-interactivity features. For example, you can make different parts of the 3-D object move in response to user requests.
- Add effects, such as fog or rain, to the object.

(2) You export a normal .DCR file from Director Shockwave Studio and place it on the Web server.

(3) The user then downloads and views the .DCR file using his or her browser and the Shockwave player (version 8.5 or higher).

This is not a trivial process, but for someone already familiar with 3-D modeling using a program like 3D Studio Max, it's a straightforward extension.

Keywords

shockwave	冲击波
manipulate	操作
animation	动画
geometric	几何
virtual	虚拟
dominate	主导
E-commerce	电子商务
merchants	商家
rotate	旋转
angle	角度
illustration	插图
obstacles	障碍
disseminate	宣传
polygons	多边形
textures	纹理
hefty	沉重
simplify	简化
wireframe	线框
Macromedia	大媒体
subdivision	细分
trivial	不重要的

19.4　Exercise 19

Multiple or single choices.

1. In previous editions of Shockwave and Director, Web artists could create only 2-D animation which can only create 2D animation, we can know that ____.
   A. classic cartoons—you see movement as a series of 2-D still images shown in a set sequence
   B. vector animation allows artists to create complex movies that have very small file sizes
   C. Your viewpoint is set by your own
   D. move with respect to one another

2. With Shockwave 3-D technology, users can become the director and move the camera, they can ____.
   A. download an object and rotate the object
   B. do something like when you play a first-person video game
   C. see the object from different perspectives
   D. download an environment and move the camera through it

3. Director Shockwave Studio helps you prepare the 3-D object for distribution on the Web, so that you can ____.
   A. make different parts of the 3-D object move in response to user requests
   B. add user-interactivity features
   C. a multi-resolution mesh or subdivision surfaces
   D. add effect like rain or cloud

True or False.

1. In the classic cartoons, you can see movement as a series of 2-D still images shown in a set sequence.

2. The newest edition of Director incorporates Intel Internet 3-D technology developed by Intel Architecture Labs.

3. Computer or game console can interpret this input and decide how to redraw the image easily when the information come from the internet.

4. The most significant benefit of 3-D is greater user involvement.

19.5 Further Reading: Computer Viruses

Computer viruses are mysterious and grab our attention. On the one hand, viruses show us how vulnerable we are. A properly engineered virus can have an amazing effect on the worldwide Internet. On the other hand, they show how sophisticated and interconnected human beings have become.

For example, the thing making big news right now is the Mydoom worm, which experts estimate infected approximately a quarter-million computers in a single day (Times Online). Back in March 1999, the Melissa virus was so powerful that it forced Microsoft and a number of other very large companies to completely turn off their e-mail systems until the virus could be contained. The ILOVEYOU virus in 2000 had a similarly devastating effect. That's pretty impressive when you consider that the Melissa and ILOVEYOU viruses are incredibly simple.

In this article, we will discuss viruses-both "traditional" viruses and the newer e-mail viruses-so that you can learn how they work and also understand how to protect yourself. Viruses in general are on the wane, but occasionally a person finds a new way to create one, and that's when they make the news.

19.5.1 Types of Infection

When you listen to the news, you hear about many different forms of electronic infection. The most common are:

- Viruses—A virus is a small piece of software that piggybacks on real programs. For example, a virus might attach itself to a program such as a spreadsheet program. Each time the spreadsheet program runs, the virus runs, too, and it has the chance to reproduce (by attaching to other programs) or wreak havoc.
- E-mail viruses—An e-mail virus moves around in e-mail messages, and usually replicates itself by automatically mailing itself to dozens of people in the victim's e-mail address book.
- Worms—A worm is a small piece of software that uses computer networks and security holes to replicate itself. A copy of the worm scans the network for another machine that has a specific security hole. It copies itself to the new machine using the security hole, and then starts replicating from there, as well.
- Trojan horses—A Trojan horse is simply a computer program. The program claims to do one thing (it may claim to be a game) but instead does damage when you run it (it may erase your hard disk). Trojan horses have no way to replicate automatically.

19.5.2 What's a "Virus"

Computer viruses are called viruses because they share some of the traits of biological viruses. A computer virus passes from computer to computer like a biological virus passes from person to person.

There are similarities at a deeper level, as well. A biological virus is not a living thing. A virus is a fragment of DNA inside a protective jacket. Unlike a cell, a virus has no way to do anything or to reproduce by itself—it is not alive. Instead, a biological virus must inject its DNA into a cell. The viral DNA then uses the cell's existing machinery to reproduce itself. In some cases, the cell fills with new viral particles until it bursts, releasing the virus. In other cases, the new virus particles bud off the cell one at a time, and the cell remains alive.

A computer virus shares some of these traits. A computer virus must piggyback on top of some other program or document in order to get executed. Once it is running, it is then able to infect other programs or documents. Obviously, the analogy between computer and biological viruses stretches things a bit, but there are enough similarities that the name sticks.

19.5.3 What's a "Worm"

A worm is a computer program that has the ability to copy itself from machine to machine. Worms normally move around and infect other machines through computer networks. Using a network, a worm can expand from a single copy incredibly quickly. For example, the Code Red worm replicated itself over 250,000 times in approximately nine hours on July 19, 2001.

A worm usually exploits some sort of security hole in a piece of software or the operating system. For example, the Slammer worm (which caused mayhem in January 2003) exploited a hole in Microsoft's SQL server. This article offers a fascinating look inside Slammer's tiny (376 byte) program.

19.5.4 Code Red

Worms use up computer time and network bandwidth when they are replicating, and they often have some sort of evil intent. A worm called Code Red made huge headlines in 2001. Experts predicted that this worm could clog the Internet so effectively that things would completely grind to a halt.

The Code Red worm slowed down Internet traffic when it began to replicate itself, but not nearly as badly as predicted. Each copy of the worm scanned the Internet for Windows NT or Windows 2000 servers that do not have the Microsoft security patch installed. Each time it found an unsecured server, the worm copied itself to that server. The new copy then scanned for other servers to infect. Depending on the number of unsecured servers, a worm could conceivably create hundreds of thousands of copies.

The Code Red worm was designed to do three things:

Replicate itself for the first 20 days of each month.

Replace Web pages on infected servers with a page that declares "Hacked by Chinese".

Launch a concerted attack on the White House Web server in an attempt to overwhelm it.

The most common version of Code Red is a variation, typically referred to as a mutated strain, of the original Ida Code Red that replicated itself on July 19, 2001. According to the National Infrastructure Protection Center:

The Ida Code Red Worm, which was first reported by eEye Digital Security, is taking advantage of known vulnerabilities in the Microsoft IIS Internet Server Application Program Interface (ISAPI) service. Un-patched systems are susceptible to a "buffer overflow" in the Idq.dll, which permits the attacker to run embedded code on the affected system. This memory resident worm, once active on a system, first attempts to spread itself by creating a sequence of random IP addresses to infect unprotected web servers. Each worm thread will then inspect the infected computer's time clock. The NIPC has determined that the trigger time for the DOS execution of the Ida Code Red Worm is at 0:00 hours, GMT on July 20, 2001. This is 8:00 PM, EST.

Upon successful infection, the worm would wait for the appointed hour and connect to the www.whitehouse.gov domain. This attack would consist of the infected systems simultaneously sending 100 connections to port 80 of www.whitehouse.gov (198.137.240.91).

The U.S. government changed the IP address of www.whitehouse.gov to circumvent that particular threat from the worm and issued a general warning about the worm, advising users of Windows NT or Windows 2000 Web servers to make sure they have installed the security patch.

19.5.5 Early Cases: Executable Viruses

Early viruses were pieces of code attached to a common program like a popular game or a popular word processor. A person might download an infected game from a bulletin board and run it. A virus like this is a small piece of code embedded in a larger, legitimate program. Any virus is designed to run first when the legitimate program gets executed. The virus loads itself into memory and looks around to see if it can find any other programs on the disk. If it can find one, it modifies it to add the virus's code to the unsuspecting program. Then the virus launches the "real program." The user really has no way to know that the virus ever ran. Unfortunately, the virus has now reproduced itself, so two programs are infected. The next time either of those programs gets executed, they infect other programs, and the cycle continues.

If one of the infected programs is given to another person on a floppy disk, or if it is uploaded to a bulletin board, then other programs get infected. This is how the virus spreads.

The spreading part is the infection phase of the virus. Viruses wouldn't be so violently despised if all they did was replicate themselves. Unfortunately, most viruses also have some sort of destructive attack phase where they do some damage. Some sort of trigger will activate the attack phase, and the virus will then "do something"—anything from printing a silly message on the screen to erasing all of your data. The trigger might be a specific date, or the number of times the virus has been replicated, or something similar.

19.5.6 Boot Sector Viruses

As virus creators got more sophisticated, they learned new tricks. One important trick was the ability to load viruses into memory so they could keep running in the background as long as the computer remained on. This gave viruses a much more effective way to replicate themselves. Another trick was the ability to infect the boot sector on floppy disks and hard disks. The boot sector is a small program that is the first part of the operating system that the computer loads. The boot sector contains a tiny program that tells the computer how to load the rest of the operating system. By putting its code in the boot sector, a virus can guarantee it gets executed. It can load itself into memory immediately, and it is able to run whenever the computer is on. Boot sector viruses can infect the boot sector of any floppy disk inserted in the machine, and on college campuses where lots of people share machines they spread like wildfire.

In general, both executable and boot sector viruses are not very threatening any more. The first reason for the decline has been the huge size of today's programs. Nearly every program you buy today comes on a compact disc. Compact discs cannot be modified, and that makes viral infection of a CD impossible. The programs are so big that the only easy way to move them around is to buy the CD. People certainly can't carry applications around on a floppy disk like they did in the 1980s, when floppies full of programs were traded like baseball cards. Boot sector viruses have also declined because operating systems now protect the boot sector.

Both boot sector viruses and executable viruses are still possible, but they are a lot harder now and they don't spread nearly as quickly as they once could. Call it "shrinking habitat," if you want to use a biological analogy. The environment of floppy disks, small programs and weak operating systems made these viruses possible in the 1980s, but that environmental niche has been largely eliminated by huge executables, unchangeable CDs and better operating system safeguards.

19.5.7 E-mail Viruses

The latest thing in the world of computer viruses is the e-mail virus, and the Melissa virus in March 1999 was spectacular. Melissa spread in Microsoft Word documents sent via e-mail, and it worked like this:

Someone created the virus as a Word document uploaded to an Internet newsgroup. Anyone who downloaded the document and opened it would trigger the virus. The virus would then send the document (and therefore itself) in an e-mail message to the first 50 people in the person's address book. The e-mail message contained a friendly note that included the person's name, so the recipient would open the document thinking it was harmless. The virus would then create 50 new messages from the recipient's machine. As a result, the Melissa virus was the fastest-spreading virus ever seen! As mentioned earlier, it forced a number of large companies to shut down their e-mail systems.

The ILOVEYOU virus, which appeared on May 4, 2000, was even simpler. It contained a piece of code as an attachment. People who double clicked on the attachment allowed the code to execute. The code sent copies of itself to everyone in the victim's address book and then started corrupting files on the victim's machine. This is as simple as a virus can get. It is really more of a Trojan horse distributed by e-mail than it is a virus.

The Melissa virus took advantage of the programming language built into Microsoft Word called VBA, or Visual Basic for Applications. It is a complete programming language and it can be programmed to do things like modify files and send e-mail messages. It also has a useful but dangerous auto-execute feature. A programmer can insert a program into a document that runs instantly whenever the document is opened. This is how the Melissa virus was programmed. Anyone who opened a document infected with Melissa would immediately activate the virus. It would send the 50 e-mails, and then infect a central file called NORMAL.DOT so that any file saved later would also contain the virus! It created a huge mess.

Microsoft applications have a feature called Macro Virus Protection built into them to prevent this sort of thing. With Macro Virus Protection turned on (the default option is ON), the auto-execute feature is disabled. So when a document tries to auto-execute viral code, a dialog pops up warning the user. Unfortunately, many people don't know what macros or macro viruses are, and when they see the dialog they ignore it, so the virus runs anyway. Many other people turn off the protection mechanism. So the Melissa virus spread despite the safeguards in place to prevent it.

In the case of the ILOVEYOU virus, the whole thing was human-powered. If a person double-clicked on the program that came as an attachment, then the program ran and did its thing. What fueled this virus was the human willingness to double-click on the executable.

19.5.8 Prevention of Virus

You can protect yourself against viruses with a few simple steps:
- If you are truly worried about traditional (as opposed to e-mail) viruses, you should be running a more secure operating system like UNIX. You never hear about viruses on these operating systems because the security features keep viruses (and unwanted human visitors) away from your hard disk.

- If you are using an unsecured operating system, then buying virus protection software is a nice safeguard.
- If you simply avoid programs from unknown sources (like the Internet), and instead stick with commercial software purchased on CDs, you eliminate almost all of the risk from traditional viruses. In addition, you should disable floppy disk booting—most computers now allow you to do this, and that will eliminate the risk of a boot sector virus coming in from a floppy disk accidentally left in the drive.
- You should make sure that Macro Virus Protection is enabled in all Microsoft applications, and you should NEVER run macros in a document unless you know what they do. There is seldom a good reason to add macros to a document, so avoiding all macros is a great policy.
- You should never double-click on an attachment that contains an executable that arrives as an e-mail attachment. Attachments that come in as Word files (.DOC), spreadsheets (.XLS), images (.GIF and .JPG), etc., are data files and they can do no damage (noting the macro virus problem in Word and Excel documents mentioned above). A file with an extension like EXE, COM or VBS is an executable, and an executable can do any sort of damage it wants. Once you run it, you have given it permission to do anything on your machine. The only defense is to never run executables that arrive via e-mail.

As shown in Figure 19.7, Open the Options dialog from the Tools menu in Microsoft Word and make sure that Macro Virus Protection is enabled.

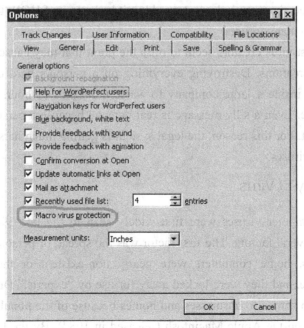

Figure 19.7 Options dialog

By following those simple steps, you can remain virus free.

19.5.9　Origins of Virus

People create viruses. A person has to write the code, test it to make sure it spreads properly and then release the virus. A person also designs the virus's attack phase, whether it's a silly message or destruction of a hard disk. So why do people do it?

There are at least three reasons. The first is the same psychology that drives vandals and arsonists. Why would someone want to bust the window on someone else's car, or spray-paint signs on buildings or burn down a beautiful forest? For some people that seems to be a thrill. If that sort of person happens to know computer programming, then he or she may funnel energy into the creation of destructive viruses.

The second reason has to do with the thrill of watching things blow up. Many people have a fascination with things like explosions and car wrecks. When you were growing up, there was probably a kid in your neighborhood who learned how to make gunpowder and then built bigger and bigger bombs until he either got bored or did some serious damage to himself. Creating a virus that spreads quickly is a little like that—it creates a bomb inside a computer, and the more computers that get infected the more "fun" the explosion.

The third reason probably involves bragging rights, or the thrill of doing it. Sort of like Mount Everest. The mountain is there, so someone is compelled to climb it. If you are a certain type of programmer and you see a security hole that could be exploited, you might simply be compelled to exploit the hole yourself before someone else beats you to it. "Sure, I could TELL someone about the hole. But wouldn't it be better to SHOW them the hole?" That sort of logic leads to many viruses.

Of course, most virus creators seem to miss the point that they cause real damage to real people with their creations. Destroying everything on a person's hard disk is real damage. Forcing the people inside a large company to waste thousands of hours cleaning up after a virus is real damage. Even a silly message is real damage because a person then has to waste time getting rid of it. For this reason, the legal system is getting much harsher in punishing the people who create viruses.

19.5.10　History of Virus

Traditional computer viruses were first widely seen in the late 1980s, and they came about because of several factors. The first factor was the spread of personal computers (PCs). Prior to the 1980s, home computers were nearly non-existent or they were toys. Real computers were rare, and they were locked away for use by "experts." During the 1980s, real computers started to spread to businesses and homes because of the popularity of the IBM PC (released in 1982) and the Apple Macintosh (released in 1984). By the late 1980s, PCs were widespread in businesses, homes and college campuses.

The second factor was the use of computer bulletin boards. People could dial up a bulletin board with a modem and download programs of all types. Games were extremely popular, and so were simple word processors, spreadsheets, etc. Bulletin boards led to the precursor of the virus known as the Trojan horse. A Trojan horse is a program that sounds really cool when you read about it. So you download it. When you run the program, however, it does something uncool like erasing your disk. So you think you are getting a neat game but it wipes out your system. Trojan horses only hit a small number of people because they are discovered quickly. Either the bulletin board owner would erase the file from the system or people would send out messages to warn one another.

The third factor that led to the creation of viruses was the floppy disk. In the 1980s, programs were small, and you could fit the operating system, a word processor (plus several other programs) and some documents onto a floppy disk or two. Many computers did not have hard disks, so you would turn on your machine and it would load the operating system and everything else off of the floppy disk.

Viruses took advantage of these three facts to create the first self-replicating programs.

Chapter 20

Kinect

Kinect (codenamed Project Natal during development) is a line of motion sensing input devices by Microsoft for Xbox 360 and Xbox One video game consoles and Windows PCs. Based around a webcam-style add-on peripheral, it enables users to control and interact with their console/computer without the need for a game controller, through a natural user interface using gestures and spoken commands. The first-generation Kinect was first introduced in November 2010 in an attempt to broaden Xbox 360's audience beyond its typical gamer base. A version for Windows was released on February 1, 2012. Kinect competes with several motion controllers on other home consoles, such as Wii Remote Plus for Wii and Wii U, PlayStation Move/PlayStation Eye for PlayStation 3, and PlayStation Camera for PlayStation 4.

Microsoft released the Kinect software development kit for Windows 7 on June 16, 2011. This SDK was meant to allow developers to write Kinecting apps in C++/CLI, C#, or Visual Basic .NET.

20.1 Technology

Kinect builds on software technology developed internally by Rare, a subsidiary of Microsoft Game Studios owned by Microsoft, and on range camera technology by Israeli developer PrimeSense, which developed a system that can interpret specific gestures, making completely hands-free control of electronic devices possible by using an infrared projector and camera and a special microchip to track the movement of objects and individuals in three dimensions. This 3D scanner system called Light Coding employs a variant of image-based 3D reconstruction.

Kinect sensor is a horizontal bar connected to a small base with a motorized pivot and is designed to be positioned lengthwise above or below the video display. The device features an "RGB camera, depth sensor and multi-array microphone running proprietary software", which provide full-body 3D motion capture, facial recognition and voice recognition capabilities. At launch, voice recognition was only made available in Japan, United Kingdom, Canada and United States. Mainland Europe received the feature later in spring 2011. Currently voice recognition is supported in Australia, Canada, France, Germany, Ireland, Italy, Japan, Mexico,

New Zealand, United Kingdom and United States. Kinect sensor's microphone array enables Xbox 360 to conduct acoustic source localization and ambient noise suppression, allowing for things such as headset-free party chat over Xbox Live.

The depth sensor consists of an infrared laser projector combined with a monochrome CMOS sensor, which captures video data in 3D under any ambient light conditions. The sensing range of the depth sensor is adjustable, and Kinect software is capable of automatically calibrating the sensor based on game play and the player's physical environment, accommodating for the presence of furniture or other obstacles.

Described by Microsoft personnel as the primary innovation of Kinect, the software technology enables advanced gesture recognition, facial recognition and voice recognition. According to information supplied to retailers, Kinect is capable of simultaneously tracking up to six people, including two active players for motion analysis with a feature extraction of 20 joints per player. However, PrimeSense has stated that the number of people the device can "see" (but not process as players) is only limited by how many will fit in the field-of-view of the camera.

As shown in Figure 20.1 and Figure 20.2, reverse engineering has determined that the Kinect's various sensors output video at a frame rate of ~9 Hz to 30 Hz depending on resolution.

Figure 20.1 This infrared image shows the laser grid Kinect uses to calculate depth

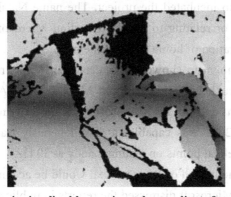

Figure 20.2 The depth map is visualized here using color gradients from white (near) to black (far)

The default RGB video stream uses 8-bit VGA resolution (640×480 pixels) with a Bayer color filter, but the hardware is capable of resolutions up to 1280×1024 (at a lower frame rate) and other colour formats such as UYVY. The monochrome depth sensing video stream is in VGA resolution (640×480 pixels) with 11-bit depth, which provides 2048 levels of sensitivity. The Kinect can also stream the view from its IR camera directly (i.e.: before it has been converted into a depth map) as 640×480 video, or 1280×1024 at a lower frame rate. The Kinect sensor has a practical ranging limit of 1.2~3.5 m (3.9~11.5 ft) distance when used with the Xbox software. The area required to play Kinect is roughly 6 m^2, although the sensor can maintain tracking through an extended range of approximately 0.7~6 m (2.3~19.7 ft). The sensor has an angular field of view of 57° horizontally and 43° vertically, while the motorized pivot is capable of tilting the sensor up to 27° either up or down. The horizontal field of the Kinect sensor at the minimum viewing distance of ~0.8 m (2.6 ft) is therefore ~87 cm (34 in), and the vertical field is ~63 cm (25 in), resulting in a resolution of just over 1.3 mm (0.051 in) per pixel. The microphone array features four microphone capsules and operates with each channel processing 16-bit audio at a sampling rate of 16 kHz.

Because the Kinect sensor's motorized tilt mechanism requires more power than the Xbox 360's USB ports can supply, the device makes use of a proprietary connector combining USB communication with additional power. Redesigned Xbox 360 S models include a special AUX port for accommodating the connector, while older models require a special power supply cable (included with the sensor) that splits the connection into separate USB and power connections; power is supplied from the mains by way of an AC adapter.

20.2 History

Kinect was first announced on June 1, 2009 at E3 2009 under the code name "Project Natal". Following in Microsoft's tradition of using cities as code names, "Project Natal" was named after the Brazilian city of Natal as a tribute to the country by Brazilian-born Microsoft director Alex Kipman, who incubated the project. The name Natal was also chosen because the word natal means "of or relating to birth", reflecting Microsoft's view of the project as "the birth of the next generation of home entertainment".

Three demos were shown to showcase Kinect when it was revealed at Microsoft's E3 2009 Media Briefing: Ricochet, Paint Party and Milo & Kate. A demo based on Burnout Paradise was also shown outside of Microsoft's media briefing. The skeletal mapping technology shown at E3 2009 was capable of simultaneously tracking four people, with a feature extraction of 48 skeletal points on a human body at 30 Hz.

It was rumored that the launch of Project Natal would be accompanied with the release of a new Xbox 360 console. Microsoft dismissed the reports in public and repeatedly emphasized that Project Natal would be fully compatible with all Xbox 360 consoles. Microsoft indicated that the company considers it to be a significant initiative, as fundamental to Xbox brand as Xbox Live, and with a launch akin to that of a new Xbox console platform. Kinect was even referred to as a "new Xbox" by Microsoft CEO Steve Ballmer at a speech for Executives' Club of Chicago. When asked if the introduction will extend the time before the next-generation console platform is launched (historically about 5 years between platforms), Microsoft corporate vice president Shane Kim reaffirmed that the company believes that the life cycle of Xbox 360 will last through 2015 (10 years).

During Kinect's development, project team members experimentally adapted numerous games to Kinect-based control schemes to help evaluate usability. Among these games were Beautiful Katamari and Space Invaders Extreme, which were demonstrated at Tokyo Game Show in September 2009. According to creative director Kudo Tsunoda, adding Kinect-based control to pre-existing games would involve significant code alterations, making it unlikely for Kinect features to be added through software updates.

Although the sensor unit was originally planned to contain a microprocessor that would perform operations such as the system's skeletal mapping, it was revealed in January 2010 that the sensor would no longer feature a dedicated processor (see Figure 20.3). Instead, processing would be handled by one of the processor cores of Xbox 360's Xenon CPU. According to Alex Kipman, Kinect system consumes about 10%~15% of Xbox 360's computing resources. However, in November, Alex Kipman made a statement that "the new motion control tech now only uses a single-digit percentage of Xbox 360's processing power, down from the previously stated 10 to 15 percent." A number of observers commented that the computational load required for Kinect makes the addition of Kinect functionality to pre-existing games through software updates even less likely, with concepts specific to Kinect more likely to be the focus for developers using the platform.

Figure 20.3 A January 2010 promotional banner indicating the expected release of Kinect (then "Project Natal") by holiday 2010

As shown in Figure 20.4, The Xbox 360 S and E models have dedicated ports for the Kinect, removing the need for an external power supply.

On March 25, 2010, Microsoft sent out a save the date flier for an event called the "World Premiere 'Project Natal' for Xbox 360 Experience" at E3 2010. The event took place on the evening of Sunday, June 13, 2010 at Galen Center and featured a performance by Cirque du Soleil. It was announced that the system would officially be called Kinect, a

Figure 20.4 Xbox 360

portmanteau of the words "kinetic" and "connect", which describe key aspects of the initiative. Microsoft also announced that the North American launch date for Kinect will be November 4, 2010. Despite previous statements dismissing speculation of a new Xbox 360 to accompany the launch of the new control system, Microsoft announced at E3 2010 that it was introducing a redesigned Xbox 360, complete with a connector port ready for Kinect. In addition, on July 20, 2010, Microsoft announced a Kinect bundle with a redesigned Xbox 360, to be available with Kinect launch.

On June 16, 2011, Microsoft announced its official release of its SDK for non-commercial use. On July 21, 2011, Microsoft announced that the first ever white Kinect sensor would be available as part of "Xbox 360 Limited Edition Kinect Star Wars Bundle", which also includes custom a Star Wars-themed console and controller, and copies of Kinect Adventures and Star Wars Kinect. Previously, all Kinect sensors had been glossy black.

On October 31, 2011, Microsoft announced launching of the commercial version of Kinect for Windows program with release of SDK to companies. David Dennis, Product Manager at Microsoft, said, "There are hundreds of organizations we are working with to help them determine what's possible with the tech".

On February 1, 2012, Microsoft released the commercial version of Kinect for Windows SDK and told that more than 300 companies from over 25 countries are working on Kinect-ready apps.

20.3 Launch

Microsoft had an advertising budget of US$500 million for the launch of Kinect, a larger sum than the investment at launch of Xbox console. The marketing campaign "You Are the Controller", aiming to reach new audiences, included advertisements on Kellogg's cereal boxes and Pepsi bottles, commercials during shows such as Dancing with the Stars and Glee as well as print ads in various magazines such as People and InStyle.

On October 19, Microsoft advertised Kinect on The Oprah Winfrey Show by giving free Xbox 360 consoles and Kinect sensors to the people in the audience. Two weeks later, Kinect bundles with Xbox 360 consoles were also given away to the audience of Late Night with Jimmy Fallon. On October 23, Microsoft held a pre-launch party for Kinect in Beverly Hills.

The party was hosted by Ashley Tisdale and was attended by soccer star David Beckham and his three sons, Cruz, Brooklyn, and Romeo. Guests were treated to sessions with Dance Central and Kinect Adventures, followed by Tisdale having a Kinect voice chat with Nick Cannon. Between November 1 and 28, Burger King gave away a free Kinect bundle "every 15 minutes".

A major event was organized on November 3 in Times Square, where singer Ne-Yo performed with hundreds of dancers in anticipation of Kinect's midnight launch. During the festivities, Microsoft gave away T-shirts and Kinect games.

Kinect was launched in North America on November 4, 2010, in Europe on November 10, 2010, in Australia, New Zealand and Singapore on November 18, 2010, and in Japan on November 20, 2010. Purchase options for the sensor peripheral include a bundle with the game Kinect Adventures and console bundles with either a 4 GB or 250 GB Xbox 360 console and Kinect Adventures.

20.4 Reception

1. Kinect for Xbox 360

Upon its release, the Kinect garnered positive opinions from reviewers and critics. IGN gave the device 7.5 out of 10, saying that "Kinect can be a tremendous amount of fun for casual players, and the creative, controller-free concept is undeniably appealing", though adding that for "$149.99, a motion-tracking camera add-on for Xbox 360 is a tough sell, especially considering that the entry level variation of Xbox 360 itself is only $199.99". Game Informer rated Kinect 8 out of 10, praising the technology but noting that the experience takes a while to get used to and that the spatial requirement may pose a barrier. Computer and Video Games called the device a technological gem and applauded the gesture and voice controls, while criticizing the launch lineup and Kinect Hub.

CNET's review pointed out how Kinect keeps players active with its full-body motion sensing but criticized the learning curve, the additional power supply needed for older Xbox 360 consoles and the space requirements. Engadget, too, listed the large space requirements as a negative, along with Kinect's launch lineup and the slowness of the hand gesture UI. The review praised the system's powerful technology and the potential of its yoga and dance games. Kotaku considered the device revolutionary upon first use but noted that games were sometimes unable to recognize gestures or had slow responses, concluding that Kinect is "not must-own yet, more like must-eventually own." TechRadar praised the voice control and saw a great deal of potential in the device whose lag and space requirements were identified as issues. Gizmodo also noted Kinect's potential and expressed curiosity in how more mainstream titles would utilize the technology. Ars Technica's review expressed concern that the core feature of Kinect, its lack of a controller, would hamper development of games beyond those that have either stationary players or control the player's movement automatically.

The mainstream press also reviewed Kinect. USA Today compared it to the futuristic control scheme seen in Minority Report, stating that "playing games feels great" and giving the device 3.5 out of 4 stars. David Pogue from The New York Times predicted players will feel a "crazy, magical, omigosh rush the first time you try the Kinect." Despite calling the motion tracking less precise than Wii's implementation, Pogue concluded that "Kinect's astonishing technology creates a completely new activity that's social, age-spanning and even athletic." The Globe and Mail titled Kinect as setting a "new standard for motion control." The slight input lag between making a physical movement and Kinect registering it was not considered a major issue with most games, and the review called Kinect "a good and innovative product," rating it 3.5 out of 4 stars.

2. Kinect for Xbox One

Although featuring improved performance over the original Kinect, its successor has been subject to mixed responses. It has been praised for its wide angle, fast response time and high quality camera. However, the Kinect's inability to understand some accents in English was criticized. Furthermore, controversies surround Microsoft's intentional tying of the sensor with the Xbox One console despite the initial requirements for the sensor being plugged in at all times having been revised since its initial announcement. There have also been a number of concerns regarding privacy. In May 2014, Microsoft announced that it would offer an Xbox One console without Kinect beginning June 9 alongside the original package; a separate Kinect sensor will be made available at a later date for users who wish to add it later.

20.5　Sales

24 million units of Kinect had been shipped by February 2013. Having sold 8 million units in its first 60 days on the market, Kinect has claimed the Guinness World Record of being the "fastest selling consumer electronics device". According to Wedbush analyst Michael Pachter, Kinect bundles accounted for about half of all Xbox 360 console sales in December 2010 and for more than two-thirds in February 2011. More than 750,000 Kinect units were sold during the week of Black Friday 2011.

20.6　Awards

- The machine learning work on human motion capture within Kinect won the 2011 MacRobert Award for engineering innovation.
- Kinect Won T3's "Gadget of the Year" award for 2011. It also won the "Gaming Gadget of the Year" prize.

- "Microsoft Kinect for Windows Software Development Kit" was ranked second in "The 10 Most Innovative Tech Products of 2011" at Popular Mechanics Breakthrough Awards ceremony in New York City.

Keywords

peripheral	外围设备
subsidiary	附加物
gestures	手势
infrared	红外线
reconstruction	重建
horizontal	横向的
motorized	机动的
pivot	枢纽
lengthwise	纵向
laser	激光
monochrome	黑白的
captures	捕获
adjustable	可调整的
accommodate	容纳
furniture	附属品
obstacles	障碍
natal	产后
demos	演示
Paradise	仙境
rumor	传闻
glossy	光滑
casual	非正式的
lineup	排队
stationary	固定的
futuristic	未来的

20.7 Exercise 20

Multiple or single choices.

1. Microsoft developed a system that can interpret specific gestures, making completely hands-free control of electronic devices possible by using _____ to track the movement of objects and individuals in three dimensions.

A. an infrared projector
   B. a camera
   C. a sensor
   D. a special microchip

2. Kinect sensor is a horizontal bar connected to a small base with a motorized pivot and is designed to be positioned lengthwise above or below the video display which can provide the following functions ____.
   A. full-body 3D motion capture
   B. fingerprint recognition
   C. facial recognition
   D. voice recognition

3. From user's opinions about Kinect, we can know that ____.
   A. games were sometimes unable to recognize gestures or had slow responses
   B. praised the voice control and saw a great deal of potential in the device whose lag and space requirements were identified as issues
   C. the experience takes a while to get used to and that the spatial requirement may pose a barrier
   D. players will feel a crazy, magical, omigosh rush the first time you try the Kinect

True or False.

1. The depth sensor consists of an infrared laser projector combined with a monochrome CMOS sensor.

2. Kinect is capable of tracking up to ten active players for motion analysis simultaneously.

3. Kinect can be controller-free which brings more fun and appeal to those casual player easily.

4. Kinect has been shown its ability to understand some accents in English accurately.

20.8 Further Reading: Software of Kinect

20.8.1 Kinect for Windows

On February 21, 2011 Microsoft announced that it would release a non-commercial

Kinect software development kit (SDK) for Windows in spring 2011, which was released for Windows 7 on June 16, 2011 in 12 countries. The SDK includes Windows 7 compatible PC drivers for Kinect device. It provides Kinect capabilities to developers to build applications with C++, C#, or Visual Basic by using Microsoft Visual Studio 2010 and includes following features:

- Raw sensor streams—Access to low-level streams from the depth sensor, color camera sensor, and four-element microphone array.
- Skeletal tracking—The capability to track the skeleton image of one or two people moving within Kinect's field of view for gesture-driven applications.
- Advanced audio capabilities—Audio processing capabilities include sophisticated acoustic noise suppression and echo cancellation, beam formation to identify the current sound source, and integration with Windows speech recognition API.
- Sample code and Documentation.

In March 2012, Craig Eisler, the general manager of Kinect for Windows, said that almost 350 companies are working with Microsoft on custom Kinect applications for Windows.

1. Version 1.5

In March 2012, Microsoft announced that next version of Kinect for Windows SDK would be available in May 2012. Kinect for Windows 1.5 was released on May 21, 2012. It adds new features, support for many new languages and debut in 19 more countries.

- Kinect for Windows 1.5 SDK would include "Kinect Studio" a new app that allows developers to record, playback, and debug clips of users interacting with applications.
- Support for new "seated" or "10-joint" skeletal system that will let apps track the head, neck, and arms of a Kinect user—whether they're sitting down or standing; which would work in default and near mode.
- Support for four new languages for speech recognition—French, Spanish, Italian, and Japanese. Additionally it would add support for regional dialects of these languages along with English.
- It would be available in Austria, Belgium, Brazil, Denmark, Finland, India, the Netherlands, Norway, Portugal, Russia, Saudi Arabia, Singapore, South Africa, Sweden, Switzerland and the United Arab Emirates in June.

2. Version 1.6, 1.7 and 1.8

Kinect for Windows SDK for the first-generation sensor was updated a few more times, with version 1.6 released October 8, 2012, version 1.7 released March 18, 2013, and version 1.8 released September 17, 2013.

3. Version 2

The second-generation Kinect for Windows, based on the same core technology as Kinect for Xbox One, including a new sensor, was first released in 2014.

20.8.2 Software

Requiring at least 190 MB of available storage space, Kinect system software allows users to operate Xbox 360 Dashboard console user interface through voice commands and hand gestures. Techniques such as voice recognition and facial recognition are employed to automatically identify users. Among the applications for Kinect is Video Kinect, which enables voice chat or video chat with other Xbox 360 users or users of Windows Live Messenger. The application can use Kinect's tracking functionality and Kinect sensor's motorized pivot to keep users in frame even as they move around. Other applications with Kinect support include ESPN, Zune Marketplace, Netflix, Hulu Plus and Last.fm. Microsoft later confirmed that all forthcoming applications would be required to have Kinect functionality for certification.

Games that require Kinect have a purple sticker on them with a white silhouette of Kinect sensor and "Requires Kinect Sensor" underneath in white text, and also come in purple packaging. Games that have optional Kinect support (meaning that Kinect is not necessary to play the game or that there are optional Kinect minigames included) feature a standard green Xbox 360 case with a purple bar underneath the header, a silhouette of Kinect sensor and "Better with Kinect Sensor" next to it in white text.

Kinect launched on November 4, 2010 with 17 titles. Third-party publishers of available and announced Kinect games include, among others, Ubisoft, Electronic Arts, LucasArts, THQ, Activision, Konami, Sega, Capcom, Namco Bandai and MTV Games. Along with retail games, there are also select Xbox Live Arcade that exclusively use Kinect peripheral.

1. Kinect Fun Labs

At E3 2011, Microsoft announced Kinect Fun Labs: a collection of various gadgets and minigames that are accessible from Xbox 360 Dashboard. These gadgets includes Build A Buddy, Air Band, Kinect Googly Eyes, Kinect Me, Bobblehead, Kinect Sparkler, Junk Fu and Avatar Kinect.

2. Open source drivers

In November 2010, Adafruit Industries offered a bounty for an open-source driver for Kinect. Microsoft initially voiced its disapproval of the bounty, stating that it "does not condone the modification of its products" and that it had "built in numerous hardware and software safeguards designed to reduce the chances of product tampering". This reaction, however, was caused by a misunderstanding within Microsoft, and the company later clarified its position, claiming that while it does not condone hacking of either the physical device or the console, the USB connection was left open by design.

The first thing to talk about is, Kinect was not actually hacked. Hacking would mean that someone got to our algorithms that sit inside of the Xbox and was able to actually use them, which hasn't happened. Or, it means that you put a device between the sensor and the Xbox for means of cheating, which also has not happened. That's what we call hacking, and that's what we have put a ton of work and effort to make sure doesn't actually occur. What has happened is someone wrote an open-source driver for PCs that essentially opens the USB connection, which we didn't protect, by design, and reads the inputs from the sensor. The sensor, again, as I talked earlier, has eyes and ears, and that's a whole bunch of noise that someone needs to take and turn into signal.

On November 10, Adafruit announced Héctor Martín as the winner, who had produced a Linux driver that allows the use of both the RGB camera and depth sensitivity functions of the device. It was later revealed that Johnny Lee, a core member of Microsoft's Kinect development team, had secretly approached Adafruit with the idea of a driver development contest and had personally financed it.

In December 2010, PrimeSense, whose depth sensing reference design Kinect is based on, released their own open source drivers along with motion tracking middleware called NITE. PrimeSense later announced that it had teamed up with Asus to develop a PC-compatible device similar to Kinect for Chinese markets, called the Wavi Xtion. The product was released in October 2011.

OpenNI is an open-source software framework that is able to read sensor data from Kinect, among other natural user interface sensors.

3. Third-party development

A demonstration of a third-party use of Kinect at Maker Faire. The visualization on the left, provided through Kinect, is of a user with a jacket featuring wearable electronic controls for VJing.

Numerous developers are researching possible applications of Kinect that go beyond the system's intended purpose of playing games. For example, Philipp Robbel of MIT combined Kinect with iRobot Create to map a room in 3D and have the robot respond to human gestures, while an MIT Media Lab team is working on a JavaScript extension for Google Chromecalled depth JS that allows users to control the browser with hand gestures. Other programmers, including Robot Locomotion Group at MIT, are using the drivers to develop a motion-controller user interface similar to the one envisioned in Minority Report. The developers of MRPT have integrated open source drivers into their libraries and provided examples of live 3D rendering and basic 3D visual SLAM. Another team has shown an application that allows Kinect users to play a virtual piano by tapping their fingers on an empty desk. Oliver Kreylos, a researcher at University of California, Davis, adopted the technology to improve live 3-dimensional videoconferencing, which NASA has shown interest in.

Alexandre Alahi from EPFL presented a video surveillance system that combines multiple Kinect devices to track groups of people even in complete darkness. Companies So touch and Evoluce have developed presentation software for Kinect that can be controlled by hand gestures; among its features is a multi-touch zoom mode. In December 2010, the free public beta of HTPC software KinEmote was launched; it allows navigation of Boxee and XBMC menus using a Kinect sensor. Soroush Falahati wrote an application that can be used to create stereoscopic 3D images with a Kinect sensor.

For a limited time in May 2011, a Topshop store in Moscow set up a Kinect kiosk that could overlay a collection of dresses onto the live video feed of customers. Through automatic tracking, position and rotation of the virtual dress were updated even as customers turned around to see the back of the outfit.

Kinect also shows compelling potential for use in medicine. Researchers at the University of Minnesota have used Kinect to measure a range of disorder symptoms in children, creating new ways of objective evaluation to detect such conditions as autism, attention-deficit disorder and obsessive-compulsive disorder. Several groups have reported using Kinect for intraoperative, review of medical imaging, allowing the surgeon to access the information without contamination. This technique is already in use at Sunnybrook Health Sciences Centre in Toronto, where doctors use it to guide imaging during cancer surgery. At least one company, GestSure Technologies, is pursuing the commercialization of such a system.

Another recent application of Kinect is for multi-touch displays. A Seattle-based company that graduated from Microsoft's Kinect Accelerator, Ubi Interactive, has developed software for the Kinect intended to work with projectors to allow touchscreen-like capabilities on various surfaces.

Kinect is used with a number of ROS robots via OpenNI.

4. Kinect for Xbox One

Xbox One consoles ship with an updated version of Kinect; the new Kinect uses a wide-angle time-of-flight camera, and processes 2 gigabits of data per second to read its environment. The new Kinect has greater accuracy with three times the fidelity over its predecessor and can track without visible light by using an active IR sensor. It has a 60% wider field of vision that can detect a user up to 3 feet from the sensor, compared to six feet for the original Kinect, and can track up to 6 skeletons at once. It can also detect a player's heart rate, facial expression, the position and orientation of 25 individual joints (including thumbs), the weight put on each limb, speed of player movements, and track gestures performed with a standard controller. Kinect's microphone is used to provide voice commands for actions such as navigation, starting games, and waking the console from sleep mode.

All Xbox One consoles were initially shipped with the Kinect sensor included—a holdover from a previously-announced, but retracted mandate requiring Kinect to be plugged into the console at all times for it to function. In June 2014, bundles without Kinect were made available, along with an updated Xbox One SDK allowing game developers to explicitly disable Kinect skeletal tracking, freeing up system resources that were previously reserved for Kinect even if it was disabled or unplugged.

A standalone Kinect for Xbox One, bundled with a digital copy of Dance Central Spotlight, was released on October 7, 2014.

附录 A 部分参考译文

第 1 章　电脑基本组件

作为大学生或从事 IT 的人，你必须了解 PC 及其组件，包括其存储设备和 I/O 设备。现在我们先来看看这些组件。

1.1　存储器

计算机中存储器的作用就是保存数据和信息，在 CPU 需要这些数据和信息时尽快地传送给 CPU。计算机用多种磁盘来存储如计算机内部的硬盘、外部的软盘和光盘等。

1. 硬盘

计算机一般有两种存储器：位于主板上的主存储器和存储在硬件上的辅助存储器。主存储器保存了所有重要的记忆用于命令计算机的操作。辅助存储器保存了所有你保存在计算机上的信息。

在硬盘里面会看到由钢制成的磨得很光亮的环形磁盘。

在这些磁盘里面，有很多磁轨或者是圆环。在硬件中，电子读写器又叫磁头，在磁盘上的轨迹上向前或向后，从磁盘上读取信息或者是写入信息。硬盘的旋转速度是 3600rpm（转/每分钟）或者更高。也就是说，一分钟之内，硬盘转了超过 3600 次。现在的硬盘能够存储的信息量非常的大，甚至超过了 20GB。

2. 软盘

当你注意软盘的时候，你会发现它是一个差不多 3.5×5 英寸的塑料盒子，里面是一片很薄的涂了微小铁粒的塑料。

这种盘很像录影带或者录音带里的盘。仔细看一下软盘的图片你会发现在它的一端有一个小金属片覆盖着一个矩形洞。那个盖子可以移动到旁边，这样就能够看到磁盘的内部。

但是千万别去碰里面的磁盘，因为可能会破坏到里面所存储的数据。在软盘的一面贴着标签，在另外一面是一个有着两个洞的银色的圆形物体。当软盘放进软驱中，软驱就钩住这两个洞使软盘旋转。这就使得里面的盘能够以 300rpm 的速度旋转起来。同时在软盘一边的银色盖子被推开，这样软驱中的磁头就能够读取软盘中的数据或者把数据写入软盘。

软盘是最小的一种存储器，它的容量只有 1.44MB。

3. 硬盘和软盘是如何工作的

从硬盘或软盘中读取或者写入数据的进程是通过电和磁来完成的。这两种类型的磁盘表面都是很容易受磁的。在磁盘驱动器中的电磁头通过在磁盘的表面上创建一系列受磁和未受磁的区域来往磁盘中记录信息。

你是否记得二进制代码是怎样用 1 和 0 来表示信息的？在磁盘中，受磁的区域用 1 表示，未受磁的区域用 0 来表示，所以所有的信息都是用二进制代码存储起来的。这就是磁头能够从磁盘的表面上读取和写入信息的原因。使磁盘和计算机始终远离磁性物质是非常重要的，因为磁铁能够消除你在磁盘中所存储的信息。

4. 光盘

CD 盘不是用电磁学的原理而是用磁盘凹陷（细微的缺口）和平面（平坦表面）来存储信息，这跟电磁学采用受磁和未受磁的区域来存储信息也是大同小异的。在光驱中是用激光来扫描光盘的某个地方是否有凹陷。根据有没有反射光就创建了能够表示数据的代码。

CD 盘通常可以存储大约 650MB 的数据。这比软盘的容量 1.44MB 要大得多。一张 DVD 盘能够比 CD 盘存储更多的数据，因为 DVD 盘可以存储两层的信息，而且它的凹陷更小甚至可以两面都存储数据。

5. 软盘的使用

你可能会感到很疑惑：如果所有的信息都安全地存储在计算机中，为什么还要把它存储在计算机外面呢？轻便的存储器之所以这么重要是有很多原因的。

软盘可以用来备份计算机中比较重要的信息以防造成信息的丢失。随机存储器在计算机关闭时会丢失所存储的所有信息，但是只读存储器可以在计算机关闭时保存信息。计算机有些时候会出现一些问题使它们发生撞击。不过并不是说它们会从桌子上掉下来摔碎在地板上。

计算机里面电路的一些事故会造成数据的丢失。甚至有一些可能造成只读存储器的数据丢失。将一些重要的信息备份在磁盘上，这样就可以在需要的时候将它们重新放回自己的计算机上。备份可以省下不少的时间并避免令人头痛的事情。

磁盘可以使信息在不同的计算机中交流。比如，在图书馆的计算机上设计一个方案，却不能够在图书馆关门前完成，那么这个方案就保留在图书馆的计算机上，应该怎么样将它带回家，并在自己的计算机上完成呢？可以把这些信息存在磁盘上，将磁盘带回家并将这些信息上传到自己的计算机上。这是多么简单的数据传输方法！

6. 光盘的使用

光盘最普遍的应用是存储软件程序(除了播放音乐)。当购买一个计算机游戏软件时，程序会告诉计算机怎么从存储的 CD 盘上运行游戏。安装了这个软件后，就把这个游戏程序送入了计算机中，一些程序完整地传送到你的计算机硬盘中。但是，有很多程序非常大，这就意味着要使用非常大的硬盘空间。为了避免占用太大的硬盘空间，那些程序就被设计成只有一部分的程序会上传到计算机中，剩余的部分就留在了软件上。所以当没有把 CD 盘放入计算机中，使得随机存储器不能读取程序需要的剩余信息时，那些程序就无法运行。

随着刻录光盘的发明（磁盘驱动器可以写入或读取光盘中的信息），CD 盘现在可以像软盘一样存储信息。使用光盘刻录，计算机中的数据可以备份在 CD 盘中。那些对于软盘太大的数据，现在可以被备份在 CD 盘中。很多人将一些音乐文件或者是家庭照片存储在 CD 盘中。

1.2 外部的硬件

仔细看一下放在面前的计算机。不要只看显示器，看一下其他的部分。知道那些是什么东西吗？知道它们是干什么用的吗？如果你已经知道了——那非常好。给自己一个大奖品吧！但是如果你不知道计算机的每一个部分，那么请接着读下去慢慢地弄明白。

1. 基础部分

让我们从一些计算机系统的核心部分开始谈起吧。看见旁边一个形状很像盒子的东西了吗？它有一个电源的开关和一个或者两个灯，有一个或者更多的地方可以让你插入磁盘。这就是装了计算机中重要组成部分的机箱。如果它是立着的，则称为立式机箱；如果是卧着的，则称为卧式机箱。

看一下计算机的后面，你将会看到机箱引出了几条电线通到了计算机的其他部分，比如到显示器。

计算机机箱可能会有一个地方可以插入软盘或者光盘。这些地方叫做软驱和光驱，软驱可以从一个非常薄而且平的方形的塑料磁盘中读取信息。

你同样可以写入信息到这些盘中并且保存下来。CD-ROM 是只读性光盘驱动的缩写，一个光盘是一个可以存储信息的非常有光泽的圆形磁盘。一个只读性光盘驱动只能够从盘中读取信息。

一些比较新的计算机用可读写光盘驱动代替了只读性光盘驱动。可读写光盘驱动器允许像读取信息一样写入一些信息到磁盘中。当然，一些比较新的计算机用 DVD 驱动器来代替只读性光盘驱动或者是可读取光盘驱动。一个 DVD 片就像一张 CD 片，但是它能够存储比 CD 片多得多的信息。你可以在这些 DVD 上看电影、听音乐或者玩一些游戏。但是有一件很重要的事情必须知道，那就是可以在 DVD 播放器中播放 CD 盘却不能在 CD 播放器中播放 DVD 盘。

2. 输入设备

有很多的途径可以让计算机获得新的信息或者输入信息到计算机中。两个最通用的方法就是键盘和鼠标。键盘上有字母、数字、标点符号和一些特殊命令的按键，按下这些按键也就代表你告诉了计算机你要它做什么操作或者是要它写出什么信息。

鼠标有一个特别的球形体，在一个鼠标垫或者是桌子上滚动它时，在计算机屏幕上的指针也会相应地移动。当按下鼠标上面的按钮时，就可以指导计算机做相应的操作。有其他类似鼠标的设备可以取代鼠标，例如，有一种跟踪球可以用手指去移动它；还有一种叫触摸屏，可以在触垫上移动手指来控制光标，压一下就相当于按了鼠标上的按钮。

一些其他类型的设备可以将一些图像输入到计算机中。一个扫描器可以将一张图片或者一个文档复制到计算机中。有很多种扫描器的类型，有一些看起来很不一样但是大部分都是一个玻璃平面上带着挡板。用一个数码相机可以将照片输入到计算机中。你可以用数码相机在远离计算机的地方拍下照片并且存储在一个很小的存储片中。然后将相机和计算机相连，这样就可以将这些照片下载到计算机中。另外一种输入设备是画图板，一个与计算机相连的很敏感的触压垫。用一种特殊的笔（千万别用钢笔或铅笔）在这块板上画图时，所画的图案将在计算机的屏幕上显示。画图板和特殊的笔可以像一个鼠标那样移动屏幕上的指针并做出相应的操作。

3. 输出设备

输出设备可以让计算机显示人们能够理解的信息。最常用的一个输出设备是显示器，它像一台电视机覆盖了计算机的屏幕。显示器可以让人们看到自己和计算机一起做的事情。

扬声器是一种可以让人们听到计算机声音的输出设备，计算机的扬声器就像立体的喇叭，它们通常由两个部分组成，而且有很多不同的类型。

打印机也是另外一个计算机系统中比较常用的设备。它可以将计算机屏幕上看到的东西打印到纸张上面。打印机有两种类型。喷墨打印机是用墨水进行打印的，它是家用计算机上最常用的打印机，它可以打印黑白的和彩色的文本。激光打印机的速度非常快，因为它是用激光进行打印的，激光打印机一般用于商业中。黑白打印在激光打印机中比较常见，当然也有一些也是用彩色打印的。

1.3 智能手机、平板电脑与笔记本电脑

智能手机、平板电脑和笔记本电脑是最流行的移动设备，下面看看这三者之间有何区别。

尽管智能手机从根本上来说是一个用来打电话和接电话的设备，但是它也可以被看成是一个微型电脑而不仅仅是一部简单的手机。它利用独立操作系统安装和运行先进、复杂的程序。大部分智能手机配有虚拟键盘，用户可以在高电容触屏方便地进行操作。

智能手机具有高速处理器、大容量内存、大屏（大约 3.5 英寸）和非常友好的操作系统。目前智能手机市场上主要的两款操作系统是苹果的 iOS 系统和谷歌的安卓系统。iOS 系统只适用于苹果的智能手机，安卓系统却是开放式的操作系统，几乎适用于其他所有制造商的智能手机。

正如它的名字，平板个人电脑是一个具有丰富功能的多媒体设备，用户能在约 10 英寸的大屏上欣赏音频和视频文件，其屏幕只比笔记本电脑的小一点。平板电脑配有虚拟键盘，能很好地完成如发邮件一类的小型打字任务。

所有的平板电脑都拥有无线功能，这意味着它能上网，同时也能玩游戏。如今，平板电脑趋于配有双摄像头，既可实现高清视频拍摄，也可实现视频通话与视频聊天。但是，由于硬件配置较低，平板电脑难以完成像多媒体任务的功能和其他复杂的操作。

在三种移动设备之中，笔记本电脑在计算和浏览网页方面最为强大。它拥有最快的处理器和最大的内存容量。笔记本电脑本质上是一本可以随身携带的个人电脑，它融合了一台电脑的所有功能。用户使用触摸屏而非鼠标，加上内置的扬声器，笔记本电脑拥

有完整的硬件套装。除此之外，笔记本电脑能够仅依靠电池工作，在不充电的情况下可运行 3~5 个小时。在 14 英寸甚至更大的显示屏上，笔记本电脑理论上能够完成一台电脑的任何操作。

注释

1. This disk is much like the tape inside a video or audio cassette. Take a look at the floppy disk pictured. At one end of it is a small metal cover with a rectangular hole in it. That cover can be moved aside to show the flexible disk inside.

译文：这种盘很像录影带或者录音带。仔细看一下软盘的图片你会发现在它的一端有一个小金属片覆盖着一个矩形洞。那个盖子可以移动到旁边，这样就能够看到磁盘的内部。

2. RAM loses its memory each time the computer is turned off, but ROM keeps information stored even when the computer is not turned on. Well, sometimes computers have problems that can cause them to crash. No, that doesn't mean they jump off the desk and smash on the floor.

译文：随机存储器在计算机关闭时会丢失所存储的所有信息，但是只读存储器可以在计算机关闭时保存信息。计算机有些时候会出现一些问题使它们发生撞击。不过并不是说它们会从桌子上掉下来摔碎在地板上。

第 2 章　计算机显示器是如何工作的

"纵横比"是什么意思？"点距"又是什么？一台显示器会消耗多少电能？CRT 和 LCD 这两种显示器的区别在哪里？"刷新率"又作何解？

在本文中，我们将会回答这些问题甚至更多。文章结束后，将会了解现有的显示器，当要购买下一台显示器时可以做出更好的决定。

2.1　基本常识

在多数情况下提到的监视器是计算机上使用最多的输出设备。在用计算机工作或娱乐的时候，显示器会以文本或图像的形式及时响应。大多数桌面显示器都使用阴极射线管（CRT），而便携式计算机如笔记本电脑，则结合液晶显示（LCD）、二极管放射线（LED）、等离子气体等其他图形显示技术。凭着轻巧的设计、更低的能耗，采用 LCD 技术的显示器开始在许多台计算机中取代古老的 CRT。

在购买显示器的时候，会做出很多的选择。这些选择将会影响显示器的作用、花费、可以浏览的信息。这些决定包括：

- 显示技术——当前主要是 CRT 和 LCD 这两种技术。
- Cable 技术——VGA 和 DVI 这两种最常见。
- 可视区域（通常是以对角线衡量的）。

- 纵横比及定位（前景或成像）。
- 最大分辨率。
- 点距。
- 刷新率。
- 色度。
- 能量消耗。

接下来我们将会对每一部分进行讨论，你将会完全了解显示器是如何工作的。

2.1.1 显示技术背景

从 1970 年代以文本为基础的计算机系统中技术未成熟的显示器开始到现在，显示器经历了很长的一段时间。让我们来看一看 IBM 在十年间的发展：

1981 年，IBM 推出彩色图形适配器（CGA），它能够显示 4 种颜色，最大分辨率为 320×200 像素。

IBM 在 1984 年推出增强型图形适配器（EGA），它能够显示 16 种不同颜色，最大分辨率增大到 640×350 像素，改善了显示的外观，并使得用显示器阅读文档更方便。

1987 年，IBM 推出视频图像阵列（VGA）显示系统。现在大部分计算机都支持 VGA 标准，许多 VGA 显示器还在使用。

IBM 在 1990 年推出扩展型图形阵列（XGA），提供 800×600 像素（1680 万色），及 1024×768 像素，65 536 色的显示器。

2.1.2 VGA 显示技术

一旦显示数据被转换为标准形式，它经过 VGA 电缆被送到显示器。

可以看到 VGA 连接器有 3 条不同的线连接红、绿、蓝色信号，有两条水平和垂直同步信号。在普通的电视机中，所有这些信号被组合成独立的复合视频信号。将各种信号分开是计算机显示器比电视具有更多像素的原因。

现今 VGA 适配器并不完全支持数字显示器，因此一个新的标准——交互式数字视频系统（DVI）产生了。

2.1.3 DVI 显示技术

因为 VGA 技术需要把信号从数字形式转化到类比（模拟）形式传输到显示器，会有一定量的信号损失。DVI 在计算机和监视器之间以数字形式保存数据，事实上消除了信号损失。

DVI 规格基于硅图像转换极低损耗微分信号（TMDS），提供高速数字接口界面。TMDS 从图形适配器取得信号，决定了显示器使用的分辨率和刷新率，并在可用的带宽上对信号进行扩展，用来优化数据在计算机和监视器之间的传送。DVI 是一项独立的技术。实质上，这意味着 DVI 将会在与其相容的显示器和显示卡上有效的运行。如果要买一台 DVI 显示器，就要确定有个能和它相连的视频适配器。

2.1.4 可视范围

有两项方法可以描述显示器的大小,即纵横比和屏幕大小。现在多数的计算机显示器,如同多数的电视机那样,有 4∶3 的纵横比。这意味着显示器的长宽比是 4∶3。另外一个比较常用的纵横比是 16∶9,在电影胶片中使用。虽然在电视刚开始发展时这个比例没有被采用,但是在如 LCD 等显示技术制造业中这个比例已经很常用。随着宽屏 DVD 电影的逐渐普及和发展,大多数的电视厂商都提供 16∶9 比例的产品。

显示器有一个发射性的表面,通常称为屏幕。屏幕通常按照对角线方向的尺度来衡量。这种对角测量系统的产生是因为早期的电视制造商想让他们的电视机的尺寸听上去让人有更深刻的印象。

现在很流行的荧屏大小是 15、17、19 和 21 英寸。笔记本计算机的屏幕通常比较小,典型的是从 12 到 15 英寸分布。很明显,显示器的大小将会直接影响分辨率。相同像素的分辨率在较小的显示器上比较清晰,而在较大的显示器上,因为相同数量的像素要在更大尺度范围分布,则会较模糊。在 21 寸显示器上 640×480 分辨率的图像看上不如在 15 英寸显示器上 640×480 分辨率的图像清晰。

2.1.5 最大分辨率和点距

分辨率指的是独立色点的数量,色点即在显示器中称为像素点的东西。分辨率典型的通过水平(行)和垂直(列)方向上的像素数量来表示,如 640×480。监视器的可视范围(在前面部分已讨论过),刷新率和点距都直接影响了显示器的最大分辨率。

2.1.6 点距

简要地说,点距是衡量像素之间有多少空间。当考虑点距的时候,记住是越小越好。将像素聚集得更集中是达到更高分辨率的基本原则。

一台显示器通常能支持与其物理尺寸大小匹配的分辨率,当然更小的也行。例如,一个有 1280 行,1024 列栅格的显示器,很明显最大可支持 1280×1024 像素的分辨率。它通常也可以支持更低的分辨率,如 1204×867 像素、800×600 像素和 640×480 像素。

2.1.7 刷新率

基于 CRT 技术的显示器中,刷新率是指图像每秒钟在显示器中显示的次数。如果 CRT 显示器有 72Hz 的分辨率,那么显示器中从顶部到底部的所有像素每秒刷新 72 次。刷新率很重要,因为它决定了闪烁程度,而且需要尽可能高的刷新率。每秒太少的刷新次数会让人感觉到闪烁,这可能导致头痛和眼睛疲劳。

电视机的刷新率比大部分计算机显示器的刷新率要低。为了帮助调整低的分辨率,会使用一种叫做隔行扫描的方法。这意味着电视机里面阴极射线管的电子枪将会从顶部到底部先扫描所有奇数的行,然后是偶数的行。磷光粉将光保持足够长的时间,让人的眼睛误以为所有的光线都是同时发射的。

显示器的刷新率取决于它要扫描的行数，这限制了可能达到的最大分辨率。很多显示器都支持多种刷新率，通常取决于所选择的分辨率。记住，在闪烁和分辨率之间有一个权衡，选择对自己来说最好的那一个。

2.1.8 颜色深度

图形适配器支持的显示模式和监视器的颜色性能这两者决定了显示器可显示的颜色种类。例如，一个可以在 SVGA 模式运行的显示器可以处理一个 24 位长的像素类型，所以它能显示 16 777 216 种颜色（一般在 1680 万左右）。通常用位深度来描述像素的位数。

在 24 位模式下，每 8 位都被分解到 3 个附加的原色，即红、绿和蓝。这个位深度也叫做真彩色，因为它可以产生 10 000 000 种可以被人眼辨别的颜色，而 16 位的显示模式只能产生 65 536 种颜色。显示器从 16 位到 24 位跳跃，是因为在 8 位增量模式下的工作对于开发者和程序员更为简单。

简单地说，颜色位深度指的是用来描述每一个像素颜色所使用的位数。位深度决定了一次可以被显示的颜色的数量。参看下面的表格可以知道不同的位深度可以产生的颜色数量。

位深	颜 色 数 量
1	2(单色)
2	4(CGA)
4	16(EGA)
8	256(VGA)
16	65,536(High Color, XGA)
24	16,777,216 (True Color, SVGA)
32	16,777,216 (True Color + Alpha Channel)

可以注意到最后一栏是 32 位的。这是数字化图像、动画和视频游戏为达到某种效果而使用的特殊图形模式。如前所述，现在销售的显示器通过标准的 VGA 连接器几乎都可以处理 24 位的彩色。

2.1.9 能量消耗

不同技术下的能耗差别很大。CRT 耗能大，一台典型的显示器大约为 110 瓦特，相比之下，LCD 平均是在 30～40 瓦特。

一个基于 CRT 技术的家用计算机，显示器占用了 80%的用电量。美国政府在 1992 年开始了能源之星计划。按照美国环保署的说法，如果计算机系统符合能源之星计划，它可以让用户在用电账单上每年节省 400 美元。同样，因为能量用法的不同，一个 LCD 显示器可能在开始时花费比较多，但从长期来看却可以省钱。

2.1.10 显示器发展趋势：平板

CRT 技术仍然是最流行的桌面显示器技术。因为标准 CRT 技术需要在发射装置和荧屏幕之间有一定的距离，使用这种显示技术的显示器看上去体积庞大。使用其他的技术有可能得到薄得多的显示器，已知的如平板显示器。

液晶显示技术是阻止光而不是产生光，发光二极管和等离子气体工作时点亮屏幕上的相应位置，该位置取决于不同栅格交叉点的电压。LCD 比 LED 和等离子气体要求的能源低得多，而且是当前笔记本计算机和便携式计算机采用的主要技术。随着平板显示器在屏幕尺寸的增长以及分辨率和供给情况的改进，它们将逐渐取代基于 CRT 技术的显示器。

注释

1. Since today's VGA adapters do not fully support the use of digital monitors, a new standard, Digital Video Interface (DVI) has been designed for this purpose.

原译：因为今天的 VGA 适配器并不完全支持数字显示器，一个新的标准——交互式数字视频（DVI）产生了。

译文：现今的 VGA 适配器并不完全支持数字显示器，因此一个新的标准——交互式数字视频系统（DVI）产生了。

分析：逻辑和用词的语气上改得更顺些。这样行文更好。

2. Obviously, the size of the display will directly affect resolution. The same pixel resolution will be sharper on a smaller monitor and fuzzier on a larger monitor because the same number of pixels is being spread out over a larger number of inches.

原译：很明显，显示器的大小将会直接影响分辨率。相同像素的分辨率在较小的显示器上比较清晰而在较大的则较模糊，因为相同数量的像素在更大尺度范围分布。

译文：很明显，显示器的大小将会直接影响分辨率。相同像素的分辨率在较小的显示器上比较清晰，而在较大的显示器上，因为相同数量的像素要在更大尺度范围分布，则会较模糊。

分析：用词上基本无改动，主要是将后面的两个短句对调，这样一来，读起来更加通顺。

3. This is a special graphics mode used by digital video, animation and video games to achieve certain effects. Nearly every monitor sold today can handle 24-bit color using a standard VGA connector, as discussed previously.

原译：这是特殊的数字化图像、动画和视频游戏为达到某种效果而使用的图形模式。像前面讨论过，现在销售的显示器通过标准的 VGA 连接器几乎都可以处理 24 位的彩色。

译文：这是数字化图像、动画和视频游戏为达到某种效果而使用的特殊图形模式。如前所述，现在销售的显示器通过标准的 VGA 连接器几乎都可以处理 24 位的彩色。

分析：首先要注意"特殊的"一词所修饰的对象是图形模式，若放前面则会产生歧义。as discussed previously 语气较简略，故翻译过来也最好用简洁一些的词语。

4. Power consumption varies greatly with different technologies. CRTs are somewhat power-hungry, at about 110 watts for a typical display, especially when compared to LCDs, which average between 30 and 40 watts.

原译：能耗随着不同的技术而不同。CRT 耗能大，每一台显示器大约 110 瓦特，相比之下，LCD 平均在 30～40 瓦特。

译文：不同技术下的能耗差别很大。CRT 耗能大，一台典型的显示器大约为 110 瓦特，相比之下，LCD 平均是在 30～40 瓦特。

分析：原译中 greatly 未体现，用译文的方式意思表达较完整。另外，后半句说的是每台显示器耗能 110 瓦特，原译说法上很奇怪，是照搬英文的语序。

5. The U.S. government initiated the Energy Star program in 1992. According to the EPA, if you use a computer system that is Energy Star compliant, it could save you approximately $400 a year on your electric bill! Similarly, because of the difference in power usage, an LCD monitor might cost more upfront but end up saving you money in the long run.

原译：美国政府在 1992 年开始了能源计划。按照美国环保署的说法，如果你的计算机系统是能量节约型的，它可以让你在用电账单上每年节省 400 美元！同样的，因为能量用法的不同，一个 LCD 显示器可能在开始时花费比较多，但经长久使用最后可以节省你的钱。

译文：美国政府在 1992 年开始了能源之星计划。按照美国环保署的说法，如果计算机系统符合能源之星计划，它可以让用户在用电账单上每年节省 400 美元！同样，因为能量用法的不同，一个 LCD 显示器可能在开始时花费比较多，但从长期来看却可以省钱。

分析：对 Energy Star 这样的专有名词，不能凭自己的理解，要多查资料，以便得到准确的中文用词。其他几个改动则着眼于用词的准确性，在此不再赘述。

6. Liquid crystal display (LCD) technology works by blocking light rather than creating it, while light-emitting diode (LED) and gas plasma work by lighting up display screen positions based on the voltages at different grid intersections.

译文：液晶显示技术是阻止光而不是产生光，发光二极管和等离子气体工作时点亮屏幕上的相应位置，该位置取决于不同栅格交叉点的电压。

分析：该句翻译的难点在于句子长，用了后置定语修饰"位置"一词，译文采用的是另起一句说明，这种化长为短的技巧在科技文翻译中很重要，要重点掌握。

第 3 章　手机如何工作

在中国和世界各地，乃至整个世界，有几十亿人在使用手机。这是一种很神奇的装置——有了它，可以同地球上任何地方的任何人进行通话。

现在的手机有着一系列难以置信的功能，一些新型手机也蜂拥而至，有了手机可以：

- 储存联系信息。
- 设置任务和待办事项。

- 设置约会和提醒。
- 使用内置计算器进行简单的数学计算。
- 收发邮件。
- 获取网络信息（新闻、娱乐和股价）。
- 玩游戏。
- 看电视。
- 发信息。
- 配置 PDA、MP3 播放器和 GPS 接收器。

但是你曾想过手机是怎样工作的吗？手机与普通电话有什么不同吗？文中将讨论手机背后的技术，带你领略它的神奇之处。如果你正想买个手机，一定要阅读《如何购买手机》来了解购买须知。

首先，手机的一个最有趣之处在于，它实际上就是个无线电收发器——一个极其复杂的无线电收发器。电话是亚历山大·格雷厄姆·贝尔（Alexander Graham Bell）于 1876 年发明的，而无线通信（通常认为是年轻的意大利人古列尔莫·马可尼（Guglielmo Marconi）于 1894 年发明）则可以追溯到 1880 年代尼古拉·特斯拉（Nikolai Tesla）发明的无线电。这两项伟大技术的最终结合，也不过就成了自然而然的事。

3.1 手机频率

在手机发明之前的黑暗年代，确实需要移动通信功能的人在汽车内安装无线电话。无线电话系统在每个城市都有一个核心天线塔，每个天线塔可能拥有 25 个可用信道。这就意味着安装在汽车内的电话，需要一个功率强大的发射器——足以传输 40 或 50 英里（约 70 千米）。也就是说，只有很少人能够使用无线电话——因为没有足够的信道。

手机单元系统的独特之处在于，其把城市划分为小的单元，这使得整个城市能够进行广泛的频率复用，数百万人可以同时使用手机。

要理解手机的复杂性，一个好的方法是拿它和 CB 对讲机或步话机进行比较。

- 全双工传输、半双工传输：步话机和 CB 对讲机都是半双工传输装置，CB 对讲机的通话双方使用的是同一频率，因此同一时间只能有一个人讲话。手机是全双工传输装置，讲话时使用一个频率，接听时使用另一个独立的频率，通话双方同一时间都可以讲话。
- 信道：一个典型的步话机有一个信道，一个 CB 对讲机有 40 个信道，而一个典型的手机可以通过 1664 个或者更多的信道进行通话。
- 覆盖范围：步话机使用 0.25 瓦的发射器能够传输大约 1 英里（1.6 千米），CB 对讲机由于功率更为强大，使用 5 瓦的发射器能够传输大约 5 英里（8 千米）。手机则是在各个划分的单元内运行。移动时，各个单元之间可以进行切换。单元划分使得手机拥有不可思议的覆盖范围。由于单元间的切换，手机用户可以驱车行驶数百英里，而一直保持通话。

在美国一个典型的手机模拟单元系统中，手机运营商在一个城市可以接收到大约 800 个频率来使用。运营商把城市划分为各个单元，每个单元大小都是约 10 平方英里

（26平方千米），就像图3.3（图略）中六边形网格中的六边形一样。

每个单元拥有一个基站，包括一个发射塔和一个装有无线电设备的小型建筑。我们以后再阐述基站，首先来看一下构成手机单元系统的"单元"。

3.2 手机信道

手机模拟单元系统中的每一个单元使用 1/7 的可用双工语音信道。也就是说，每个单元都在使用 1/7 的可用信道，这样每个单元都有一个独特的频率，而不会出现冲突：

- 城市中一个典型的手机运营商有 832 个无线电频率可供使用。
- 每个手机每次通话使用两个频率（手机为双工通道），因此每个运营商拥有 395 个典型语音信道（另外 42 个信道用来做控制信道，后面再做阐述）。

因此，每个单元有大约 56 个可用语音信道。换句话说，在任一单元中，56 个人可以通过手机同时进行通话。手机模拟单元系统被称为第一代移动通信技术，或者叫做 1G。数字传输方式（2G）出现后，增加了可用信道数量。比如，基于 TDMA 技术（本文后面将做阐述）的数字系统，每个单元有大约 168 个可用信道，可支持的通话数量是模拟系统的 3 倍。

手机中有低功率的发射器，许多手机有两种信号强度，即 0.6 瓦和 3 瓦（相比之下，大多数 CB 对讲机传输功率为 4 瓦）。基站也是低功率传输，低功率发射器拥有两大优势：

- 基站的传输和单元内的手机相距不远，因此如图 3.3（图略）所示，紫色单元可以复用同一的 56 个频率，这一频率可以在整个城市广泛复用。
- 手机通常都是由电池供电，功耗相对较低。功耗越低，电池也就越小，可以握在手中的手机也就有了可能。

单元系统要求在任何大小的城市都要有大量基站。典型的大城市会有数百个发射塔。但由于手机用户很多，每个用户付出的成本也就较低。每个城市中的每个运营商还运营着一个中心站点，叫做移动电话交换站（MTSO）。这一站点负责把所有手机链接到通常的陆基电话系统，并且控制区域内的所有基站。

3.3 模拟手机

1983 年，美国联邦通信委员会（FCC）通过了模拟手机标准 AMPS（高级移动电话系统），并首先在芝加哥使用。AMPS 为模拟手机提供从 824 兆赫（MHz）到 894 兆赫（MHz）之间的一系列频率。为促进竞争和保持低价，美国政府要求每个市场要有两个运营商：A 和 B。其中一个运营商通常被叫做本地交换运营商（LEC），这是对本地电话公司的另一种称呼。

A 和 B 运营商分别分配到 832 个频率：790 个用来传输语音，42 个用来传输数据。一个信道有两个频率（一个传输，另一个接收）。模拟语音信道使用的频率频宽是典型的 30 千赫。把 30 千赫作为标准尺寸，可以使语音质量与有限电话相媲美。

每个语音信道的传输和接受频率以 45 千赫为界，而不互相干扰。每个运营商有 395 个语音信道和 21 个数据信道，用作像登记和呼叫这样的内部事务活动。

AMPS 的一个版本窄带高级移动服务（NAMPS）纳入了一些数字技术，使系统传输的通话数量比原来的版本高出3倍。尽管使用了数字技术，NAMPS仍然是模拟系统。AMPS和NAMPS只在800兆赫的频宽下运行，并不提供许多数字单元服务的常见特征，如邮箱和网页浏览。

3.4 数字时代的到来

数字手机是第二代移动通信技术（2G），同模拟手机一样使用无线电技术，但是方式不同。模拟系统没有充分利用手机和单元网络之间的信号，模拟信号不像真正的数字信号那样容易压缩和控制。许多电缆公司都在转向数字信号，因为那样的话，在一定频宽的情况下就能传输更多信道。数字系统的效率之高令人称奇。

数字手机把语音转换为二进制信息（1和0）（转换过程详情参阅《模拟数字记录如何工作》），然后进行压缩。压缩后，3~10个数字手机通话占据的空间，相当于一个模拟手机通话占据的空间。

许多数字手机系统依靠移频键控（FSK）来与AMPS交换数据。FSK使用两种频率，即1和0，两者之间迅速交替，在信号发射塔和手机之间传递数字信息。利用智能调制和编码体系，把模拟信息转换为数字信息，然后压缩，再转换回模拟信息，同时保持可接受的语音质量。所有这一切要求数字手机必须拥有很多处理能力。

我们来看一下数字手机的内部情况。

3.5 数字手机的内部

手机的每立方英寸都十分复杂，它是人们每天使用的最复杂的装置之一。现代数字手机每秒可以处理数以百万计的计算，来压缩和解压缩语音流。

如果把基本的数字手机拆开，就会发现里面的零部件很少，主要包含：

- 手机核心的神奇的电路板。
- 天线。
- 液晶显示器（LCD）。
- 键盘（类似于电视遥控器上的键盘）。
- 麦克风。
- 扬声器。
- 电池。

电路板是这一系统的心脏。下图是一个典型的诺基亚数字手机电路板：

电路板正面

电路板背面

从上图中可以看到数个计算机芯片，让我们来讨论一下这些芯片的作用。模拟至数字（A-to-D）转换芯片把发出的音频信号从模拟信号转化为数字信号，数字至模拟（D-to-A）转换芯片把接收的信号从数字信号转换为模拟信号。在光盘如何工作中，可以了解到更多关于数字和模拟信号的转换以及它们对于数字音频的重要性。设计高度用户化的数字信号处理器（DSP）是要进行高速信号操纵计算。

这一微处理器负责键盘和显示器的所有内部事务，处理和基站之间命令和控制信号，并协调电路板上的其他功能。

只读存储器（ROM）和闪存芯片存储手机的运行系统和用户特征（比如电话簿）。无线电频率（RF）和电源部分负责电源管理和充电，并处理数百个调频（FM）频道。最后，射频放大器（RF amplifiers）负责天线发出和接收的信号。

随着手机的特征的增加，显示器的尺寸也扩大了很多。现在的手机大部分都有内置的电话簿、计算器和游戏。许多手机内置了许多种 PDA 和网页浏览器。

一些手机在内部闪存中存储像 SID 和 MIN 代码的信息，还有一些使用像 SM 卡（SmartMedia）那样的外卡。

手机的扬声器和麦克风都很小，它们中的大部分产生的声音却不可思议。如下图所示，扬声器如硬币般大小，麦克风比它旁边的手表电池大不了多少。这里的手表电池是用来为手机的内部时钟芯片供电。

手机扬声器、麦克风和后备电池

令人惊奇的是，所有这些功能，仅仅在 30 年前还能铺满办公大楼的一整块地板，而现在则可以舒服地握在手中。

注释

1. Someone using a cell phone can drive hundreds of miles and maintain a conversation the entire time because of the cellular approach.

译文：由于单元间的切换，手机用户可以驱车行驶数百英里，而一直保持通话。

2. If you are thinking about buying a cell phone, be sure to check out How Buying a Cell Phone Works to learn what you should know before making a purchase.

译文：如果你正想买个手机，一定要阅读《如何购买手机》来了解购机须知。

3. In this article, we will discuss the technology behind cell phones so that you can see

how amazing they really are.

译文：本文将讨论手机背后的技术，带你领略它的神奇之处。

第 4 章　数码相机基础知识

4.1　数码相机的基本原理

假设要照一张照片并通过电子邮件将照片传给朋友，则需要以计算机能够识别的语言来代表图片——比特和字节。实质上，数码图片只是一长串能够表示小色点（像素）的 1 和 0，它们一起能够构成图片。

如果想使照片成形，则有两个选择：
- 可以用传统胶片相机来拍照，用化学方式处理胶片，将图片印在相纸上，然后使用数码扫描仪来进行样本印刷（把光线模式记录成一系列的像素值）。
- 可以直接将主题上充裕的原始光作为样本，并立即将光图像分割成一系列的像素值，也就是说，可以使用数码相机。

数码相机将光聚焦到半导体上来创建数字图像。

就像传统相机一样，数码相机有多个能集中光产生图像的棱镜。然而，它没有将光集中在一张胶片上，而是将光集中在能够电子地拍摄光的半导体设备上。计算机能够将电子信息分割成数字数据。这个过程也直接使得数码相机拥有了很多有趣的特性。

在下面的几个部分中，我们将准确地探寻数码相机是如何做到这一切的。

4.2　CCD 和 CMOS：不用胶片的相机

数码相机没有使用胶片，取而代之的是能够将光转换为电荷的传感器。

大多数的数码相机使用的图像传感器都是电荷耦合元件。然而，也有一些相机使用的是互补金属氧化物半导体技术。CCD 和 CMOS 图像传感器都可以将光转换为电子。如果读过《太阳能电池是如何工作的》这本书，就能明白在文章所提到的技术中有一种能使转换发生。我们可以用一种简单的方法来思考传感器，那就是想想几千或者几百万小太阳能电池的二维阵列。

一旦传感器将光转换为电子，它将读取图像中的每个电池的值（堆积电荷）。这是两种传感器运作的区别所在：
- CCD 通过芯片传输电荷，并且在排列的一角读取电荷，再通过测量每个图片位置的电荷量，将测量转换为二进制形式，并借由模拟至数字转换器将每个像素值转换为数码值。
- CMOS 设备在每个像素上使用晶体管来增强电荷，并使用较多的传统电线来传导电荷。

这两个传感器的区别也使得二者各有优劣：
- CCD 传感器能够制造高质量低噪点的图像。
- CMOS 传感器则通常易有噪点。

因为 CMOS 传感器上的每个像素旁都有几个晶体管，而 CMOS 芯片的感光度要比 CCD 的低。许多光子碰撞的是晶体管而不是光敏二极管。

按理说，CMOS 传感器用电少，而 CCD 使用的进程用电多，CCD 所用的电是相等的 CMOS 所用的电的 100 倍。

CCD 传感器长期以来被大量生产，所以变得越来越成熟。它们趋向有高质量的像素，并且越来越多。

尽管两者之间有着很大的区别，它们却在相机中起到同样的作用——将光转化为电。为了了解数码相机是如何工作的，可以把两种传感器当成同一设备。

4.3 数码相机分辨率

相机所能捕捉的细节量就叫做分辨率，它是以像素为单位衡量的。相机的像素越大，其所能抓住的细节越多，也就可以在不模糊又呈现颗粒状的条件下拍出越大的照片。

一些典型的分辨率有：

256×256——很廉价的相机的分辨率。此分辨率非常低以至于画面质量通常都不好。其总像素为 65 000。

640×480——这是大多数低端相机的分辨率。此分辨率适合电邮图片或者在网上发布的图片。

1216×912——这是百万像素的图像尺寸，即 1 109 000 个像素，适合打印图片。

1600×1200——接近 200 万的像素，属于"高分辨率"。可以以同样的分辨率打印出 4×5 英寸的和专业照片实验室拿到的照片同质量的照片。

2240×1680——400 万像素相机所有（现行标准）可以打印出更大的照片，达到 16×20 英寸的高质量的照片。

4064×2704——顶级的数码相机，在这个分辨率拍摄的照片有 111 万像素。在此设置下，能制造出 13.5×9 英寸的照片而不失其画面质量。

下面来看看相机是如何给图像增加色彩的：

你可能注意到了像素值，最大的分辨率却很难估算。比如，一台 210 万像素的相机能够拍摄出分辨率为 1600×1200，或者 1 920 000 像素的图片。但是"210 万像素"意味着至少有 210 万个像素。

这并非四舍五入或者二进制数学弄虚作假所造成的错误。这是因为 CCD 需要容纳电路来使 ADC 估量电荷。电路被着成黑色，这样才不会吸收光扭曲图像。

4.4 捕捉色彩

不幸的是，每个图素都是色盲。它只保留了到达其表面的光的总强度的轨迹。为了得到一张彩色的图像，大多数传感器使用过滤将光视为 3 种基本色彩的结合体。一旦记录了这 3 种色彩，相机就可将它们组合起来创作成完整的光谱。

用相机来记录 3 种色彩有几种方法。最高质量的相机使用的是 3 种不同的传感器，每个都有不同的过滤。电子束分裂器将光分给不同的传感器。想象一下，光像水流入管道一样进入相机。使用电子束分裂期就像是将同量的水分到 3 个管道中。每个传感器都得到同样的图像，但是因为过滤器，每个传感器只对基本色彩之一作出反应。

如图 4.5（图略）所示，我们可以看到分割图像的过程。

这种方法的好处在于相机在每个像素位置上记录这 3 种色彩。缺点是，使用这种方法的相机一般体积庞大且价格昂贵。

另一种方法是在单个传感器前旋转红、蓝、绿光的过滤器。传感器接二连三地快速记录 3 种不同的图像。这种方法在每个像素位置上提供了 3 种色彩的信息。然而，因为 3 种图像不是在同一精确时间进行拍摄的，所以就需要在进行 3 次读入时将相机和相片的目标保持静止。这对手持相机以及抓拍摄影则不现实。

这两种方法都适合在专业摄影室相机里使用。但是，对于随意抓拍，它们却显得不现实。

4.5 数码摄影基础知识

数码摄影较传统胶片摄影有众多优点。数码照片方便，可以立刻呈现拍摄结果，不需要胶片和冲洗的费用，适合软件编辑与网络上传。

然而，胶片拍摄在摄影世界里总有一席地位。数码相机几乎完全占据了相机消费市场。就在 5 年之前，买一台与胶片机视觉效果相当的数码相机要花费 1000 美元。然而，现在价格大幅下降，相机质量也提高了。如今，500 美元以内的相机质量几乎达到专业水平，除了最便宜的数码相机以外，其他价位的相机都能制作出不错的图像。

数码相机也有很多附加功能，包括图像稳定、机上图像编辑、色彩校正功能、自动包围式曝光与突发模式。很多这样的功能都能通过图像编辑软件实现，所以将它们设置在相机中是多余的（通常是低级的）。突发模式、微距模式以及图像稳定是最有用的性能，但是挑选最适合自己的相机的最佳方法是浏览众多数码摄影杂志中的任意一本，或者浏览任意能够提供不同相机的对比以及用户评论的网站。

如果拥有或者打算购买数码相机，这篇文章将使你了解数码相机，并帮助你更好地利用它。

4.6 百万像素等级

500 万像素的图像质量接近胶片机图像质量。数码相机决定图像质量的最基本的特性就是它的百万像素等级。这个数字就是一张照片内相机传感器能抓到的信息量。高像素等级的相机拍摄的照片更加细腻，因而照片更大。这些照片，特别是在打印成大尺码时，效果会更好。

图 4.6（图略）展示了一台数码相机。

下面我们来了解一下数码相机的设置。

4.7 数码相机的设置与模式

一台体面的数码相机，加一点练习，可以让任何人在相机设置在全自动模式下拍出质量不错的照片。甚至可以拍一堆马马虎虎的照片，通过图像编辑照片来让它们看起来不错。但是为了能够借相机之力来拍摄真正美丽的照片，就需要了解手动设置下的一些东西。请记住，低端相机可能没有手动调节设置。

在相机上改变设置时，你需要找到合适的曝光量以及照明条件。曝光量是指在拍摄照片时进入相机传感器的光的量。通常，为了使得相机传感器抓住的图像与眼睛所看到的一致，就要设置曝光量。相机在全自动模式下尽力完成这一任务，但是相机很容易受骗，过程有些慢，所以需要人工设置来制作更好的照片。

对相机越来越熟悉的时候，就能通过设置曝光量来达到不同效果。有些时候自动模式会更好——事情突然发生，只有几秒的时间来进行拍摄。这时只需设置在自动模式上进行拍照。

可以通过调整两个不同的设置来调整曝光量，即光圈和快门速度。光圈是镜头口径的直径——大一些的光圈意味着更多的光能进入相机。光圈以 f 值衡量，高一些的 f 值意味着光圈小一些。光圈设置也会影响景深——照片中清晰对焦的部分。光圈越小（高一些的 f 值），景深越大。用足够小的光圈可以将前景中的人和人后面 20 英尺的汽车清晰对焦。光圈越大，景深越小。这通常被用于近距离拍摄的画像。

图 4.7 为数码相机设置示例。

下来我们将仔细地研究一下快门速度。

4.8 快门速度

快门速度指快门保持打开，允许光线进入快门的时间量。最快的快门速度是 1/2000 秒，相机设置允许的慢快门速度为 1 秒。1/60 秒的快门速度可以使你在手执相机时拍的照片不模糊。有些摄影家强制将他们的快门保持打开更长一段时间，以此来制造各种奇特的效果。将快门打开的相机对准夜空，若干个小时之后就能够产生一系列星辰的轨迹照片，从中可以发现随着地球的转动，星星好像划过天空。

实践以及经验是发现各种各样的照片最佳光圈和快门速度搭配的最好的途径。慢快门速度能够容纳更多的光，同时很难拍到清楚地照片。任何动作（主题的动作或者相机的动作）都会使照片模糊。有时候你或许想要这种效果，但是为了能够将运动中的主题拍摄得清楚，快门速度必须快。

很多的相机都有能够设置光圈优先或者快门优先的半自动模式。你可以通过设置光圈或者快门速度（这取决于哪种优先已被启动）来满足所需设置。相机估量正确的设置并以此来适应照明条件。相机也可能会有各种模式供你选择，例如，运动模式或者户外模式。这些都对光圈以及快门进行了预先设置。同样，经验会让你知道什么条件下该用哪种模式。

注释

1. You can directly sample the original light that bounces off your subject, immediately breaking that light pattern down into a series of pixel values—in other words, you can use a

digital camera.

译文：可以直接将主题上充裕的原始光作为样本，并立即将光图像分割成一系列的像素值。就是说，可以使用数码相机。

2. Aperture is measured in f-stops. Higher f-stop numbers mean a smaller aperture. The aperture setting also affects depth of field, the amount of the photograph that is in focus. Smaller apertures (higher f-stops) give longer depth of field.

译文：光圈以 f 值衡量。高一些的 f 值意味着光圈小一些。光圈设置也会影响景深——照片中清晰对焦的部分。光圈越小（高一些的 f 值），景深越大。

3. Because each pixel on a CMOS sensor has several transistors located next to it, the light sensitivity of a CMOS chip is lower. Many of the photons hit the transistors instead of the photodiode.

译文：因为 CMOS 传感器上的每个像素旁都有几个晶体管，所以 CMOS 芯片的感光度要比 CCD 的低。许多光子碰撞的是晶体管而不是光敏二极管。

4. The amount of detail that the camera can capture is called the resolution, and it is measured in pixels. The more pixels a camera has, the more detail it can capture and the larger pictures can be without becoming blurry or "grainy."

译文：相机所能捕捉的细节量就叫做分辨率，它是以像素为单位衡量的。相机的像素越大，所能抓住的细节越多，也就可以在不模糊不呈现颗粒状的条件下拍出越大的照片。

5. A beam splitter directs light to the different sensors. Think of the light entering the camera as water flowing through a pipe. Using a beam splitter would be like dividing an identical amount of water into three different pipes.

译文：电子束分裂器将光分给不同的传感器。想象一下，光像水流入管道一样进入相机。使用电子束分裂期就像是将同量的水分到 3 个管道中。

6. Exposure is the amount of light hitting the camera's sensor when you take a photo. Generally, you will want the exposure set so that the image captured by the camera's sensor closely matches what you see with your eyes.

译文：曝光量是指拍摄照片时进入相机传感器的光的量。通常，为了使得相机传感器抓住的图像与眼睛所看到的一致，就要设置曝光量。

第 5 章 位和字节是怎样工作的

如果你用一台计算机超过 5 分钟，那么你应该已经知道位和字节这两个词。随机存取存储器和磁盘的容量都是用字节来衡量的，就像用文件浏览器来查看文件的大小一样。

你可能听过这样的一则广告：这台计算机拥有 32 位的奔腾处理器和 64MB 的内存以及 2.1GB 的硬盘空间。在这篇文章中，我们将讨论位和字节，对位和字节有个完整的认识。

5.1 十进制数

了解位的最好方法就是跟阿拉伯数字做比较。一个阿拉伯数字就是能表示数值（hold numerical）0～9 数字中的任何一个。阿拉伯数字通常组合成一组来形成大的数值。比如，6357 有 4 个数字，它被理解为 7 在个位上，5 在十位上，3 在百位上，6 在千位上，所以如果想清楚地表达，可以这样写：

(6*1000)+(3*100)+(5*10)+(7*1)=6000+300+50+7=6357

另一种表示方法，可以用 10 的幂。假设我们用 "^" 符号来隔开底数和幂（如 10 的平方写成 10^2），则可以表示为：

(6*10^3)+(3*10^2)+(5*10^1)+(7*10^0)=6000+300+50+7=6357

从这个表达式中可以看到，每个阿拉伯数字对于 10 的下一级更高的幂是一个占位符，第一个数字开始于 10 的 0 次幂。

这样感觉相当不错——我们每天都跟十进制数打交道。该数字系统灵活的一点就在于让你不再需要一个数位表示十个不同的数值。以 10 为基数的数字系统的出现似乎是因为我们有 10 个手指，但是如果人们进化成 8 个手指，那么大概就应该有以 8 为基数的数字系统了。我们应该能有以任何数为基数的数字系统。事实上，我们有很多的理由需要以不同进制解决各种不同情形的问题。

5.2 位

计算机是以 2 为基数的数字系统来运算操作的，也就是大家所知道的二进制数字系统（就像以 10 为基数的数字系统称为十进制数）。计算机使用 2 为基数的数字系统是因为二进制更容易用当前电子技术来实现。也可以构造以 10 为基数来操作的计算机，但它们将是极其昂贵的。另一方面，以 2 为基数的计算机相对便宜得多。

因此，计算机用二进制来替代十进制。"位"就是"二进制位"的缩写。尽管十进制数字有从 0～9 十个可能的取值，位只有两个可能的取值：0 和 1。因此二进制只由 0 和 1 组成，如 1011。我们怎么判断二进制数 1011 表示的数值呢？可以像上面计算 6357 那样来计算它，但是我们要用 2 来代替 10，方法如下：

(1*2^3)+(0*2^2)+(1*2^1)+(1*2^0)=8+0+2+1=11

可以看到，在二进制数字中，各个位以 2 为底数的升幂排列，这使得二进制的计算非常的简单。从 0 到 20 的十进制与二进制的对比如下：

```
0=    0              11=  1011
1=    1              12=  1100
2=    10             13=  1101
3=    11             14=  1110
4=    100            15=  1111
5=    101            16=  10000
6=    110            17=  10001
```

```
7=    111                18= 10010
8=    1000               19= 10011
9=    1001               20= 10100
10= 1010
```

观察这个序列就会发现，0 和 1 在二进制和十进制系统中都是一样的。在二进制中从数字 2 开始发生变化。如果一个位数值是 1，加上 1，则这个位就变为 0，同时它的下一位增加 1。从 15 到 16 的转变实现中，连续进了 4 位，从 1111 到 10000。

5.3 字节

位很少在计算机中单独出现。它们几乎总是每 8 位组成一组，这些组就叫做字节。为什么一个字节要有 8 个位呢？就如一个类似的问题："为什么 12 个蛋是一打呢？" 8 个位为一个字节是人们在过去的 50 多年里，经过了反复试验决定的。

8 个位一个字节，可以如下表示从 0～255 这 256 个数值：

```
0=00000000
1=00000001
2=00000010
    ...
254=11111110
255=11111111
```

光盘的内容是用两个字节或者 16 位为单位（sample）来存储的，而每个单位的存储范围是 0～65 535，如下：

```
0=0000000000000000
1=0000000000000001
2=0000000000000010
    ...
65534=1111111111111110
65535=1111111111111111
```

5.4 字节：ASCII 码

在一个文本文件中，每个字符经常都用字节来存储。在 ASCII 码字符集中，从 0～127 每个二进制位的值都被赋予一个特定的字符。大部分计算机都扩充 ASCII 码字符集，使它达到一个字节可以表示的 256 个字符的范围。更高的 128 个字符表示特别的东西，如一些外语中的字符。

以下可以看到 127 个标准的 ASCII 码。计算机用这些编码存储文本文件在磁盘和内存中。举个例子，如果在 Windows 95/98 用记事本创建一个文本文件，对文件内容 "Four score and seven years ago" 中的每个字符，记事本将会用存储器中的一个字节来存储（包括用一个字节来存储单词间的空格符——ASCII 码值 32）。当记事本在磁盘中存储句子于文件中时，文件中每个字符和空格也占用一个字节。

试做一下这个实验：在记事本中打开一个新的文件，里面插入这个句子 "Four score and seven years ago"。用文件名 getty.txt 将这个文件保存在磁盘中。然后用资源管理器查看这个文件的大小，将会发现这个文件在磁盘中占据 30 个字节的空间：每个字符占一个字节。如果在这个句子的末尾加入另外一个单词，然后重新保存一下，则这个文件的大小立即增加相应的字节，每个字符占一个字节。

如果像计算机一样看一个文件，将会发现每个字节包含的不是一个字母，而是一个数——与这个字母相应的 ASCII 码值。所以在磁盘中，表示文件的数字是这样子的：

F　o　u　r　　a　n　d　　s　e　v　e　n
70　111　117　114　32　97　110　100　32　115　101　118　101　110

通过 ASCII 码表可以发现，每个字符和 ASCII 码存在着一一对应的关系。以 32 表示空格——在 ASCII 码表中 32 就表示空格。当我们要进行技术处理的时候，我们可以把这些十进制转化为二进制（如 32＝00100000）——这正是计算机的处理过程。

5.5 标准的 ASCII 码字符集

前 32 个数值（0～31）代码表示像回车和换行这样的字符。空格符是第 33 个数值，接着是标点符号、阿拉伯数字、大写字母、小写字母。如表 5.1 所示。

表 5.1 ASCII 码字符集

0	NUL	16	DLE	32		48	0	64	@	80	P	96	`	112	p
1	SOH	17	DC1	33	!	49	1	65	A	81	Q	97	a	113	q
2	STX	18	DC2	34	"	50	2	66	B	82	R	98	b	114	r
3	ETX	19	DC3	35	#	51	3	67	C	83	S	99	c	115	s
4	EOT	20	DC4	36	$	52	4	68	D	84	T	100	d	116	t
5	ENQ	21	NAK	37	%	53	5	69	E	85	U	101	e	117	u
6	ACK	22	SYN	38	&	54	6	70	F	86	V	102	f	118	v
7	BEL	23	ETB	39	'	55	7	71	G	87	W	103	g	119	w
8	BS	24	CAN	40	(	56	8	72	H	88	X	104	h	120	x
9	TAB	25	EM	41	)	57	9	73	I	89	Y	105	i	121	y
10	LF	26	SUB	42	*	58	:	74	J	90	Z	106	j	122	z
11	VT	27	ESC	43	+	59	;	75	K	91	[	107	k	123	{
12	FF	28	FS	44	,	60	<	76	L	92	\	108	l	124	|
13	CR	29	GS	45	-	61	=	77	M	93	]	109	m	125	}
14	SO	30	RS	46	.	62	>	78	N	94	^	110	n	126	~
15	SI	31	US	47	/	63	?	79	O	95	_	111	o	127	DEL

5.6 海量字节

开始谈论大量字节时，会碰到像 kilo、mege、giga 这样的前缀，如 kilobyte、megabyte、

gigabyte（也缩写为 K、M 和 G，如 Kbytes、Mbytes、Gbytes，或者 KB、MB、GB）。表 5.2 说明了它们间的级数关系。

表5.2 字节单位

名称	缩写	大小
Kilo	K	$2^{10}=1,024$
Mega	M	$2^{20}=1,048,576$
Giga	G	$2^{30}=1,073,741,824$
Tera	T	$2^{40}=1,099,511,627,776$
Peta	P	$2^{50}=1,125,899,906,842,624$
Exa	E	$2^{60}=1,152,921,504,606,846,976$
Zetta	Z	$2^{70}=1,180,591,620,717,411,303,424$
Yotta	Y	$2^{80}=1,208,925,819,614,629,174,706,176$

从表 5.2 中可以看到，kilo 大约等于 10^3，mega 大约等于 10^6，giga 大约为 10^9 等。所以当有人说"这台计算机有 2GB 的硬盘驱动"，他的意思是说，这个硬盘驱动可以存储 2GB 字节或者说接近 20 亿字节，或者确切地说是 2 147 483 648 个字节。你需要 2GB 的空间吗？如果想一想，一个 CD 能够保存 650MB，那么你就会发现只要用 3 个 CD 的数据就会占满整个空间了。T 字节的数据库现在已经非常普遍了，目前在五角大楼的里面还存在着少数 P 字节的数据库。

5.7 二进制计算

除了每个位上只能有 0 和 1 两个数值外，二进制计算就像十进制计算一样。为了对二进制计算有个感性的认识，让我们从十进制加法开始，看看它是怎么计算的。假设我们要计算 452＋751：

$$\begin{array}{r} 452 \\ +751 \\ \hline 1203 \end{array}$$

把这两个数值加起来，从右开始 2＋1＝3，没问题。接着，5＋5＝10，所以你保存 0，并且进 1 给下一位。接着，4＋7＋1（进位值）＝12，所以保存 2 进 1，最后 0＋0＋1＝1，所以答案是 1203。

二进制加法与此类似：

$$\begin{array}{r} 010 \\ +111 \\ \hline 1001 \end{array}$$

从右边开始，第一位 0＋1＝1，没有进位。第二位 1＋1＝10，保存 0 进 1。第三位 0＋1＋1（进位值）＝10，保存 0 进 1。最后一位，0＋0＋1＝1。所以答案是 1001。如果把每个数都转化成十进制，会发现这是正确的：2＋7＝9。

5.8 迅速回顾

总结整篇文章，以下是我们关于位和二进制所学的：
- 位是二进制的阿拉伯数字，一位能保存数值 0 或 1。
- 每个字节由 8 个位组成。
- 二进制计算就像十进制一样，但是每个位上只有 0 或 1 两个数值。

没有更多的可以讲了——位和字节就是这么简单。

注释

1. That should all feel pretty comfortable—we work with decimal digits every day. The neat thing about number systems is that there is nothing that forces you to have 10 different values in a digit.

译文：这样感觉相当不错——我们每天都跟十进制数打交道。该数字系统灵巧的一点就在让你不再需要一个数位表示十个不同的数值。

2. In the ASCII character set, each binary value between 0 and 127 is given a specific character. Most computers extend the ASCII character set to use the full range of 256 characters available in a byte. The upper 128 characters handle special things like accented characters from common foreign languages.

译文：在 ASCII 码字符集中，从 0～127 每个二进制位的值都被赋予一个特定的字符。大部分计算机都扩充 ASCII 码字符集，使它达到一个字节可以表示的 256 个字符的范围。更高的 128 个字符表示特别的东西，如一些外语中的字符。

第 6 章 微处理器概述

我们所用的计算机是用微处理器进行工作的。微处理器是任何正常计算机的核心，无论是台式机、服务器还是笔记本计算机。我们所用的微处理器可能是 Pentium、K6、PowerPC、Sparc 或任何其他品牌和类型的微处理器，但它们都以大致相同的方式进行相似的工作。

如果想知道计算机中的微处理器干些什么，想知道不同微处理器的差别，请往下读。

6.1 微处理器简史

微处理器也称为 CPU 或中央处理单元，是单个芯片上制成的完整计算引擎。第一个微处理器是 1971 年引入的 Intel 4004。Intel 4004 不太强大，只能进行加法和减法，一次只能处理 4 位。但是，它的妙处在于一切都在一个芯片上。在 4004 之前，工程师建立计算机时要用一组芯片或离散组件（一个个晶体管）。4004 装备在最早的便携式电子计算器上。

家用计算机使用的第一个微处理器是 Intel 8080，这是单个芯片上的完整 8 位计算机，是 1974 年引入的。第一个真正在市场上掀起浪花的微处理器是 Intel 8088，于 1979 年推出，装备到 IBM PC 中（最早在 1982 年左右出现）。如果熟悉 PC 市场及其历史，你一定知道 PC 从 8088 转到 80286，到 80386、80486，直到奔腾机、奔腾 II、奔腾 III 和奔腾 4。所有这些微处理器都是 Intel 公司制造的，都是在 8088 基本设计上改进而成的。奔腾 4 可以执行原先 Intel 8088 上运行的任何代码，但速度快近 5000 倍。

6.2 微处理器的进程

表 6.1 有助于了解多年来 Intel 公司推出的不同微处理器间的差别。

表 6.1 微处理器间的差别

名称	日期	晶体管个数	微米	时钟速度	数据宽度	MIPS
8080	1974	6,000	6	2MHz	8 位	0.64
8088	1979	29,000	3	5MHz	16 位 8 位总线	0.33
80286	1982	134,000	1.5	6MHz	16 位	1
80386	1985	275,000	1.5	16MHz	32 位	5
80486	1989	1,200,000	1	25MHz	32 位	20
Pentium	1993	3,100,000	0.8	60MHz	32 位 64 位总线	100
Pentium II	1997	7,500,000	0.35	233MHz	32 位 64 位总线	~300
Pentium III	1999	9,500,000	0.25	450MHz	32 位 64 位总线	~510
Pentium 4	2000	42,000,000	0.18	1.5GHz	32 位 64 位总线	~1,700

表 6.1 中的信息说明：

- 日期指处理器首次推出的年份。许多处理器在原发布日期后的多年内以更高的时钟速度重新推出。
- 晶体管指芯片上的晶体管个数。可以看到，单个芯片上的晶体管个数多年来稳步上升。
- 微米指芯片上最细线的宽度（微米）。作为比较，人的头发粗细为 100 微米。随着芯片上的特性及其不断下降，晶体管个数不断增加。
- 时钟速度是芯片的最大时钟速度，时钟速度的更详细含义将在下节介绍。
- 数据宽度指 ALU 的宽度。8 位 ALU 可以进行两个 8 位数的加法、减法、乘法等。32 位 ALU 可以处理 32 位数。8 位 ALU 要执行 4 个指令才能将两个 32 位数相加，而 32 位 ALU 只要用一个指令。许多情况下，外部数据总线的宽度与 ALU 宽度相同，但也不一定。8088 使用 16 位 ALU 和 8 位总线，而现代奔腾机对 32 位 ALU 一次取 64 位数据。

- MIPS 指每秒百万指令数，是 CPU 性能的大致指标。现代 CPU 可以进行许多不同工作，使 MIPS 指标的意义不那么大，但还是可以用这个值大致看到 CPU 的相对强弱。

从表 6.1 中可以看出，一般来说，时钟速度与 MIPS 存在一定关系。最大时钟速度是制造工艺与芯片延迟的函数、晶体管个数与 MIPS 也存在一定关系。例如，8088 的时钟速度为 5MHz，但只有 0.33MIPS（大约每 15 个时钟循环一个指令）。现代处理器通常可以执行每个时钟循环两个指令。这个改进与芯片上的晶体管个数直接相关，请见下节介绍。

6.3 微处理器内部

要了解微处理器如何工作，最好看看其内部，了解建立微处理器的逻辑。在这个过程中，我们还可以学习汇编语言（微处理器的自然语言）和工程师为处理器提速的许多项目。

微处理器执行一组指令，告诉机器要干什么。根据指令，微处理器可以执行 3 个基本任务：

- 微处理器用 ALU（算术/逻辑单元）。可以进行加、减、乘、除等数学运算。现代微处理器包含完整的浮点处理器，可以对大浮点数进行相当复杂的运算。
- 微处理器可以将数据从一个内存地址移到另一个内存地址。
- 微处理器可以进行决策和根据决策跳转到新的指令集。

微处理器可以进行非常复杂的工作，上面是其 3 个基本工作。图 6.4（图略）显示了具有 3 个功能的相当简单的微处理器。

这是最简单的微处理器，这个微处理器包括：

- 地址总线（8 位、16 位或 32 位），将一个地址发给内存。
- 数据总线（8 位、16 位或 32 位），将数据发给内存或从内存接收数据。
- RD（读）和 WR（写）线，告诉内存要设置或读取指定地址。
- 时钟线，让时钟脉冲将处理器序列化。
- 复位线，将程序计数器复位为 0（或某个值），重新开始执行。

本例中假设地址总线和和数据总线均为 8 位。

下面是这个简单微处理器的组件：

- 寄存器 A、B、C 是由触发器组成的简单锁存。
- 地址锁存与寄存器 A、B、C 相似。
- 程序计数器也是个锁存，但还能够在指示下递增 1，还能够在指示下复位为 0。
- ALU 可以是简单的 8 位加法器，也可以对 8 位值进行加、减、乘、除。这里假设为后者。
- 测试寄存器是个特殊锁存，可以保存 ALU 中比较所得的值。ALU 通常可以比较两个数，确定其是否相等、一个大于另一个，等等。测试寄存器通常还可以保存加法器上一阶段的进位。它把这些值保存在触发器中，然后指令译码器可以用这个值进行决策。

- 图中（图略）有 6 个标为 "3-State" 的框，是三态缓冲器。三态缓冲器可以传入 1、0 或切断输出（就像开关从连接输出端的线路上切断输出线）。三态缓冲器使一个线路可以连接多个输出，但只有其中一个实际在线路上驱动 0 或 1。
- 指令寄存器和指令译码器负数控制所有其他组件。

尽管图 6.4 中（图略）没有显示，但指令译码器还有下列功能的控制线：

- 让寄存器 A 锁存当前在数据总线上的值。
- 让寄存器 B 锁存当前在数据总线上的值。
- 让寄存器 C 锁存当前在数据总线上的值。
- 让程序计数器寄存器锁存当前在数据总线上的值。
- 让地址寄存器锁存当前在数据总线上的值。
- 指令寄存器锁存当前在数据总线上的值。
- 让程序计数器递增。
- 让程序计数器复位为 0。
- 激活 6 个三态缓冲器（6 条线）。
- 告诉 ALU 进行什么运算。
- 让测试寄存器锁存 ALU 的测试位。
- 激活 RD 线。
- 激活 WD 线。

指令译码器接收来自测试寄存器与时钟线的位和来自指令寄存器的位。

6.4 微处理器指令

即使上例所示的简单微处理器也可以执行大量指令集。指令集合实现为位模式，每个位模式装入测试寄存器时具有不同含义。人们不太容易记住位模式，因此定义了一组短字，表示不同的位模式。这个短字集合称为处理器的汇编语言。汇编器可以将短字方便地变成位模式，然后把汇编器的输出放进内存中，使微处理器执行。

这是设计者在我们的示例中在简单微处理器中创建的一组汇编语言指令。

- LOADA mem——从内存地址装入寄存器 A。
- LOADB mem——从内存地址装入寄存器 B。
- CONB con——将常量值装入寄存器 B。
- SAVEB mem——将寄存器 B 保存到内存地址。
- SAVEC mem——将寄存器 C 保存到内存地址。
- ADD——A 和 B 相加，结果保存到 C。
- SUB——A 和 B 相减，结果保存到 C。
- MUL——A 和 B 相乘，结果保存到 C。
- DIV——A 和 B 相除，结果保存到 C。
- COM——比较 A 和 B，结果保存到测试寄存器中。
- JUMP addr——跳到一个地址。
- JEQ addr——如果相等，则跳到一个地址。

- JNEQ addr——如果不相等，则跳到一个地址。
- JG addr——如果大于，则跳到一个地址。
- JGE addr——如果大于或等于，则跳到一个地址。
- JL addr——如果小于，则跳到一个地址。
- JLE addr——如果小于或等于，则跳到一个地址。
- STOP——停止执行。

下列简单 C 语言代码计算 5 的阶乘（5 的阶乘为 5!=5*4*3*2*1=120）：

```
a=1;
f=1;
while (a <=5)
{
  f=f*a;
  a=a+1;
}
```

程序执行结束时，变量 f 包含 5 的阶乘值。

C 语言编译器将这个简单 C 语言代码变成汇编语言。没这个处理器的 RAM 从地址 128 开始，ROM（包含汇编语言程序）从地址 0 开始，则这个简单微处理器的汇编语言如下：

```
//Assume a is at address 128
//Assume F is at address 129
0  CONB 1   //a=1;
1  SAVEB 128
2  CONB 1   //f=1;
3  SAVEB 129
4  LOADA 128 //if a > 5 the jump to 17
5  CONB 5
6  COM
7  JG 17
8  LOADA 129 //f=f a;
9  LOADB 128
10 MUL
11 SAVEC 129
12 LOADA 128 //a=a+1;
13 CONB 1
14 ADD
15 SAVEC 128
16 JUMP 4   //loop back to if
17 STOP
```

问题是：所有这些指令在 ROM 中是什么样子？每个汇编语言指令要表示为二进制数。为了简单起见，假设每个汇编语言指令指定唯一数字如下：

- LOADA-1
- LOADB-2
- CONB-3
- SAVEB-4
- SAVEC mem-5
- ADD-6
- SUB-7
- MUL-8
- DIV-9
- COM-10
- JUMP addr-11
- JEQ addr-12
- JNEQ addr-13
- JG addr-14
- JGE addr-15
- JL addr-16
- JLE addr-17
- STOP-18

数字称为算子。在 ROM 中，这个小程序如下：

```
//Assume a is at address 128
//Assume F is at address 129
Addr opcode/value
0    3          //CONB 1
1    1
2    4          //SAVEB 128
3    128
4    3          //CONB 1
5    1
6    4          //SAVEB 129
7    129
8    1          //LOADA 128
9    128
10   3          //CONB 5
11   5
12   10         //COM
13   14         //JG 17
14   31
15   1          //LOADA 129
16   129
17   2          //LOADB 128
```

```
18  128
19  8    //MUL
20  5    //SAVEC 129
21  129
22  1    //LOADA 128
23  128
24  3    //CONB 1
25  1
26  6    //ADD
27  5    //SAVEC 128
28  128
29  11   //JUMP 4
30  8
31  18   //STOP
```

可以看出，7 行 C 语言代码变成 17 行汇编语言，在 ROM 中变成 31 个字节。

6.5 译码微处理器指令

指令译码器要将每个算子变成一组信号，驱动微处理器中的不同组件。下面以 ADD 指令为例，看看要干什么：

（1）第一个时钟周期，要实际装入指令。因此，指令译码器要：

- 激活程序计数器的三态缓冲器。
- 激活 RD 线。
- 激活数据输入三态缓冲器。
- 将指令锁存到指令寄存器。

（2）第一个时钟周期，译码 ADD 指令，很简单：

- 将 ALU 运算设置为加法。
- 将 ALU 输出锁存到 C 寄存器。

（3）第三个时钟周期，程序计数器递增（理论上也可以和第一个时钟周期重叠）。

每个指令可以分解为这样一组顺序操作，按正确顺序操纵微处理器的组件。有些指令（如这个 ADD 指令）可能要两个或三个时钟周期，而有些可能要五六个时钟周期。

6.6 微处理器的性能

晶体管个数对处理器性能有巨大影响。前面曾介绍过，8088 之类处理器的典型指令要用 15 个时钟周期来执行。由于乘法器的设计，使 8088 上进行一个 16 位乘法要大约 80 个时钟周期。增加晶体管就可以得到更强大的乘法器，达到单周期速度。

增加晶体管要利用管道技术。管道体系结构中，指令执行是重叠的。因此，尽管执行每个指令要 5 个时钟周期，但 5 个指令可以同时在不同执行阶段。这样，好像每个时钟周期完成一个指令。

大多数现代处理器具有多个指令编码器，各有自己的管道。这样就支持多个指令流，即每个时钟周期可以完成多个指令。这个技术的实现相当复杂，因此会采用大量晶体管。

6.7 微处理器的趋势

微处理器的设计趋势主要是完全 32 位 ALU，内置快速浮点处理器和多个指令流管道执行。处理器设计中的最新情况是 64 位 ALU。还有一种趋势是用特殊指令（如 MAX 指令）使某个运算特别高效，以及在处理器芯片上增加硬件虚拟内存支持和 L1 缓存。所有这些趋势都要增加晶体管个数，从而出现了当今几百万个晶体管的微处理器。这些处理器每秒可执行大约 10 亿条指令。

6.8 64 位处理器

64 位处理器从 1992 年就已经出现，21 世纪已开始成为主流。Intel 与 AMD 公司都引入了 64 位芯片，同 Mac G5 支持 64 位处理器。64 位处理器具有 64 位 ALU、64 位寄存器、64 位总线，等等。

之所以需要 64 位处理器，是因为其扩大地址空间。32 位芯片通常限于最多 2GB 或 4GB 的 RAM 存取，听起来好像很多，因为大多数家用计算机目前只用 256～512MB 的 RAM。但是，4GB 的局限对服务器和运行大型数据库的机器可能造成严重问题。按照目前的趋势发展下去，就连家用计算机也很快会撞击到 2GB 或 4GB 的极限。64 位芯片没有这些限制，因为 64 位 RAM 地址空间在可以预见的将来实际上是无限的，因为 2^{64} 字节内存是 10^{15} 吉字节内存数量级的。

除了 64 位地址总线与宽度和主板上的高速数据总线，64 位机器还提供了硬盘驱动器、显示卡等的高速 I/O（输入/输出）速度。这些特性可以大大提高系统性能。

注释

1. A tri-state buffer can pass a 1, a 0 or it can essentially disconnect its output.

译文：三态缓冲器可以传入 1、0 或切断输出。

2. The collection of instructions is implemented as bit patterns, each one of which has a different meaning when loaded into the instruction register.

译文：指令集合实现为位模式，每个位模式装入测试寄存器时具有不同含义。

3. An assembler can translate the words into their bit patterns very easily, and then the output of the assembler is placed in memory for the microprocessor to execute.

译文：汇编器可以将短字方便地变成位模式，然后把汇编器的输出放进内存中，使微处理器执行。

4. Every instruction can be broken down as a set of sequenced operations like these that manipulate the components of the microprocessor in the proper order.

译文：每个指令可以分解为这样一组顺序操作，按正确顺序操纵微处理器的组件。

5. The trend in processor design has primarily been toward full 32-bit ALUs with fast

floating point processors built in and pipelined execution with multiple instruction streams.

译文：微处理器设计趋势主要是完全 32 位 ALU，内置快速浮点处理器和多个指令流管道执行。

第 7 章 应 用 软 件

7.1 什么是软件

让我们从定义开始，程序是引导计算机通过一个过程的一组指令。每条指令都告诉机器完成它的基本功能中的一个，如加、减、乘、除、比较、复制、请求输入或请求输出等。我们知道在每个机器周期内处理器取出并执行一条指令。一条典型的指令包含一个操作码，它规定了存放要被处理的数据的存储单元地址或寄存器编号。

例如：指令 ADD 3, 4, 告诉一台假想的计算机把寄存器 R3 和 R4 的内容相加。

由于计算机指令组的功能有限，甚至一个简单的逻辑操作也需要若干条指令。例如，假设两个数据的值存放在主存中，在很多机器上，为了把它们相加，首先要把两个数值加载到寄存器中，然后把寄存器的内容相加，结果再送回到主存中。这样就要用 4 条指令：LOAD、LOAD、ADD 和 STORE。如果为两个数相加就需要 4 条指令的话，可以想象在一个完整的程序中会需要多少条指令。计算机是由存放在自己主存中的程序控制的。如果程序员必须用机器语言编写程序的话，那么几乎没有人能做程序员了。

7.2 编程语言

7.2.1 汇编语言

用汇编语言编程是一种选择。程序员为每一条机器指令写一条助记符指令。AR（把寄存器的内容相加）比等效的二进制操作数 000110101 容易记忆；字长 L（装入）也比 01011000 容易记忆。操作数用标志符 A、B 和 C，取代了主存的地址编号，这样也简化了代码。

遗憾的是，没有能直接执行汇编语言指令的计算机。写助记符代码可以简化程序员的工作，但是计算机是二进制的机器，要求二进制指令，因此翻译是很必要的。汇编程序读入程序的源代码，把源语句翻译成二进制数，产生目标模块。因为目标模块是程序源代码机器级的版本，所以它能被装入主存中并加以执行。

汇编语言程序设计员为每条机器指令写一条助记符指令。因为汇编指令和机器指令之间存在一一对应的关系，汇编语言是依赖于机器的，所以为一种型号计算机编写的程序不能在另一种型号的机器上执行。在一台给定的机器上，汇编语言能产生最直接有效的程序，因此经常用它编写操作系统或其他系统软件。

然而，当它应用于应用程序时，依赖于机器的属性是要为效率付出高代价的，所以应用程序很少用汇编语言来写。

7.2.2 编译程序和解释程序

为了把两个数相加，计算机需要 4 条机器指令，因为这是计算机工作的方法。人类不必像计算机一样地思考，程序员只简单地指明做加法，采用另一些指令就行了。例如，一种方法就是把加法看成一个代数表达式：

$$C=A+B$$

为什么不允许程序员用类似于代数表达式的形式去写语句，然后将这些源代码读入程序中，让程序自己产生机器代码呢？肯定行。这就是编译器所要完成的。

许多编程语言，包括 FORTRAN、BASIC、Pascal、PL/1 和 ALGOL 都是基于代数表达式的。面向商业的最常用的 COBOL 语言要求语句类似于简短的英语句子。然而必须注意的是，不管采用什么语言，目标是相同的。程序员编源码程序，汇编程序接受助记符源码程序，并产生机器目标模块；FORTRAN 编译程序接受 FORTRAN 源码，也产生机器目标模块；COBOL 编译程序接受 COBOL 源码，产生同上的目标模块。

汇编程序和编译程序之间有何区别呢？对于汇编程序，每一个源语句都转换成一条机器指令。但对于编译程序，一个源语句可以转换成好几条机器指令。还有一种选择是采用解释程序。汇编程序或编译程序都是读入一个完整的源程序，并产生一个完整的目标模块。而解释程序不同，它每次只对一个源语句操作，读入它，把它转换成机器指令，执行被转换成的二进制指令，然后继续下一条源语句。编译程序和解释程序两者都产生机器指令，但过程是不同的。

每种语言都有它自己的语法、标点和拼写规则。例如，Pascal 源程序对 COBOL 编译程序或者 BASIC 解释程序是无意义的。但是所有的这些语言都支持编写程序。无论用什么语言，程序员的目标是相同的：确定一系列步骤引导计算机通过某个过程。

7.2.3 非过程语言

用传统的汇编语言、编译语言和解释语言，程序员确定了一个确切地告诉计算机如何去解决问题的程序。

然而，用现代非过程语言，有时称第四代语言或说明语言，程序员只简单地确定问题的逻辑关系，而让语言变换程序推算出如何解决问题。商业上使用的非过程语言有 Prolog、Focus、Lotusl-2-3 等。

7.3 程序库

设想程序员编写一个庞大程序，当源语句被输入时，可用编辑程序处理它并存放在磁盘上。因为大程序不可能一次就全部写入，所以程序员最终要停止输入，并把磁盘从驱动器中取出。以后当又开始工作时，重新插入磁盘，在原来的语句后面输入新的源语句。同一磁盘可以保存其他一些源程序甚至其他程序员所写的一些程序。这就是一个源语句库的很好的例子。

最终，完成了源程序的输入和编译，得到的目标模块可以直接装入到主存，但更多的时候它存放在目标模块库中。因为目标模块是二进制的机器级的子程序，所以由汇编程序产生的目标模块和 FORTRAN 编译程序（或任何其他编译程序）产生的目标模块根本没有差别。因此，用不同的源语言产生的目标模块可存放在同一个程序库内。

有些目标模块能装入到主存中，并被执行。而另一些模块，包括了对一些子程序的调用，这些子程序不属于目标模块的一部分。例如，一个计算机仿真扑克牌的程序，如果以前有人编写了一段很优秀的发牌子程序，那么重新用那个子程序是有意义的。设想扑克牌游戏新程序已被写入、编译完并存放在目标模块库中，发牌子程序也存放在同一个库中了。在新程序被装入之前，这两个程序必须合并在一起，形成一个装配模块。目标模块是由源模块转换成的机器语言程序，它可以包括其他子程序的调用。装配模块是完整的可执行的程序，该程序包含了所有相应的子程序。把目标模块组成一个装配模块是联接编辑程序模块或装入程序模块的任务。

电视游戏、电子表格软件、文字处理软件、数据库、记账子程序和其他一些商业软件包以装配模块的形式存放在磁盘上，通常可买到。如果在源代码、目标模块和装配模块之间进行选择的话，大多数人会发现，装配模块较易于使用，因为它不必再经过翻译计算机就能执行它。然而，装配模块是很难改变的。如果程序员想修改或定制一个软件包，那么源代码很重要。

7.4 程序开发过程

程序如何修改软件包呢？更一般地说，职业的程序员如何编写源程序呢？编程序不完全是科学，还涉及一点艺术，因此不同的程序员用不同的方法编程序是没什么奇怪的。然而大多数程序员都会在编写代码前认真定义问题，然后计划他们的解决方案。让我们简单地研究一下程序的开发过程。

7.4.1 问题的定义

程序开发的第一步是问题定义，这看起来是常识，但常常有些程序还不清楚为什么需要这么做就编出程序来了。对一个未能正确定义的问题所得到的结果，即使是一个伟大的结果，也是无用的。

因为人们需要信息才有程序，因此程序员应在明确了所期望的信息后才能开始编写。接着，要确定产生那个信息的算法或规则。给出了所要求的输出信息和算法，要输入的数据也就可以确定了。结果是，一个清晰的问题定义能让程序员明确程序到底要完成什么。

顺便插一句，程序通常是在系统意义上被确定的。

7.4.2 计划编制

算法确定了必须做的事情，接下来的任务是决定如何去做。目的是要用计算机能理解的术语去陈述问题的解。由于计算机只能完成算术运算、比较、复制和请求输入输出等操作，因此程序员会受到这些基本操作的限制。一个良好的起点是先确定求解问题的一个小型方案，然后通过实际求解的算法甚至是在规模上有限的一个算法，使程序员能明确编程问题所需的步骤。

程序员可用大量的工具，把一个问题的解法转换成计算机术语。例如，流程图就形象地表示了程序的逻辑结构。在把它转换成源码之前，程序员能够用伪码"画"出程序逻辑结构的草图。较复杂的程序经常由两个或更多的程序员编写成，或者包含大量的逻辑关系。这些程序通常被分割成一些较小的单功能模块，能独立地对它们编码。一个好程序员在开始写源程序之前会规划好每一个模块的内容，认真地定义模块之间的相互关系。正像一个承包人在开始建房之前，准备详细的蓝图，程序员在开始写源程序之前也要制定出一个详细的编程计划。

7.4.3 编写程序

在实施阶段，程序员把问题的解法转换成用某种编程语言写成的一系列源语句，每一种编程语言都有它自己的语法、标点和拼写规则，而且学习一种新语言要花一定的时间，因此写指令基本上是机械性工作。编程的真正秘密不是简单地编写指令，而是要思考接下来该写什么指令，这就需要逻辑性。幸运的是，知道如何编程序并不是使用计算机的必备条件。

7.4.4 调试和文档编写

程序一旦编成，程序员就必须调试它。首先要纠正它的一些无意识的错误，譬如错误的标点或拼写等。编译程序或解释程序通常能发现这些错误，较困难的是发现并纠正逻辑错误，这类错误是由于采用错误的指令而引起的。仅仅指令正确是不够的，程序必须是顺序正确的指令组。再重复一次，精心的编程设计是关键，优良的计划编制可简化程序的调试。

程序的文档由图表、注释和其他解释或阐明编程的说明材料组成，在程序调试阶段，文档是极其宝贵的，也是有效维护程序的基础。最有用的是出现在程序中的注释，这些注释列举并解释了逻辑关系。

7.4.5 维护

程序一调好，维护就开始了。由于不可能彻底地测试许多大程序。在调试阶段中，一些 bug 可能没能被发现，几个月甚至几年以后才暴露出来，修复这些 bug 是非常重要的维护任务。更为重要的维护是修改程序使它能适应现时应用。例如，由于所得税经常变化，工资程序必须经常修改。认真的计划、完整的文档资料和优良的程序设计是维护的关键。

7.5 编写你自己的程序

能演奏一种乐器并不是欣赏音乐的基础。相似地，知道如何编程序也不是使用计算机的必备条件。大多数计算机用户不会编程。同样，正如一种乐器的初步知识能增长你对音乐的欣赏水平一样,知道了如何编程序能使你更有效地利用计算机。当然,如果你希望进入计算机行业,那么编程知识是基础,这就不用说了。有些人会发现编程很容易,但有些人又会发现编程极其困难,关键在于实践。学习编程的唯一方法就是实际编程。

注释

1. Why not allow a programmer to write statements in a form similar to algebraic expressions, read those source statements into a program, and let the program generate the necessary machine-level code. That's exactly what happens with a compiler.

原译:为什么不许程序员用类似于代数表达式的形式去写语句呢?又为什么不能把这些源语句编在程序中,并让程序产生机器码呢?肯定可以。这就是编译程序所要完成的。

译文:为什么不允许程序员用类似于代数表达式的形式去写语句,然后将这些源代码读入程序中,让程序自己产生机器代码呢?肯定行。这就是编译器所要完成的。

分析:译文较简洁,原译把不该扩展的也扩展成两句了。其他用词上没有很大出入。

2. A good starting point is solving a small version of the problem; by actually solving the algorithm, even on a limited scale, the programmer can gain a good sense of the steps required to program it.

原译:一个良好的起点总是首先确定求解问题的一个小型方案,通过实际求解的这一算法甚至是在规模上有限的一个算法,使程序员获得编程问题所需的步骤。

译文:一个良好的起点是先确定求解问题的一个小型方案,然后通过实际求解的算法甚至是在规模上有限的一个算法,使程序员能明确编程问题所需的步骤。

分析:句子结构并不复杂,是两个句子,这里把它们合并了。词组 gain a good sense of 表示理解、明白某事,这里译为明确。

3. For example, flowcharts can programmer can "draft" the logic before converting it to source code.

译文:例如,流程图就形象地表示了程序的逻辑结构。在把它转换成源码之前,程序员能够用伪码"画"出程序逻辑结构的草图。

分析:此句难点在一些专门术语上的翻译,要通过耐心查证得到。

4. Of course, it goes without saying that if you hope to earn your living as a computer professional, a knowledge of programming is essential.

译文:如果你希望进入计算机行业,那么编程知识是基础,这就不用说了。

分析:这里 it goes without saying that 是一个比较口语的东西,指很基本的常识,甚至都不用说了。

第 8 章 编 译 器

编译器是把用某种计算机程序语言(源程序语言)编写的计算机程序转变为用另一种计算机程序语言(输出语言或目标语言)编写的等价的计算机程序。

8.1 绪论及其发展历史

大多数编译器是把用高级程序语言编写的源程序翻译成能被计算机或虚拟计算机直接执行的目标代码或机器语言。然而，把一种低级程序语言翻译成高级的也是有可能的。我们通常所知的反编译器可以重构那些产生低级语言程序的高级语言程序。编译器也能把一种高级语言翻译成另一种高级语言（交叉编译），或者有时会翻译成某种还需进一步处理的中间语言。这些有时被认为是层译器。

典型的编译器输出所谓的目标代码，它主要包含了扩充有关入口指针以及外部调用（对不在目标程序中的函数来说）的名字和位置的机器代码。一系列的目标文件，它们不一定都来自同一个编译器，但其格式相同，并且可以连接在一起生成一个用户能直接运行的最终可执行程序。

很多实验性的编译器是在19世纪50年代被开发出来的。一般认为，第一个完善的编译器是IBM公司John Backus带领的FORTRAN小组在1957年开发出的。COBOL语言是一种早期的在多重体系结构上被编译的语言，时间是1960年。此后，编译思想迅速流行，而且大多数的编译设计理论是在19世纪60年代被揭示出来的。

编译器本身就是用某些可执行的语言编写的计算机程序。早期的编译器是用汇编语言编写的。第一个自我集成编译器（self-hosting complier）是Hart和Levin于1962年在MIT为Lisp公司创造的，它能把自身的源代码编译成某种高级语言。20世纪70年代早期，自从Pascal和C利用自身语言编写出了编译器后，用高级语言来写编译器得到了进一步的发展。建立一个自我集成编译器是个需要逐步解决的难题——首先这种面向某种语言的编译器必须要么能被用另一种不同的语言编写的编译器编译，要么通过在解释器中运行编译器来编译，如Hart和Levin's Lisp的编译器。

20世纪90年代期间，人们开发了大量的面向各种各样的语言的免费编译器和编译器开发工具，这两个既属于GNU工程的一部分，又属于其他开放源码的自主行为。其中一些代码质量很高，并且其免费源代码为任何对现代编译理论有兴趣的人提供了很好的学习机会。

8.2 编译器类型

编译器产生的代码必须能在它本身运行所依赖的计算机和操作系统上运行。这样的编译器有时被称为本地编码编译器（native-code complier）。还有另一种，它产生的代码能在一个不同的平台上运行。这样的编译器通常被称为交叉编译器。交叉编译器在新的平台产生时非常的有用。一个源对源编译器是一种把某种高级语言作为输入然后输出另一种高级语言的编译器。例如，一个自动并行编译器经常把一种高级程序语言作为输入然后翻译代码并注释上相应的注解（例如OpenMP）或者语言结构（例如Fortran's DOALL声明）。

（1）一遍编译器，像Pascal的早期编译器，只编译一次，因此速度很快。

（2）线性代码编译器，像大多数的前向执行器。这样的编译器可看成是数据库查找程序。它只不过是把源程序中给定的字符串用给定的二进制代码替换了。这样的二进制代码是可变的。实际上，一些前向编译器还能够编译无须操作系统作为载体的程序。

（3）增量编译器，如大部分 Lisp 系统。个别的功能程序能在一个包含解释功能的运行环境中被编译。增量编译技术要追溯到 1962 年的第一个 Lisp 编译器，它仍然在 Common Lisp 系统中使用。

（4）段编译器能实现编译成理论机制化的汇编语言，像一些 Prolog 执行器。这样的 Prolog 机器也被认为是 Warren 抽象机（或叫 WAM）。Java 字节码编译器和 Python（还有很多）也都是这种类型的。

（5）即时编译，被用在 Smalltalk 和 Java 系统中。应用程序被转换成在执行前转换成本地机器代码的字节码。

（6）可重定义编译器能轻易地被改造成能产生适用于不同 CPU 体系的代码。这样产生出的目标代码通常都比那些单个 CPU 专用编译器译出的代码质量差。可重定义编译器通常也是交叉编译器，GCC 就是一个实例。

（7）并行编译器能把一系列输入程序转变成适合在并行计算机体系中有效执行的代码形式。

8.3 编译语言与解释语言

很多人都把高级程序语言分成编译型和解释型。然而，一种语言对其自身是编译型或解释型的要求并无多大关系。编译和解释是语言的执行，不是语言本身。这种分类思想通常体现在最流行、应用最广的语言上，例如 BASIC 被认为是一种解释型语言，C 语言被当作是编译型语言，尽管存在着 BASIC 语言编译器和 C 语言解释器。但是，有例外，一些语言规范采用了编译理论（就像 C 语言），或者清楚地说明了执行器必须包含编译器（就像 Common Lisp）。

8.4 编译器的设计

过去为了节省空间，编译器被分成很多通道。在这个上下文环境中，通道是指编译器贯穿待编译源代码进行一次编译并由此建立编译器的内部数据的整个过程的一次运行（就像扩展符表和其他的辅助数据）。当每个通道都结束时，编译器能释放该通道的内部数据空间。这种"多通道"技术是当时很常用的编译技术，但也是由于主机的主存空间相对于源代码和数据来说比较小。

很多最新的编译器都采用了"双端"设计技术。前端把源代码翻译成中间语言。第二个是后端，它把中间语言经过处理形成输出语言。前端和后端应该用不同的通道来执行，或者前端调用后端作为子程序，并把中间语言传递给后端。

这种方法减轻了把处理语义、检测错误等的前端工作和主要产生正确有效输出的后端工作分离开来的复杂性。它也同样有允许使用多语源的单一后端的优点，同样也有允许对于不同对象使用不同后端的优点。

通常，如果前端和后端都能够对由前端送达后端的中间语言进行操作，那它们就可以共同处理代码的最优化和出错检测。这样可以让很多编译器（前后端的联合）重用大量的经常性的代码分析和最优化工作。

有些语言，由于其语言的设计以及变量和对象的特定的声明规则，还有可执行程序被提及或使用前的声明等方面的原因，能够单通道编译。Pascal 编程语言就是因为有这种能力而众所周知，而且实际上很多的 Pascal 语言的编译器就是用 Pascal 语言写的，因为这种语言有很多严格的规范并且拥有用单通道编译 Pascal 程序的能力。

8.5 编译器前端

编译器前端本身是由多个阶段组成的，每个都含有丰富的形式语言理论。

（1）词法分析：把源代码分成很多小块（符号或者终结符），每个都代表一个单一的语言单位，例如一个关键词，标识符或者符号名。这种符号语言是典型的正规语言，因此一个由规则表达式组建的有限状态自动机能识别它。这阶段也叫做罗列或者扫描。

（2）语法分析：按造句法确定源代码的结构。它只注重结构。换句话说，它定义了代码中符号的顺序并推出代码层次结构。这个阶段也叫做分析。

（3）语义分析将识别出程序代码的含义并开始做好输出的准备。在这个阶段，进行了类型校验并指出了大多数的编译错误。

（4）中间语言的产生：一个与源程序等价的中间语言程序被创建。

8.6 编译的后端

当应用程序只需要编译器前端时，就像静态语言测试工具，一个真正的编译器把由前端产生的中间语言传递给后端，后端再产生一个功能等价的用输出语言编写的程序。这是多步实现的。

（1）编译分析：这是把用由输入源文件产生出的中间语言编写的程序的信息收集起来的过程。典型的分析有可变的定义—使用和使用—定义链、数据关系分析、别名分析等。精确分析是编译最优化的基础。调用图和控制流程图在分析阶段中也经常使用的。

（2）优化：中间语言形式被转变成功能一样但执行较快（或者所占空间较小）的形式。普遍的优化是嵌入式扩展、消除死代码、常量传播、循环转换、注册分配或者甚至是自动并行。

（3）代码生成：优化转变过的中间语言被翻译成输出语言，通常是本地系统的机器语言。这包括资源和存储空间的确定，比如确定哪些变量适合寄存器和主存以及选择和定制与其寻址模式适合的机器指令。

注释

1. However, translation from a low level language to a high level one is also possible; this is normally known as a decompiler if it is reconstructing a high level language program which (could have) generated the low level language program.

原译：然而，把一种低级程序语言翻译成高级的也是有可能的；这就是通常所说的经重构某种高级语言而产生低级程序语言的反编译器。

译文：然而，把一种低级程序语言翻译成高级的也是有可能的；我们通常所知的反编译器可以重构那些产生低级语言程序的高级语言程序。

2. A pass in this context is a run of the compiler through the source code of the program to be compiled, resulting in the building up of the internal data of the compiler (such as the evolving symbol table and other assisting data).

译文：在这个上下文环境中，通道是编译器贯穿待编译源代码进行一次编译并由此建立编译器的内部数据的整个过程的一次运行（就像扩展符表和其他的辅助数据）。

分析：in this context 有上下文环境的意思，但这里指的是上文说的 pass；整句的翻译很到位。注意句子主干讲的是 A pass is a run。

3. This approach mitigates complexity separating the concerns of the front end, which typically revolve around language semantics, error checking, and the like, from the concerns of the back end, which concentrates on producing output that is both efficient and correct.

译文：这种方法减轻了把处理语义、检测错误等的前端工作和主要产生正确有效输出的后端工作分离开来的复杂性。

分析：译文十分精练地翻译了这个长句。原文中有多重从句，增加了翻译的难度。但只要主干抓住就好办了：This approach mitigates complexity。后面部分都是修饰 complexity。

4. Certain languages, due to the design of the language and certain rules placed on the declaration of variables and other objects used, and the predeclaration of executable procedures prior to reference or use, are capable of being compiled in a single pass.

译文：有些语言，由于其语言的设计以及变量和对象的特定的声明规则，还有可执行程序被提及或使用前的声明等方面的原因，能够单通道编译。

分析：本文长难句较多。此句主干是 Certain languages are capable of being compiled in a single pass，中间是插入说明的部分。由于句子结构比较正常，语序不用变化。

第 9 章　Java 是如何工作的

你是否曾经想知道计算机程序是怎么工作的？是否想过自己如何去写一个计算机程序？不管你是一个 14 岁很想学怎么写自己的第一个游戏的小孩，还是一个 70 岁已经有 20 年的编程爱好的老人，这篇文章都很适合你。本文将通过告诉你们怎么用 Java 语言编程来教会你们计算机程序是如何工作的。

为了向你们介绍计算机程序，我从开始就做如下几点假设：

- 假设你对计算机程序一点也不了解。如果你已有所了解，本文的第一部分对你来说应该没什么问题，你可以跳过，直到遇到不懂的地方。
- 假设你对你所用的计算机有一定的了解，也就是说，假定你已经知道如何编辑、复制、删除和重命名文件，会在系统中查找信息，等等。
- 简单地说，假设你的计算机安装的是 Windows 95、Windows 98、Windows 2000、

Windows NT 或 Windows XP。对于使用其他操作系统的人，通过本文也能很容易地理解这方面的概念。
- 假设你有学习的渴望。

Java 编程所需的所有工具可以很容易在网上找到，并且免费使用。同时网上还有大量的免费的教学材料。因此在读完本篇文章后，你可以做更多的学习以提高你的技能。学习 Java 编程可以不花费一分钱在编译器、开发环境、教学材料等方面。一旦你学会了 Java，其他语言就很容易学会，所以这是一个好的开始。

说了这么多，我们可以正式开始了。

9.1 一些术语

记住我是假设你对编程一无所知，明白以下的一些术语会给你带来方便。
- 计算机程序（computer program）：一个程序就是一个指令集，它将精确地告诉计算机要做什么。这些指令可能是执行一串数字的相加，或者是比较两个数的大小，并根据结果做出决策，或是其他的指令。但是一个计算机程序只不过是一个计算机指令序列，就如菜谱是为厨师准备的一个指令集合，曲谱是音乐家用的指令集合。计算机精确地遵循你的指令，在这过程中做一些有用的事情——就像计算处理一个账本或是在屏幕上显示一个游戏或是实现一个文档处理。
- 编程语言（programming language）：为了让计算机识别你给出的指令，这些指令要用一种计算机能理解的语言——计算机语言来编写。有很多的编程语言，如 Fortran、Cobol、Basic、Pascal、C、C++、Java、Perl 等，就像有很多种口语一样，它们用不同的方式表达大约相同的意思。
- 编译器（compiler）：编译器把我们能够理解的计算机语言（如 Java）翻译成机器可以执行的形式。你很可能已经看到你机器上的 EXE 文件了。这些 EXE 文件就是编译器生成的结果。它们包含机器可以识别的可执行程序。

为了开始用 Java 编程语言编写计算机程序，你需要一个 Java 语言编译器。接下来的部分将引导你如何下载和安装编译器。一旦有了编译器，你就可以开始了。这个过程将会花费几个小时，大部分时间将被下载几个大的文件占去。同时还需要有 40MB 的可用空间（在你开始前要确保有足够的可用空间）。

9.2 下载 Java 编译器

为了在机器上得到 Java 开发环境（在开发环境下开发计算机程序），需要完成以下几个步骤：

下载 Java 开发环境软件包（包含编译器和其他的工具）。

（1）下载 Java 文档。

（2）假如计算机上还没有 WinZip（或其他功能相同的软件），则需要下载并且安装它们。

（3）安装 Java 开发环境。

（4）安装文档。
（5）设置一些环境变量。
（6）彻底检测每一部分。

在开始之前，建立一个临时文件夹来装将要下载的文件会使事情变得容易，我们将它称为下载文件夹。

第 1 步：下载 Java 开发环境包

到 http://java.sun.com/j2se/1.4.2/download.html 单击 Download J2SE SDK 链接，下载 SDK 软件包，将会看到一个许可协议，单击"接受"按钮。根据所使用的操作系统选择相应的文件下载到 Download 文件夹。这是个大文件，如果你是拨号上网的，可能会花几个小时时间。接下来的两个文件也很大。

第 2 步：下载 Java 帮助文档

根据所使用的操作系统单击下载 SDK 1.4.1 帮助文档。

第 3 步：下载和安装 WinZip

假如你没有 WinZip 或其他的类似的工具，就到 http://www.winzip.com/ 下载。运行 EXE 文件开始安装。我们一会儿在安装帮助文档时要用它。

第 4 步：安装开发包

运行在第一步下载的 j2sdk-1_4_1.exe，它将会解压并且自动安装开发包。

第 5 步：安装帮助文档

请阅读安装说明，它会教你移动文档到一个相同的文件夹，这些文件夹包含了刚刚安装的开发包。解压文档，它会自动解压到正确的位置。

第 6 步：设置好环境

就像在另一页文章所介绍的那样，你需要改变一下路径变量。这个很容易做到。打开 MS-DOS 提示界面，输入 PATH 命令看看当前环境变量的设置情况。用记事本打开 autoexec.bat，按说明上说的指定值改变 PATH 变量。

第 7 步：测试

打开另一个 MS-DOS 窗口并且输入 javac。假如一切设置正确，将会看到两行文字，它告诉你怎么用 javac。这意味着一切就绪了。假如看到的是 Bad Command or File Name，这说明还没有安装好。重新看安装说明，找出什么地方弄错了。确保 PATH 的设置正确无误。回到刚才的窗口看它显示的文字，直到问题全部解决。

现在你将为你的机器能够编译 Java 程序而感到自豪，可以开始写软件了。

值得一提的是，所解压的其中一个叫 demo 的文件夹里包含有很小精简的程序例子，这些例子可以直接运行。所以可以先找到那个文件夹，并去执行一些例子看看。它们中的大多是可以发出声音的，所以要打开喇叭。为了执行这些程序，还要去找像 example1.html 的文件并用浏览器打开它。

9.3 您的第一个程序

你的第一个程序短小可爱，它将会创建一个绘图区域并在其上画出一条对角线。为了建立它，需要完成下面的工作：

打开记事本并输入（或剪切、粘贴）程序代码。

（1）保存程序代码。

（2）用 Java 编译器去编译这个程序以产生 java.applet 程序。

（3）修改出现的各种错误。

（4）创建一个 HTML 网页来装入刚才的 java.applet 程序。

（5）运行 java applet 程序。

这里是我们用到的示例程序。

```
import java.awt.Graphics;

public class FirstApplet extends java.applet.Applet
{
public void paint(Graphics g)
  {
    g.drawLine(0, 0, 200, 200);
  }
}
```

第 1 步：输入程序代码

新建一个文件夹来存放你所写的程序，打开记事本（或者其他可以创建 TXT 文档的文本编辑器）。输入或剪切并粘贴程序到记事本窗口内。这个很重要：当输入程序时，大小写要匹配。这意味着你必须准确地输入大小写字母。回顾上述的程序员事项，如果没有准确地输入，程序就会出错。

第 2 步：保存文件

以文件名 FirstApplet.java 把文件保存在第 1 步创建的文件夹里。文件名大小写要匹配。确保 F 和 A 是大写，其他的是小写。

第 3 步：编译程序

打开 MS-DOS 窗口。改变目录（cd）到包含 FirstApplet.java 的目录。输入：

```
Javac FirstApplet.java
```

注意大小写。程序要么运行，这样窗口上不会显示任何东西，要么就是出错。假如没有什么错误，一个名为 FirstApplet.class 的文件将会被创建在 FirstApplet.java 旁边。

第 4 步：修正错误

假如存在错误，修正它们。把你的程序和上述的比较，确保完全一样。再编译直到没有错误为止。假如 Javac 不能工作，再看看前面的部分并且修改安装。

第 5 步：建立一个 HTML 网页

创建一个 HTML 网页以容纳 Applet 小程序。打开记事本，并输入以下内容：

```
<html>
<body>
<applet code=FirstApplet.class width=200 height=200>
```

```
</applet>
</body>
</html>
```

在同一个文件夹里以 applet.htm 为文件名保存这个文件。
第 6 步：运行 Applet 小程序。
MS-DOS 窗口中输入：

```
appletviewer applet.htm
```

应该可以看到一条从左上角到右下角的斜线。

稍微拉大一点小程序浏览器以看到整条线。也可以用现代浏览器，如 Netscape Navigator 或 Microsoft Internet Explorer 装载 HTML 网页，并看到基本上相同的东西。

到目前为止，你已经成功地创建了第一个程序。

9.4 理解刚刚是怎么回事

到底是怎么回事呢？首先，你写了几行很简单的 Java Applet 代码。一个 Applet 是一个可以在浏览器里运行的 Java 程序。与之相对的是 Java 应用程序，它是一个在你机器上独立运行的程序（Java 应用程序相对复杂一点，因而没有那么流行，所以我们从 Applet 讲起）。用 Javac 编译 Applet 程序，然后建立一个很简单的网页以"装入（hold）" Applet。我们用 Appletviewer 运行它，也可以简单地用浏览器运行它。

这个程序本身只有 8 行代码：

```
import java.awt.Graphics;
public class FirstApplet extends java.applet.Applet
{
public void paint(Graphics g)
  {
    g.drawLine(0, 0, 200, 200);
  }
}
```

这可是你所能创建的最简单的 Java Applet 小程序了。要完全了解它，必须学会更多，特别是在面向对象技术领域。既然已经假设你没有任何编程经验，现在希望你做的是要集中注意力在这一行代码：

```
g.drawLine(0, 0, 200, 200);
```

这一行代码在程序中起主要作用。由它来画出斜线，程序的其他代码只是为这一行提供框架，这里可以先暂时忽略那些框框。这是要计算机从左上角坐标（0, 0）到右下角坐标（200, 200）画一条线。计算机就按照我们所说的去做了。这是计算机程序的本质。（同样，在 HTML 网页中，在第 5 步设置 Applet 的窗口高和宽都是 200。）

在这个程序里，调用了一个名为 drawLine 方法（也叫函数），并且传递给它 4 个参数（0, 0, 200, 200）。这行代码以分号结束，这分号就像句子末尾的句号。这行代码以 g 开始，这表示要调用的函数 drawLine 来自一个特定的对象叫做 g（也就是在上一行代码中的类 Graphics，在本文的后面将会更详细地介绍类和类的方法）。

一个函数就像一个命令，它告诉计算机要做某件事情。在这个例子中，drawLine 告诉计算机在两个指定点（0, 0）和（200, 200）画一条线。可以认为窗口的左上角坐标是 (0, 0)，X 轴向右延伸，Y 轴向左延伸。屏幕在的每个点（像素）就是一个增量。

试着用不同的数为 4 个参数赋值。改变一个或两个参数并保存，重新用 Javac 编译并用 Appletviewer 重新运行，看看有什么变化。

除了 drawLine 还有其他可用的函数吗？可以通过看帮助文档中的 Graphics 类来找到答案。安装 Java 开发包并解压帮助文档时，会有一个叫做 java.awt.Graphics.html 的网页在机器上生成，就是这个文件对 Graphics 类做了详细的说明。在"我的电脑"上，它精确的路径是 D:\\jdk1.1.7\\docs\\api\\java.awt.Graphics.html。而在你的计算机上可能会有一点不同，但是会很相近，取决于你把它们安装到哪里了。找到这个文件并打开它。在这个网页的上方有一部分叫"函数索引"。这是这个类中所有函数的一个列表。drawLine 是其中的一个函数，你可以看到还有很多其他函数。你可以画其他的东西：

- 直线。
- 圆弧。
- 椭圆。
- 多边形。
- 矩形。
- 字符串。
- 字符。

看一下这个文档并且努力去试验其中的一些函数，看看可能会发生什么。比如，尝试一下以下代码：

```
g.drawLine(0, 0, 200, 200);
g.drawRect(0, 0, 200, 200);
g.drawLine(200, 0, 0, 200);
```

它们将会画出一个正方形和两条斜线（确保窗口足够大以看到全部的结果），试着画其他的形状，并用 setColor 函数改变其颜色。例如：

```
import java.awt.Graphics;
import java.awt.Color;
public class FirstApplet extends java.applet.Applet
{
public void paint(Graphics g)
  {
    g.setColor(Color.red);
    g.fillRect(0, 0, 200, 200);
    g.setColor(Color.black);
```

```
        g.drawLine(0, 0, 200, 200);
        g.drawLine(200, 0, 0, 200);
    }
}
```

注意这里在第二行增加了一行代码，程序的输出如图 9.3 所示（图略）。

现在你的脑子里可能会有这么个疑问：它是怎么知道要用 Color.red 而不仅仅是 red，并且它是怎么知道要加上第二行 import 语句的？你得通过例子来学习，因为我只是举一个例子，以说明怎么调用 setColor 这个函数。现在你懂得了当你想要改变颜色时可以用 Color 函数。后面跟上你需要的颜色的名称作为 setColor 函数的参数。而且你还要在程序的顶部增加一条正确的 import 语句。假如查找 setColor，将有一个链接，它会告诉你一些关于 Color 类的信息。这个类中有一个有效的颜色名字列表和创建新颜色的函数（未命名的）。通过阅读这个信息，把它记在脑子里，现在就可以知道怎么在 Java 里改变颜色了。学习新的技术并在下次编程时还能记住——这是成为一个程序员的基本要求。你可以通过阅读例子（就像现在这样）或是看文档，或是看示例的源代码（在 demo 文件夹里）来学习方法。假如你乐于探索、学习和记忆，那么你将会喜欢上编程。

注释

1. It should be relatively straightforward for people running other operating systems to map the concepts over to those.

原译：对于使用其他操作系统的人，通过上边具体操作系统的例子，也会容易理解这方面的概念。

译文：对于使用其他操作系统的人，通过本文也能相对容易地理解这方面的概念。

分析：此句不难，但注意 map the concepts 的翻译。还有就是原译中漏了"相对"一词，这在程度上是有差别的。

2. The computer follows your instructions exactly and in the process does something useful—like balancing a checkbook or displaying a game on the screen or implementing a word processor.

原译：计算机精确地遵循你的指令，并且在这过程中做一些有用的事情——就像平衡一个账本或是在屏幕上显示一个游戏或是实现一个文档处理。

译文：计算机精确地遵循你的指令，在这过程中做一些有用的事情——就像计算处理一个账本或是在屏幕上显示一个游戏或是实现一个文档处理。

分析：有一个词需要强调一下，就是 balance，原意是平衡，在会计中有 balance sheet 一词，指的是资产负债表，所以这里的 balance 可以翻译成作账或说计算处理。只说"平衡"，有些令人费解。

3. Either it will work, in which case nothing will be printed to the window, or there will be errors. If there are no errors, a file named FirstApplet.class will be created in the directory right next to FirstApplet.java.

译文：程序要么运行，这样窗口上不会显示任何东西，要么就是出错。假如没有什么错误,一个名为 FirstApplet.class 的文件将会被创建在 FirstApplet.java 旁边。

第 10 章 数据库与 VC++

  1992 年 11 月到 1995 年年底之间,微软推出了一系列新的与 Windows 相关的数据库产品:Access 7、为 Windows 设计的 Visual FoxPro 3.0 和 Visual C++ 4.0。Microsoft 宣称 Access 是"任何人都会使用的数据库",并且 Access 1.x 以低廉的价格一经发行,在 90 天内便卖出了 750 000 套。Windows 平台的 FoxPro 是针对 FoxPro 开发者和预期的 Borland 的 dBase 的忠实使用者。Access 和 FoxPro 的目标都是和 Borland 的 Paradox 争夺市场,Paradox 是在 Access 1.0 的零售版发行后不久出现的。Access(在 Windows 95 面世时升级到了 7.0 版本)、Visual FoxPro 和 Paradox 都被归类为桌面数据库系统。

  自从推出以来,Visual Basic 比 Visual C++ 能更好地支持数据库界面。然而 C/C++ 程序员只有依靠 Visual C++ 4 才能真正有一个 Microsoft Jet 发动机界面。桌面数据库系统是使用私人(本地)的数据库文件,或者索引机构(或者二者兼而有之)的应用软件,并且可以在中等性能的 PC 上运行(如 8MB 内存的 80486 机)。桌面数据库系统也可以包括一种用来设计本地数据库结构的编程语言。桌面数据库系统还包括一种专门为本地数据库结构设计的程序设计语言。

  编著本书的第一版时,Microsoft 已经售出超过 400 万套的 Access,包括 1.0、2.0、3.0 这 3 个版本。从 1995 年 6 月中旬到 11 月中旬,Microsoft 发行 Windows 95 和 32 位的"为 Windows 95 设计"的 Access 7.0、Visual FoxPro 3.0、Visual Basic 4.0 和 Visual C++ 4.0,并同时发行了 32 位的 Microsoft Office 95。Microsoft 想确保 Windows 95 早期使用者能使用 32 位的应用程序。

  Visual C++ 是 Microsoft 应用最广泛和最强大的编程语言。Microsoft 的 Visual C++ 的最初目的是提供一种强大而方便的平台,让程序员可在 Windows 下创建他们自己的 Windows 应用程序。Microsoft 通过 Visual C++ 1.0 实现这个目标。很多有经验的程序员丢弃 DOS 下的 C、C++ 和 Pascal 而选择 Visual C++,因为在 Windows 的图形界面接口环境中,用 Visual C++ 能够比传统的编程语言更快地开发 Windows 应用软件。Microsoft 通过改进接口和扩充 MFC C++ 库强化了 Visual C++ 1.5,而 Visual C++ 2.x 将程序员带入 32 位的应用程序世界。Visual C++ 4.0 则将编程界面、类型库和各种特性都提升到一个新的高度。随着 Visual C++ 的推出,增加了一系列数据库特征。Visual C++ 4.0 除了支持 ODBC,还支持 DAO(Data Access Objects),并极大地扩充了其他的支持,比如增加了 OLE 自定义控件的容器支持。一些独立的公司还为 Visual C++ 创造了许多有用的工具、类型库及附加属性,其中大多数都应用于数据库应用。今后,将会有更多为程序员设计的 OLE 自定义控件。

  1993 年早期,一份 Microsoft 市场调查表明超过 70%的 Windows 应用程序或多或少与数据库有关。1995 年 10 月,Microsoft 的 Visual C++产品经理在 Boston 的演讲中强调 40%至 60%的 Visual C++ 应用程序都是以数据库为导向的。Visual C++ 也可以被认为是一个很受欢迎的数据库应用程序开发工具。甚至在 Visual C++ 4.0 推出之前,C 和 C++

因为它的数据存取对象（CRecordset、CDatabase 和 CRecordView）极大地增强了数据库的功能性，就已是 Windows 数据库市场上主要的但并未被意识到的竞争者。如今 Visual C++ 4.0 的出现使 Visual C 成为 Visual Basic 在数据库发展平台的舞台上的强大竞争者。市场调查公司对 Visual C++在 Windows 数据库种类定位的失误导致了 1993 年及以后数据库市场统计的重大扭曲。

本章描述了 Visual C++ 在数据库应用程序发展中的角色并且说明了 Visual C++、OLE（对象链接和嵌入）自动化、ODBC（开放数据库连通）、DAO 和 MFC 如何适应 Microsoft 的策略来保持在 Windows 应用程序中的主导地位。本章还讨论了 Visual C++ 在数据库应用程序的优缺点并且让你简要了解一下本书其他章节将要讨论的内容。

这是一个 32 位的、"为 Windows 95 设计的"世界，因此本书聚焦于用 Visual C++ 4.0 的 32 位的应用程序开发。

选择 Visual C++作为你的数据库开发平台

Visual C++ 保持了数据库连通性和数据操作特性，使得编程语言有一个成熟的数据库开发环境。Microsoft 在 Visual C++ 中增加了新的数据存取特征，使 Visual C++ 在桌面数据库市场上成为 Visual Basic 的直接竞争对手，并使 Visual C++ 更好地支持 Access、FoxPro 和 Paradox。Visual C++ 相对于其他数据库竞争者的主要优势在于它简单、方便，易扩张性强。

- 32 位的 Microsoft Jet 3.0 数据库引擎相对于 16 位的 Jet 2 有了充分的改良。Jet 3.0 是多线程的，最少可有 3 个线程（可以进入 Windows 95 或者 Windows NT 的注册表在可用范围内增加线程的数目）。全面优化和代码的调整也加快了 Jet 3.0 queries 的执行。
- Visual C++ 内置的 MFC 类，连同 AppWizard，可让你很少，甚至不用 Visual C++ 代码就能快速地建立一个表格来显示数据库信息。
- Visual C++ 比较灵活，不会把程序员限定在一个特定的应用程序结构中，正如 Access 的多文档界面（MDI）。同样，你也没有必要按照 DOCMD 说明去操作现行的开放数据库。
- Visual C++ 4.0 数据库前端末端明显比 Access 需要更少的资源。大多数 Visual C++ 4.0 数据库应用程序在有 8MB 内存的计算机的 Windows 95 和有 16MB 以上内存计算机的 Windows NT3.51+平台下运行得很好。Microsoft 宣称 Access 95 在 Windows 95 平台下需要 12MB 内存，但是除了一些琐碎的 Access 95 应用程序之外，大都需要 16MB 内存才能实现足够的性能。一个典型的 Visual C++ 数据库前端末端程序可能至少要在有 12MB 内存的 Windows 平台下才能良好运行。同样的程序在后来有同样内存，但版本为 Windows NT 的平台下就能很好地运行。
- OLE 自定义控件，尽管它不是在其他所有的数据库开发平台上都可用，但却可以很容易地用它来为你的 Visual C++应用程序添加许多新的特征。第三方的开发人员可以开发自定义插件来扩展 Visual C++的数据存储控制指令。自定义控件可以自定义控件可以采用 32 位环境下的 OLE 自定义控件形式。

虽然并没有明显地列于 Microsoft 的特征列表中，但选择 Visual C++作为主要数据库开发平台的最大好处在于，它有一个非常巨大的工具队列，对 ODBC 的支持，还可用来开发数据库。

另外一个选择 Visual C++进行数据库应用程序开发的原因是其 OLE 的兼容性。Visual C++是兼容 OLE 的最好的数据库开发环境。

OLE 自动控制对大多数 Visual C++数据库开发人员来说可能是 OLE 特征中最重要的。OLE 自动控制允许用 Visual C++数据库应用程序控制其他支持 OLE 的服务器的应用程序的操作，应用程序不需要包括 Visual Basic 作为 OLE 自动控制的原程序（服务器）。与 Windows 95 配套的 Word 及其后继版本可以使用传统的 Word Basic 宏语言句法来支持 OLE 自动控制。

Windows 数据库战争并没有结束（可能永远也不会结束），Microsoft 依靠 Visual Basic、Access、Visual FoxPro、SQL Server 多方面进行攻击，并且 Visual C++逼得桌面数据库管理的竞争者只剩抵挡之力。作为一个团队，Microsoft 为 Windows 开发的数据库应用程序，加上辅助的产品，如 ODBC、DAO 和 Access Jet 数据库引擎，有了目前任何其他软件发布者无法匹敌的深度和广度。

注释

1. FoxPro for Windows targeted existing FoxPro developers and prospective users of Borland's long-promised dBase for Windows. Both Access and FoxPro targeted the market for Borland's Paradox for Windows, which emerged shortly after the retail release of Access 1.0. Access (upgraded to version 7.0 with the introduction of Windows 95), Visual FoxPro, and Paradox for Windows are categorized as desktop databases.

译文：Windows 平台的 FoxPro 是针对 FoxPro 开发者和预期的 Borland 的 dBase 的忠实使用者。Access 和 FoxPro 的目标都是和 Borland 的 Paradox 争夺市场，Paradox 是在 Access 1.0 的零售版发行后不久出现的。Access（在 Windows 95 面世时升级到了 7.0 版本），Visual FoxPro 和 Paradox 都被归类为桌面数据库系统。

分析：这段话是由几个并列句组成的，结构都差不多，译文采用了概括翻译的办法，在说法上有所变化，较为灵活。

2. Microsoft's original objective for Visual C++ was to provide a powerful and flexible platform that programmers could use to create their own Windows applications while running under Windows.

原译：Microsoft 的 Visual C++ 的最初的目的是提供一种强大的并且方便的平台，程序员可以用来创建他们自己的 Windows 应用程序运行在 Windows 下。

译文：Microsoft 的 Visual C++ 的最初的目的是提供一种强大而方便的平台，让程序员可在 Windows 下创建他们自己的 Windows 应用程序。

分析：关于英语与汉语语序的问题已经强调过多次，中文中的状语（本句中的"在 Windows 下"）一般是在谓语之前的。其他一些细节注意一下就更好了。

3. Most 32-bit Visual C++ 4.0 database applications run fine under Windows 95 with

PCs having 8MB of RAM and under Windows NT 3.51+ in the 16M range. Microsoft says Access 95 requires 12MB of RAM under Windows 95, but you need 16MB to achieve adequate performance of all but trivial Access 95 applications.

原译：大多数 Visual C++ 4.0 数据库应用程序在有 8MB 内存的 Windows 95 或者有 16MB 内存的 Windows NT 3.51+PC 平台下运行得很好。Microsoft 宣称 Access 95 在 Windows 95 平台下要求 12MB 内存，但是需要 16MB 内存来达到最好的性能，除了一些琐碎的 Access 95 应用程序。

译文：大多数 Visual C++ 4.0 数据库应用程序在有 8MB 内存的计算机的 Windows 95 和有 16MB 以上内存计算机的 Windows NT 3.51+平台下运行得很好。Microsoft 宣称 Access 95 在 Windows 95 平台下需要 12MB 内存，但是除了一些琐碎的 Access 95 应用程序之外，大都需要 16MB 内存才能实现足够的性能。

4. The Windows database war wasn't over (heck, it may never be over), but Microsoft's multipronged attack with Visual Basic, Access, Visual FoxPro, SQL Server, and Visual C++ is forcing competing publishers of desktop database managers into their defensive trenches.

原译：Windows 数据库战争并没有结束（可能永远也不会结束），但 Microsoft 依靠 Visual Basic、Access、Visual FoxPro、SQL Server 多方面的攻击，并且 Visual C++ 强迫桌面数据库经理的竞争发行者（竞争者）进入防御的战壕。

译文：Windows 数据库战争并没有结束（可能永远也不会结束），Microsoft 依靠 Visual Basic、Access、Visual FoxPro、SQL Server 多方面进行攻击，并且 Visual C++ 逼得桌面数据库管理的竞争者只剩抵挡之力。

分析：原文写得比较生动，原译中对最后的比喻却很遗憾地采用了直译，显得有些生硬。译文中就把其意味体现出来了。

第 11 章 人 工 智 能

人工智能，也称为机器智能，被定义为人造（也就是人工）系统展示的智能。这个术语经常被应用于通用计算机，而且在科学研究领域深入到了 AI 的理论和实际应用。

11.1 概述

人工智能的概念可以看成两个方面："什么是人工的本质？"和"什么是智能？"第一个问题相对比较容易回答，然而就必然会引出什么是可以人造的问题。比如，某种类型的系统的局限性，例如古典的计算系统，现有制造工序的局限性、人类智力的局限性，在各方面约束了人造的能力。

第二个问题引出了关于意识、自我和精神（包括无意识的）的基础存在问题。而且也引出了关于人类表现出的智能的本质的问题，因为人类的智能行为是复杂的而且总是难以学习或理解。对动物的研究和对一些不是现有的简单模型的人工系统的研究经常被认为是高度相关的。

以下将讨论一些独特类型的人工智能以及它们的区分、历史发展、支持者、反对者和在这个领域的应用研究。最后将提供一些描述虚拟和现实的 AI 的资料。

11.2 强人工智能和弱人工智能

人工智能的一个流行的早期定义是由约翰·麦卡锡（John McCarthy）在 1956 年的达特茅斯会议（Dartmouth Conference）上提出的：制造一种机器可以像人表现出智能的行为一样行动。它重提了艾伦·图灵（Alan Turing）"计算机和智能（Computing machinery and intelligence）"（1950 年 10 月）的主张。但是这个定义似乎忽略了强人工智能的可能性（见下）。另一个人工智能的定义是来源于人工设计的智能。对人工智能的定义大多可分类为关于：系统"像人一样思考""像人一样行动""理性地思考"或"理性地行动"。

11.2.1 强人工智能

强人工智能的研究用来创造一些计算机为基础、能真正推理和解决问题的人工智能，一个强类型的人工智能被认为是有知觉的，有自我意识的。理论上可分为两类强人工智能：

类人的人工智能，即计算机像人的头脑一样思考和推理。

非类人的人工智能，即计算机产生了完全和人不一样的意识，使用和人完全不一样的思考和推论方式。

11.2.2 弱人工智能

弱人工智能的研究用于创造一些基于计算机，但只能在有限的领域能推理和解决问题的人工智能，这些机器在某些方面表现出智能，但并不是真正拥有智能或感觉。图灵测试是这方面最经典的测试。

弱人工智能有很多领域，其中有一个就是自然语言。很多弱人工智能的领域都有专用的软件和编程语言。例如，最近似人的自然语言 chatterbot A.L.I.C.E.使用专用的编程语言 AIML，和命名为 Alicebots 的各种复制品。Jabberwacky 近似于强 AI，因为它通过与用户的单独交互作用为基础学习如何交谈。

迄今为止，在这个领域大部分的工作已在以预先确定的规则为基础的计算机智能模拟实现。强人工智能方面的研究则处于停滞不前的状态。从如何确定目标的基础上看，弱人工智能方面已经取得了一定的成就。

半开玩笑地说，弱人工智能可以看作"在这方面没有好的解决办法的计算机科学问题"。一旦一个分支学科有了成效，它就从人工智能分离出来并拥有新的名字，例如，模式识别、图像处理、神经网络、自然语言处理、机器人技术和博弈论。虽然所有的这些学科的基础都已经作为人工智能的一部分被稳固建立，它们之间现在还是被认为有所区别。

11.2.3 强人工智能的哲学争论和支持

"强人工智能"一词最初是由约翰·希尔勒针对数字计算机和其他信息处理机器提出的，他把强 AI 定义为：

对强人工智能来说，计算机不仅是用来研究人的思维的一种工具；相反，只要运行适当的程序，计算机本身就是有思维的。（J. Searle in Minds Brains and Programs. The Behavioral and Brain Sciences, vol. 3, 1980）

不同于更广义的一元论和二元论的争论，希尔勒与多数人争论的要点是：如果一台机器的唯一工作原理就是对编码数据进行转换，那么这台机器是不是有思维的？（无论什么类型的计算机，包括生物计算机，都会包含一个思想。）

希尔勒在他的中文房间的例子中陈述，信息处理机携带的编码数据是对某些事情的描述。这些编码的数据离开他描述的事物将会变得毫无意义。基于这一点，希尔勒指出对信息处理机自身并没有意义或理解。因此希尔勒宣称即使机器通过了图灵测试，也不一定真的像人一样有思维和意识。

一些哲学家坚持认为，如果弱人工智能是被承认的，那么强人工智能也是可实现的。Daniel C. Dennett 在 Consciousness Explained 中认为，没有了思想或灵魂，人也不过是一台机器而已，那么为什么提到关于智能或思维时，人这个机器可以处于一种凌驾于其他可能的机器之上的特权位置？Simon Blackburn 在其哲学入门教材 Think 里指出：一个人看起来可能是"智能"的，但这并不能真正说明这个人就真的是智能的。然而，如果讨论只限制在强人工智能而不是人造意识，这就可确定人的思维活动的特性不会发生在信息处理计算机里。

强人工智能包括以下几个关于思想和头脑的设想：

思想是软件，一个应用 Church-Turing 理论的有限状态机器。

头脑是纯硬件（只是服从了经典计算机的规则）。

第一个设想是很有问题的，因为老格言说所有计算机只是美其名的算盘。其实用球和木头构造任何类型的信息处理器是完全可能的，尽管这样的装置会很慢而且很可能失败，它可以做任何现代计算机可以做的事。这说明，信息处理器等于说：思想可以以在木制通道里的滚珠的形式存在。

一些人（包括 Roger Penrose）直接攻击图灵理论的适用性，他们提出停机问题是某些信息系统解决不了的，而用人的思想就可以解决。

最后，强人工智能的关键是信息处理器能否包含所有的头脑的内容，比如说意识。尽管如此，弱人工智能是和强人工智能无关的，而且毫无疑问的是，很多当今的计算机能做的事，如算术运算、数据库搜索等，在一个世纪前是被认为是智能的。

11.3 人工智能理论的发展史

人工智能大部分的初期研究来源于近似于心理学的实验，而且强调的是什么会被称为语言上的智能（最好的例子就是图灵测试）。

近似于人工智能但并不是以语言上的智能为中心的，包括机器人学和集中智能近似，他们以动态环境处理或一致同意决策判定为中心，而且从生物学和政治方面寻求智能行为的组织模式。

人工智能技术也来源于动物的学习，特别是更容易用机器人模拟的昆虫和有比较复杂认识的动物，比如说和人很相像可是计划和认识能力比较弱的猿。人工智能的研究者认为对这些比人类简单的动物应该可以更加简单地模拟，可是还没有令人满意的计算模型。

开创性的论文发展了机器智能的概念，包括 A Logical Calculus of the Ideas Immanent in Nervous Activity (1943)，Warren McCulloch and Walter Pitts 编写；On Computing Machinery and Intelligence (1950)，Alan Turing 编写；Man-Computer Symbiosis，J.C.R. Licklider 编写。

还有一些否认机器在逻辑和哲学方面的智能的可能性的早期文章，比如说 Minds, Machines and Gödel（1961），John Lucas 编写。

随着人工智能研究上实际技术的进步，人工智能的提倡者认为反对者在计算机国际象棋和语音识别这些以前被用来否认人工智能造诣的事情上反复改变他们的观点。他们指出这种目标的交换有效地把智能定义为"人可以做而机器不能做的"。

John von Neumann 在一次演讲上（1948 年），作为对机器是无法思考这个评论的回应，他预言："你们坚持说有些事是机器不能做的。如果你能准确告诉我什么是机器不能做的，我总可以做出正好可以做这件事的机器！"Von Neumann 大概是在影射声明所有有效的程序都可以被计算机模拟的 Church-Turing 理论。

1969 年 McCarthy 和 Hayes 从他们的论文《关于人工智能立场的一些哲学问题》开始了关于框架问题的讨论。

11.4 实验的人工智能研究

人工智能于 20 世纪 50 年代开始进入实验的领域，其先驱包括在 Carnegie-Mellon 大学建立第一个人工智能实验室的 Allen Newell 和 Herbert Simon 以及在 1959 年建立 MIT 人工智能实验室的 McCarthy 和 Marvin Minsky。他们都出席了 1956 年由刚提到的 McCarthy、Minsky、IBM 的 Nathan Rochester 和 Claude Shannon 组织的 Dartmouth 大学夏季人工智能会议。

历史上，有两个主要的人工智能研究类型——"简洁的"和"不简洁的"。"简洁的"也叫经典或符号型的人工智能研究，大体上讲，包括抽象概念的符号处理，而且是大多数专家系统中使用的方法论。与之对应的是"不简洁的"或者说"有联系的"，它最广为人知的例子是神经系统网络，它试着通过建立系统，并用一些自动的进程而不是系统地设计一些东西来完成这一工作。这两种方法在人工智能的历史上出现时间都比较早，在 20 世纪 60 到 70 年代，不简洁的方法并不被重视，可是在 20 世纪 80 年代当简洁的方法在时间上的局限性变得越来越明显的时候，它又被重新重视起来。然而，使用这两种主要途径的现代方法的严格的限制也越来越明显。

在 20 世纪 80 年代美国的 Defense Advanced Research Projects Agency（高级防御研究计划机构）和日本的 fifth generation computer systems project（第五代电脑系统计划）都在人工智能研究方面有很大的投资。尽管有某些人工智能的从业者冠冕堂皇地承诺会尽快得出结果，投资工作还是失败了，因此在 20 世纪 80 年代后期，政府机构的投资相对大大减少了，导致了一段被称为"人工智能的冬天"的普遍低迷时期。在接下来的 10 年，尽管纯人工智能的研究还在较低的水平上继续，很多人工智能的研究人员转向相关较小目标的领域，比如，机器学习、机器人学和电脑视觉。

人工智能技术的实际应用

在制造像人一样的智能方面的发展已经减缓的同时，很多分支学科却发展起来了。最显著的例子是原本为人工智能研究发明，现在却应用于非人工智能工作的 LISP 和 Prolog 语言。黑客文化首先从人工智能实验室中开始，特别是麻省理工学院人工智能实验室，在不同时期产生了很多名人，如 McCarthy、Minsky、Seymour Papert（在那里发明了 Logo）、Terry Winograd（在研究了 SHRDLU 之后放弃了人工智能）。

很多其他有用的系统的建立都利用了某些曾经是人工智能研究的活跃领域的技术。包括以下例子：

（1）深蓝国际象棋比赛计算机，在 1997 年的一场著名的比赛中击败了卡斯帕罗夫。

（2）InfoTame，KGB 为了自动排列成千上万页截获的情报而研发的文本分析搜索引擎。

（3）模糊逻辑（Fuzzy logic），一种用于不确定情况下的逻辑，已经在工业控制系统中得到广泛的应用。

（4）专家系统（Expert systems）在一些区域企业中使用。

（5）机器翻译系统，比如 SYSTRAN 已经得到广泛的应用，尽管效果不如人工翻译的好。

（6）神经网络技术在多种工作中已经得到了广泛的应用，从侵入检测系统到电脑游戏。

（7）光学字符识别系统可以把任何打字机打出的欧洲语系的稿件转化为文本。

（8）手写识别在许多 PDA（私人数字助理）中得到。

（9）应用语音识别可用于商业并已得到广泛使用。

（10）计算机代数系统（Computer algebra systems），如 Mathematica 和 Macsyma，都是非常普遍的。

（11）机器视觉（Machine vision）系统应用于很多工业应用软件，涵盖了从硬件识别（hardware verification）到安全系统（security systems）。

（12）人工智能计划编制方法在第一次海湾战争期间被用于美国军队的自动计划部署。如果用人工来做，这个工作将耗费几个月的时间和数百万美元，而且 DARPA 认为在这一次行动上节约的钱比他们在过去 50 年对人工智能研究的总支出还多。

人工智能代替人的专业判断的梦想曾多次被提到，不管是在这个领域的历史上，还是一些科幻小说，或当今的采用专家系统来辅助或代替专业人员判断的工程和医学等专门领域。

注释

1. The first question is relatively easy to answer, although it also necessarily leads to an examination of what it is possible to manufacture. For example, the limitations of certain types of systems, such as classical computational systems, or of available manufacturing processes, or of human intellect, may all place constraints on what can be manufactured.

译文：第一个问题相对比较容易回答，然而就必然会引出什么是可以人造的问题。比如说，某种类型的系统的局限性，例如古典的计算系统，现有制造工序的局限性、人类智力的局限性，在各方面约束了人造的能力。

分析：有的读者可能会这样翻译：第一个问题相对比较容易回答，尽管会导致一个关于什么是可以人造的实验。比如说，确信类型的系统的局限性，像古典的计算系统，或关于可用的制造业进程，或关于人类的智力，在很多地方约束了什么是可以人造的。翻译中不少词都是照搬字典的意思，不能变通，例如，certain 此处为"某种"，而非"确信"；manufacuring processes 则应该是制造工序。还有就是后句结构的理解也有值得注意的地方，应该把主干抓住，主语是三者的局限性，而翻译漏了后两个。

2. Weak artificial intelligence research deals with the creation of some form of computer-based artificial intelligence that can reason and solve problems only in a limited domain; such a machine would, in some ways, act as if it were intelligent, but it would not possess true intelligence or sentience.

原译：弱人工智能的研究涉及以计算机为基础但只能在有限的领域推理和解决问题的人工智能类型，例如，机器领域，在某些方面它表现出智能，但是不会拥有真正的智能或感觉。图灵测试是这方面最经典的测试。

译文：弱人工智能的研究用于创造一些基于计算机，但只能在有限的领域能推理和解决问题的人工智能，这些机器在某些方面表现出智能，但并不会真正拥有智能或感觉。图灵测试是这方面最经典的测试。

分析：此段的关键在翻译"deals with the creation of"，原译中似乎漏译了，只是意译。译文运用了将名词动词化的技巧。

3. When viewed with a moderate dose of cynicism, weak artificial intelligence can be viewed as "the set of computer science problems without good solutions at this point." Once a sub-discipline results in useful work, it is carved out of artificial intelligence and given its own name. Examples of this are pattern recognition, image processing, neural networks, natural language processing, robotics and game theory.

原译：简单地说，弱人工智能可以被看作"在这方面没有好的解决办法的计算机科学问题"。一旦一个学科分支导致了有用的工作，它就从人工智能分离开来而且赋予它自己的名字，例如，模式识别、图像处理、神经网络、自然语言处理、机器人技术和博弈论。

译文：半开玩笑地说，弱人工智能可以看作"在这方面没有好的解决办法的计算机科学问题"。一旦一个分支学科有了成效，它就从人工智能分离出来并拥有新的名字，例如，模式识别、图像处理、神经网络、自然语言处理、机器人技术和博弈论。

分析：此句的难点在"When viewed with a moderate dose of cynicism"及"it is carved out of artificial intelligence and given its own name"。原译有些生硬，而且对第一个难点没有正确理解。这里其实是说弱人工智能没有好的 solution，本身有些讽刺意味，故用"半开玩笑地说"会比较好。

4. Searle and most others involved in this debate are addressing the problem of whether a

machine that works solely through the transformation of encoded data could be a mind, not the wider issue of Monism versus Dualism.

译文：不同于更广义的一元论和二元论的争论，希尔勒与多数人争论的要点是：如果一台机器的唯一工作原理就是对编码数据进行转换，那么这台机器是不是有思维的？

分析：本句要注意的是把后半句调到前面先说，这样才符合中文的习惯。另外将"Searle and most others involved in this debate are addressing the problem"直接采用意译也使得译文简洁许多。

5. This means that the proposition that information processors can be minds is equivalent to proposing that minds can exist as devices made of rolling balls in wooden channels.

原译：这说明信息处理器是思想的设想可以等同于思想可以以滚珠和木棍的形式存在的说法。

译文：这说明，信息处理器是思想的设想等于说：思想可以以在木制通道里的滚珠的形式存在。

分析：此句较长，原译虽然把握到该句子的结构，然而不甚通顺，感觉有些拗口。译文再次使用名词动词化的技巧，将"is equivalent to proposing"译为"等于说"。

6. The failure of the work funded at the time to produce immediate results, despite the grandiose promises of some AI practitioners, led to correspondingly large cutbacks in funding by government agencies in the late 1980s, leading to a general downturn in activity in the field known as AI winter.

译文：尽管有某些人工智能的从业者冠冕堂皇地承诺会尽快得出结果，投资工作还是失败了，因此在20世纪80年代后期，政府机构的投资相对大幅减少了，导致了一段被称为"人工智能的冬天"的普遍低迷时期。

分析：此段就翻译得比较好。逻辑关系的处理得当，一些词语也翻译得比较准确、恰当。

第12章 机器学习

机器学习是计算机科学的子领域。它是由人工智能中的模式识别和计算学习理论衍化而来。Arthur Samuel 于1959年将机器学习定义为"一个可以在没有精确指令的情况下赋予计算机学习能力的研究领域"。机器学习理论主要是分析和设计一些可以从数据中学习和预测的算法。这些算法通过样本输入来建立模型，并以此来做出以数据为导向的预测或决定，而不是严格地执行固定的程序。

机器学习与致力于用计算机做预测的计算统计学关系密切，两者经常会有交叉与重合。机器学习与最优化也有着紧密的联系，后者不仅提供了理论方法还提供了应用的方向。机器学习被应用于精确算法无法处理的大规模计算。具体的应用有：垃圾邮件分类、光学文字识别、搜索引擎和计算机视觉。有时会将机器学习与数据挖掘合并到一起，后者更强调探索性的数据分析，即无监督学习。

在数据分析领域，机器学习被视为一种设计依靠自身预测的复杂模型和算法的方法。

研究员、数据学家、工程师以及分析师可以依靠这些分析模型"得出可依靠、可重复的决策和结果"并通过数据历史的关系和趋势来发现"未观测到的见解"。

机器学习可以看作一个尝试从现实世界的观察中推演出某种模式并得到某种见解的工具和方法的集合。比如说，如果你想要依据一套房子的房间的数目、洗手间的数目、房子面积以及占地尺寸来预测这套房子的价格，你可以用一个简单的机器学习算法（比如线性回归）去从现有已知的真实房子的估值的集合学习，然后便可以预测那些房价未知的房子了。

支持向量机是一种分类器，它可以将输入空间通过线性边界分割成两个区域，如图 12.1 所示（图略），它已经学会怎么区分黑圈和白圈。

12.1 综述

Tom M. Mitchell 给出了一个广泛引用并且更正式的定义："在度量 P 下，如果一个计算机程序在任务 T 中的表现通过经验 E 提升了，则该程序被称为从经验 E 中学习到了在某类任务 T 和表现的度量 P 下的知识。"这个定义之所以出名是因为它通过基本流程来定义了机器学习而不是经验性的词汇，因此之后 Alan Turing 在其论文"计算机器与智能"中提到的问题"机器可以思考吗？"被替换为"机器可以做我们（可以思考的实体）做的吗？"。

图 12.1 为支持向量机的一个示例。

12.1.1 任务和问题的种类

依据学习的本质"信号"或者"反馈"是否可以在学习系统中得到，机器学习的任务通常被分为三大类。它们分别是：

- 监督学习——为了学习到一个从输入到输出的一般映射，会由"老师"来提供输入样本和这些样本目标的输出供计算机学习。
- 无监督学习——学习算法中没有任何标签，程序需要自己找出输入数据的结构。无监督学习本身可以是一个目标（发现数据中隐藏的模式）也可以是达到某种目的的方法（特征学习）。
- 强化学习——在没有老师明确地告诉它是否已接近其目标的情况下，计算机程序在一个动态的环境中必须执行一个特定的目标（如驾驶车辆）。另一个例子是通过和对手对战学习如何玩游戏。

介于监督学习和非监督学习之间的是半监督学习。训练过程中，训练者给了一个不完备的训练信号：缺失了一些（通常是许多）目标的输出。转导学习是一种满足在学习的时候，除了目标缺失了一部分整个问题实例是已知的这一原则的具体情况。

在其他类别的机器学习问题中，自学习指的是基于先前的经验学会自己的学习倾向。发展学习（即发育机器人学）通过自主探索和与人类老师的社会互动以及使用如主动学习，汽车协同效应和模仿等的指导机制来生成自己关于学习情况的序列（也称为课程）以此来逐渐地获得一系列新的技能。

当我们考虑一个系统的期望的输出值时，另一种关于机器学习任务的分类就产生了。

在分类中，输入的样本被分为两个或更多的类，同时学习者必须产生一个可以分配看不见的输入到一个（或多标记分类）或多个类。这类问题通常是用监督学习的方式加以解决。垃圾邮件过滤是一个分类的例子，输入的样本是电子邮件（或其他）信息而类是"垃圾邮件"和"不是垃圾邮件"。

在回归中，也是一个监督学习的问题，输出是连续的而非离散的。

在聚类中，输入集被分成组。分类不同，聚类得到的组在事先是未知的，因此这通常是一个无监督学习。

密度估计会发现输入样本在某些空间下的分布情况。

降维可以通过将输入样本映射到低维空间来简化它们。主题模型是一个相关的问题，其中程序会被给出一个人类语言的列表文件，它的任务是找出哪些文档包含了相似的主题。

12.1.2 历史以及与其他领域的关系

作为一门科学，机器学习的发展源于对人工智能的探索。早在早期人工智能成为一门学术科目时，一些研究者就对机器学习数据很感兴趣。他们尝试用各种符号方法，以及当时被称为"神经网络"的方法，来解决这个问题，这些大多数是感知器和其他一些模型，后来发现是统计的广义线性模型的再造。概率推理也被采用了，尤其是在自动化的医疗诊断中。

然而，对逻辑的、以知识为基础的方法的越发重视导致了人工智能和机器学习之间的分歧。概率体系受到了关于数据采集及其代表性的理论与实际问题的困扰。直到1980年，专家体系已经主导了人工智能，而统计学却得不到认可。符号/知识学习方面的工作一直在人工智能领域中继续，产生了归纳逻辑程序设计，但更偏于统计的研究现在不在人工智能领域，而是在模式识别和信息检索中。神经网络研究大约也在同一时间被人工智能和计算科学领域抛弃。这个研究同样也在人工智能/计算科学领域外继续，作为"连接机制"，由来自其他领域的研究者（包括 Hopfield、Rumelhart 和 Hinton）进行。他们的主要成就是20世纪80年代中期反向传播算法的再造。

机器学习，作为一个独立的领域，在20世纪90年代开始蓬勃发展。这个领域的研究目标从实现人工智能转变为解决实际的可解的问题。它把研究焦点从由人工智能中得来的符号方法上转移开了，并转向由统计和概率论中借鉴来的方法和模型。它同时也得益于数字化信息的日益普及，以及通过网络传播信息的可能。

机器学习和数据挖掘经常采用相同的方法并且有明显的交叉。它们大致可以区分如下：

机器学习重点在于预测，以从训练数据中学习到的已知性质为基础。

数据挖掘重点在于发现数据中（以前）未知的性质。这是数据库中知识发掘的分析步骤。

这两个领域在许多方面有交叉：数据挖掘使用很多机器学习的方法，但往往与头脑中的目标略有不同。另一方面，机器学习也采用数据挖掘的方法像"无监督学习"或作为预处理步骤来提高学习的准确性。这两个研究团体的很多混淆（他们确实通常有单独的会议和各自的期刊，ECML PKDD 是一个主要的特例）来自他们工作中的基本假设：在机器学习中，性能通常由已知知识的再现能力进行评价，而在知识发掘和数据挖掘（KDD）中，主要任务是发现以前未知的知识。以已知的知识进行评价，一个未知的（无监督）方法会轻易被一个有监督方法超过，而在一个典型的 KDD 任务中，由于无法获得训练数据，有监督方法并不能使用。

机器学习与优化也有着密切的联系：很多机器学习的问题在一个训练集上可以表示为某个损失函数的最小化问题。损失函数表示训练模型的预测结果和实际问题实例之间的差距（比如，在分类问题中，某人想给实例分配标签，且模型被训练来正确预测一组实例预先分配的标签）。两个领域的差别源于整体的目标：优化算法可以最小化一个训练集上的损失，而机器学习和最小化未知样本上的损失有关。

12.1.3 理论

学习的一个主要目的是泛化经验。这里泛化指的是一个学习机器在学习了训练集后正确处理新的、未曾见过的样本或任务的能力。训练样本有着一些在总体上是未知的概率分布（样本被认为可以是代表事件的全空间），而学习者必须建立一个能在这个空间内使它在新的例子下产生足够准确的预测的一般模型。

机器学习算法及其性能的计算分析是理论计算机科学的一个分支，被称为计算学习理论。因为训练集的有限和未来的不确定，学习理论通常无法保证算法的性能。相反，性能的概率限定是很常见的。方差权衡就是一种量化泛化误差的方法。

模型用现有的例子训练并正确地预测未知例子的程度称为泛化。对最佳泛化来说，假设的复杂性应该匹配数据背后函数的复杂性。如果假设是没有函数复杂，便是欠拟合。随后增加了复杂性，训练误差就会减小。但是如果假设过于复杂，便是过度拟合。随后我们应该找到使得训练误差最小的假设。

除了性能界限之外，计算学习理论还研究学习的时间复杂度和可行性。在计算学习理论中，如果一个计算可以在多项式时间内完成，则该计算被认为是可行的。有两类时间复杂度的结果。积极的结果表明，一类明确的函数可以在多项式时间内被学习。而消极的结果表明，一类明确的函数无法在多项式时间内被学习。

尽管使用术语不同，机器学习理论和统计推断仍有很多相似之处。

12.1.4 方法

1. 决策树学习

决策树学习用决策树作为可以将一项观测值映射到该项目标值的结论的预测模型。

2. 关联式规则学习

关联式规则学习是一种用来在大型数据库中发现变量间关联的方法。

3. 人工神经网络

人工神经网络学习算法，通常称为"神经网络"，是一种参考生物神经网络的结构和功能的学习算法。该算法是由一组相互关联的人工神经元构成，其处理信息使用联结主义的方法来计算。现代神经网络是非线性统计数据建模的工具。它们通常用来模拟输入和输出之间的复杂关系，发现数据中的模式，或在观察到的变量联合概率分布未知的情况下，描述变量之间的统计结构。

4. 归纳逻辑编程

归纳逻辑编程是一种（机器）学习的方法。具体表现在把逻辑编程运用到已知事实、规则设定和提出的假定上。对于一个已设定的规则，和一些作为已知逻辑的事实，归纳逻辑编程可以获得一个符合所有积极（positive）样本并且不符合所有消极（negative）样本的假定。归纳逻辑编程与任何一种编程语言都息息相关，而不仅仅是函数式编程之类的逻辑编程。

5. 支持向量机

支持向量机是一种用来分类和回归的监督学习的方法。对于给定的训练样本，训练数据会被标记为两类，支持向量机训练算法建立了一个可以将新样本归为两类中某一类的模型。

6. 聚类

聚类分析是将一组观测值分配到子集（称为簇），因此根据一些固定的准则或标准在相同的簇中的观测值是相似的，而反之亦然。不同的聚类技术会在数据结构上做出不同的假设，通常由一些相似性度量并且评估，例如内部的紧凑性（在同一集群的成员的相似性）和不同集群之间的分离性。其他的方法都是基于估计的密度和图的连通性的。聚类是一种无监督学习方法，统计数据分析的常用技术。

7. 贝叶斯网络

贝叶斯网络、信念网络或有向无环图形模型是通过一个有向无环图（DAG）表示一组随机变量及条件独立的概率图形化模型。例如，一个贝叶斯网络可以代表疾病和症状之间的概率关系。在知道症状的情况下，该网络可以被用来计算各种疾病存在的概率。存在可以执行推理和学习的有效算法。

8. 强化学习

强化学习关注在某个环境中个体应该如何采取行动以最大限度地利用一些概念的长期回报。强化学习算法试图找到一个可以将不同区域映射到应该在这些区域上采取的行动的决策。强化学习与监督学习问题的不同之处在于：从未出现过正确的一对输入与输出，也没有次优的行动被明确纠正过。

9. 表征学习

一些学习算法（大部分是无监督学习算法）在训练的过程中希望发现输入样本更好的表示。经典的例子有主成分分析和聚类分析。表征学习算法经常会尝试在保留输入样本的原有信息的基础上用另一种表示来让这些信息更有效。该方法经常作为分类或者预测前预处理的步骤。它能重构那些分布未知的输入样本，虽然这一方法对于那些分布不合情理的结构并不是足够可信。

流型学习算法尝试在约束下学习低秩表示。稀疏表示算法尝试在约束下学习稀疏表示（矩阵中包含许多 0）。多线性子空间学习算法尝试不将多维度的数据变为高维向量而是直接考虑从张量的角度来考虑。深度学习算法通过高维度发现多层次的表示方法或者某种等级的特征。有人认为，智能机器是可以学习到用于表示对于观察到的数据的潜在因素变化的机器。

10. 相似性和度量学习

在这个问题中，计算机会被给予成对的相似和不相似的例子。它需要从中学习到一个可以预测新的样本是否相似的相似性方程（或者是一个度量方程）。这种方法有时候会用于评价系统中。

11. 稀疏字典学习

在该方法中，数据被表示为基底的线性组合，同时系数矩阵被认为是稀疏的。令 x 是一个 d 维的数据，D 是 d×n 阶的矩阵，其每一列代表了一个基底。r 是 D 表示 x 对应的系数矩阵。数学上来说，稀疏字典学习方法是在 r 稀疏的情况下用来解决 x ≈ Dr 问题的。一般来说，n 的取值会大于 d 以保证稀疏表示的自由性。

学习一个稀疏表示的字典是一个严格的 NP 问题而且也难以给出十分准确的解。一个比较流行的发启发式方法是 K-SVD 方法。

稀疏字典学习已经被应用在一些环境下。在分类当中被应用于决定之前不可见的数据属于哪一类的问题。假设有一个每个分类都已知的字典，那么一个新的数据会被对应字典分配到最好的稀疏表示的那一类。稀疏字典学习也被应用于图像去噪。该方法的关键思想是一个未被污染的图像块可以通过图像字典来稀疏表示，但噪声却不能。

12. 遗传算法

遗传算法（Genetic Algorithm）是模拟达尔文生物进化论的自然选择和遗传学机理（通过基因的变异与组合交叉产生新的遗传算子）的生物进化过程的计算模型，是一种通过模拟自然进化过程搜索最优解的方法。遗传算法曾在 20 世纪 80、90 年代被用于机器学习。反过来，机器学习也被用于改进遗传算法。

注释

1. Machine learning explores the study and construction of algorithms that can learn from and make predictions on data. Such algorithms operate by building a model from example inputs in order to make data-driven predictions or decisions, rather than following strictly static program instructions.

译文：机器学习理论主要是分析和设计一些可以从数据中学习和预测的算法。这些算法通过样本输入来建立模型，并以此来做出以数据为导向的预测或决定，而不是严格地执行固定的程序。

分析：英语句法要求一个句子只能有一个谓语动词（并列谓语必须有连词），其他动词必须变成非谓语形式。而汉语中使用动词比英语多，也比较灵活，有时一个短句子就用了好几个动词。因此英语中名词、形容词、介词和副词均可转译为汉语动词。原文中第一句使用了较多的名词，例如 study、construction。因此在翻译的时候将其转译为分析和设计两个动词。

2. Machine learning is closely related to and often overlaps with computational statistics; a discipline which also focuses in prediction-making through the use of computers. It has strong ties to mathematical optimization, which delivers methods, theory and application

domains to the field.

译文：机器学习与致力于用计算机做预测的计算统计学关系密切，两者经常会有交叉与重合。机器学习与最优化也有着紧密的联系，后者不仅为其提供了理论方法还提供了应用的方向。

分析：It 作人称代词时，通常用来代替刚提到过的一个具体事物。翻译时一般有三种译法：一是仍译成代词"它"；二是重复译法；三是省略译法。英语中在同一句中或同一段内除非出于强调语气的需要或避免引起误解，一般不重复名词，所以代词用得比较多；而汉语中，代词用得比较少。因此在本段中第二句的 it 为了不与前一句说到的计算统计学相混淆，便采用了重复译法。

3. Within the field of data analytics, machine learning is a method used to devise complex models and algorithms that lend themselves to prediction. These analytical models allow researchers, data scientists, engineers, and analysts to "produce reliable, repeatable decisions and results" and uncover "hidden insights" through learning from historical relationships and trends in the data.

译文：在数据分析领域，机器学习被视为一种设计依靠自身预测的复杂模型和算法的方法。研究员、数据学家、工程师以及分析师可以依靠这些分析模型"得出可依靠、可重复的决策和结果"并通过数据历史的关系和趋势来发现"未观测到的见解"。

分析：英语中被动语态使用广泛，尤其科技作品使用更多。汉语虽然也有被动语态，但使用面窄得多。因此在翻译科技作品时，英语被动语态一般翻译成汉语主动语态。只是在特别强调被动动作或者特别突出被动者时才译成汉语被动句。本段第一句便是把 used to 的被动翻译出来。

4. Tom M. Mitchell provided a widely quoted, more formal definition: "A computer program is said to learn from experience E with respect to some class of tasks T and performance measure P if its performance at tasks in T, as measured by P, improves with experience E". This definition is notable for its defining machine learning in fundamentally operational rather than cognitive terms, thus following Alan Turing's proposal in his paper "Computing Machinery and Intelligence" that the question "Can machines think?" be replaced with the question "Can machines do what we (as thinking entities) can do?"

译文：Tom M. Mitchell 给出了一个广泛引用并且更正式的定义："在度量 P 下，如果一个计算机程序在任务 T 中的表现通过经验 E 提升了，则该程序被称为从经验 E 中学习到了在某类任务 T 和表现的度量 P 下的知识。"这个定义之所以出名是因为它通过基本流程来定义了机器学习而不是经验性的词汇，因此之后 Alan Turing 在其论文"计算机器与智能"中提到的问题"机器可以思考吗？"被替换为"机器可以做我们（可以思考的实体）做的吗？"。

分析：语序是指句中各成分的恰当位置。汉语是分析性语言，不是靠词形变化，而是靠语序表示各成分之间的关系，因此语序是严格的、固定的。英语是正在演变的综合性语言，语序上比较灵活。此外，两种语言在语序上的差别跟民族习惯及思维逻辑也有关系。语序上的错误会造成概念上的错误。英语中，主语+谓语+宾语，这样的语序与汉语基本相同。但定语和状语则有不同，例如本段中的第一句中描述 Tom M. Mitchell 给出的定义时，有很多定语与状语。在翻译中，对语序做出了较大改动。

5. Machine learning is a subfield of computer science that evolved from the study of pattern recognition and computational learning theory in artificial intelligence. In 1959, Arthur Samuel defined machine learning as a "Field of study that gives computers the ability to learn without being explicitly programmed".

译文：机器学习是计算机科学的子领域。它是由人工智能中的模式识别和计算学习理论衍化而来。Arthur Samuel 于 1959 年将机器学习定义为"一个可以在没有精确指令的情况下赋予计算机学习能力的研究领域"。

分析：限制性定语从句是句中不可缺少的部分，如果除去它整个句子就可能失去意义。因此在翻译时通常省去关系代词，把定语从句放在其所修饰的名词之前，如同本段对第一句中机器学习的翻译。

第 13 章 DSL 是如何工作的

13.1 概述

当连接到因特网的时候，可以通过一个调制解调器来连接，或通过经过办公室中的区域网连接，或通过一个电缆调制解调器连接，或通过数字用户线路（DSL）连接。DSL 是一个高速的连接，它和一般电话使用的是一样的电线。

DSL 有以下这些优点：
- 能够既与因特网保持连接，又能使用电话线打电话。
- 它的速度比一般的调制解调器高许多（1.5 Mbps 与 56 kbps）。
- DSL 不需要新的配线，只需使用已有的电话线。
- DSL 公司通常会在安装时提供调制解调器。

但它也有以下缺点：
- 当比较接近供应商的服务站时，DSL 将会工作得更棒。
- 在因特网上，它接收数据速度比发送数据的速度快。
- 并不是到处都有这样的服务。

在这篇文章中，我们解释 DSL 如何设法通过标准电话线发送和接收更多的信息——并且能让你在线时还能打电话。

13.2 电话线

在美国正常安装电话，电话公司需要在你的家里安装一对铜线。铜电线提供比你打电话所需的更多的带宽——它们拥有非常大的带宽，或者比传播声音所需的频率范围更广的频率。DSL 开发这"额外的容量"使得信息顺利传送的同时不影响人们通过电话交谈。整个的计划是基于为特殊任务匹配特定的频率的。

为了了解 DSL，首先需要知道一些关于一个普通的电话线的知识——普通老式电话业务，电话专业人士称为 POTS。POTS 充分利用公司电话线及设备的方法之一是限制开关、电话及其他设备的频率。人们在普通对话中声音的频率是从 0~3400Hz 的。这是频率的一个很小的范围。比如，大多数立体声扬声器的范围就有 20~20 000Hz。而且电线本身在大部分的情况下都有处理高达百万赫兹的频率的潜能。如此小部分地使用电线的总带宽已成为历史——请记住，电话系统通过一对铜线连接到每家每户已有一个世纪的历史了。由于频率被限制，电话系统只需一个较小的空间就能容纳很多的电话线，而不用担心电话线之间会有冲突。传送数字数据而不是模拟数据的现代仪器能安全地使用电话线更多的容量。

13.3 非对称数字用户线路

大多数的家庭和生意人使用非对称数字用户线路（ADSL）来连接。ADSL 在一条线路中划分可用频率，假设大多数用户查看或下载比发送或上传需要更多信息。如果从因特网到使用者的连接比从使用者返回到因特网连接速度快 3 到 4 倍，那么使用者将会得到最大的好处（大部分的时候）。

其他类型的 DSL 包括：

- 高速数字用户线路（VDSL）——这是一个快速的连接，但是只在短距离内工作。
- 对称数字用户线路（SDSL）——它主要用于小型业务，不允许同时使用电话，但是接收和发送数据的速度是相同的。
- 可调整的 DSL（RADSL）——这是 ADSL 的一种变种，但是调制解调器可根据电话线的长度和质量调整连接的速度。

13.4 距离限制

你到底能得到多少好处取决于你和提供 ADSL 服务的公司的中央办公室的距离。ADSL 是一种对距离敏感的技术：随着连接距离增加，信号质量将降低，连接速度也变慢。ADSL 服务的最大范围是 18 000 英尺（5460 米），但是为了服务器速度和质量能更好，ADSL 的提供者通常会对速度有个更小范围的限制。在最远端的 ADSL 用户的速度将会远低于承诺的最大速度，而靠近中央的用户速度将更快，并且在将来可能会非常的快。ADSL 技术为大约 6000 英尺（1820 米）远的用户提供的下行速率是 8 Mbps（因特网到用户），上行速率是 640 kbps。实际上现在普遍提供的最好的是下行速率 1.5 Mbps，上行速率 64~640 kbps。

你可能有疑问，如果距离是 DSL 的限制，为什么它不也是对电话的限制？原因在于电话公司用来推进声音信号的一个小放大器，叫加感线圈。很不幸，这些加感线圈与 ADSL 的信号不兼容，因此，在你的电话和电话公司的中央办公室之间的加感线圈将使你无法从 ADSL 接收信息。其他可能使你无法从 ADSL 接收信息的原因有：

- 桥接抽头——这些是你与中央办公室之间的扩展，将服务扩展给其他客户。如果你不注意，这些普通电话服务里的桥接抽头，可能使回路的总长度超过服务提供者的限制的距离。
- 海底光缆——如果你的电话回路有一部分是通过海底光缆的，那么 ADSL 信号将不能从模拟量转到数据量再转变到模拟量。
- 距离——即使你知道你的中央办公室在哪里（如果你不知道也不用感到惊讶——电话公司不会做广告，告诉大家自己的位置），在地图上不会指示出信号是如何在你的房子和服务台之间传送的。

13.5 分离信号（CAP）

目前有两种相互竞争、互不兼容的 ADSL 标准：一种是官方的 ANSI（T1.143）标准，称为离散多音频（DMT）。由于仪器制造商的问题，现在大部分 ADSL 仪器都是使用 DMT 的。另一种是更早更容易执行的无载波调幅/调相（CAP）技术，它在早期 ADSL 的安装上使用较多。

CAP 将电话线上的信号分为 3 个波段：声音交谈从 0~4000Hz 波段，如同所有 TOPS 回路中一样；上行频道（从使用者到服务器）从 25Hz~160kHz 波段；下行频道（从服务器到使用者）从 240Hz 开始，上限由很多因素决定（线长度、线噪音、某家电话公司的使用者数量），但最大不超过 1.5MHz。这个将 3 个频道分开的系统，将一条线上的不同频道的冲突或不同线上的信号冲突，都减少到最小。

13.6 分离信号（DMT）

DMT 也把信号分为单独的频道，但不是分为两个非常宽广的频道分别用来传送上行和下行的数据。相反地，DMT 把数据分为 247 个单独的频道，每个 4kHz。

你可以把它想象成电话公司把你的铜线分为 247 条 4kHz 的线，并且每条附加一个调制解调器，这样就等价于有 247 个调制解调器连接到你的计算机。每个频道都被监控，如果某条的信号很弱，那么信号就被转移到另一条上。这个系统不停地将信号在不同的频道间转换，直到找到发送和接收的最好频道。另外，一些较低的频道（从 8kHz 开始），都是双向使用的：上行和下行。监控并挑出双向频道的信息，保持 247 条的质量，而不是给不同质量的线更大的弹性，使得 DMT 变得比 CAP 更具有灵活性。

13.7 DSL 设备

ADSL 使用两个仪器：一个在客户终端，另一个在因特网服务提供商、电话公司或者其他 DSL 的提供者上。在客户那有一个 DSL 无线电收发机，它还能提供其他的服务。DSL 服务提供者有一个数字用户线接入复用器（DSLAM）来接收用户的连接。

在下一段文章中，我们来看一看这两个仪器。

13.7.1 DSL 设备：无线电收发器

大多数的用户称他们的 DSL 无线电收发机为"DSL 调制解调器"。电话公司的工程师或因特网服务提供商（ISP）则称它为 ATU-R。

不管怎么叫，它都是用户计算机或网络连接到 DSL 线的点。

虽然大多数用户安装了 USB 或 10 base-T Ethernet connections，无线电收发机能以多种方式连接到客户的仪器上。网际服务和电话公司销售大部分的 ADSL 无线电收发机是简单的无线电收发机,而公司用的设备可在同一平台上联结网络路由器、网络开关或其他网络设备。

13.7.2　DSL 设备：DSLAM

通路提供者的 DSLAM 是使得 DSL 实现的仪器。DSLAM 连接许多客户并把他们通过一个单独的、高容量的连接连到因特网。DSLAM 一般都有弹性,可支持多个单一中央办公室的多种类型的 DSL,以及同种类型的 DSL 中的多种协议和调制——比如 CAP 和 DMT。除此之外, DSLAM 还有一些附加功能,如路由、用户的动态地址分配。

经过 ADSL 和 cable modems 的 DSLAM 有一个最主要的不同。由于 cable modems 的使用者通常是一个邻近地区共享一个网络回路,用户的增加将意味着一些场合的性能会降低。而 ADSL 为每个使用者返回 DSLAM 提供一个专门的连接,这意味着使用者的增加不会使性能降低——直到总用户数在单个高速因特网连接上面达到饱和。那时,服务提供者将升级,为所有连接到 DSLAM 的用户提供更多的性能。

注释

1. The copper wires have lots of room for carrying more than your phone conversations—they are capable of handling a much greater bandwidth, or range of frequencies, than that demanded for voice.

译文:铜电线提供比你打电话所需的更多的带宽——它们拥有非常大的带宽,或者比传播声音所需的频率范围更广的频率。

分析:次句的翻译关键在抓住主干,并且注意一些在英文中省略,但中文又不得不交代的意思。例如,demanded for voice 字面上省略了传输声音的意思。

2. Other types of DSL include:
- Very high bit-rate DSL (VDSL)—This is a fast connection, but works only over a short distance.
- Symmetric DSL (SDSL)—This connection, used mainly by small businesses, doesn't allow you to use the phone at the same time, but the speed of receiving and sending data is the same.
- Rate-adaptive DSL (RADSL)—This is a variation of ADSL, but the modem can adjust the speed of the connection depending on the length and quality of the line.

原译:

其他类型的 DSL 包括:
- 高速数字用户线路(VDSL)——这是一个快速的连接,但是只在短距离内工作。
- 对称数字用户线路(SDSL)——它被小生意使用,不允许你同时使用电话,但是接收和发送数据的速度是相同的。
- 估价-适应的 DSL(RADSL)——这是 ADSL 的一种变化,但是调制解调器可根据电话线的长度和质量调整连接的速度。

译文：

其他类型的 DSL 包括：

- 高速数字用户线路（VDSL）——这是一个快速的连接，但是只在短距离内工作。
- 对称数字用户线路（SDSL）——它主要用于小型业务，不允许同时使用电话，但是接收和发送数据的速度是相同的。
- 可调整的 DSL（RADSL）——这是 ADSL 的一种变种，但是调制解调器可根据电话线的长度和质量调整连接的速度。

分析：这里谈 3 个词语的分析。by small businesses 翻译为小型业务为好，说小生意所指范围太窄；rate-adaptable 其实就是可调整，不用想太多；还有 variation 一般的意思是各种各样的，但可不要被迷惑，这里是"变种"的意思。

3. At the extremes of the distance limits, ADSL customers may see speeds far below the promised maximums, while customers nearer the central office have faster connections and may see extremely high speeds in the future. ADSL technology can provide maximum downstream (Internet to customer) speeds of up to 8 megabits per second (Mbps) at a distance of about 6,000 feet (1,820 meters), and upstream speeds of up to 640 kilobits per second (kbps).

译文：在最远端的 ADSL 用户的速度将会远低于承诺的最大速度，而靠近中央的用户速度将更快，并且在将来可能会非常的快。ADSL 技术为大约 6000 英尺（1820 米）远的用户提供的下行速率是 8 Mbps（因特网到用户），上行速率是 640 kbps。

分析：此段翻译得比较好，一些用词比较到位，At the extremes of the distance limits 采用了意译的方法译为"最远端"。

第 14 章 Internet 基础结构

Internet 的妙处之一是不真正属于任何人，而是全球大小网络的集合。这些网络用多种不同方式连接起来，构成所谓 Internet 的单一实体。事实上，这个名称就是指互联网络。

Internet 始于 1969 年，从 4 台主机增长到几千万台主机。但是，虽然 Internet 不真正属于任何人，但仍然以不同方式得到监管和维护。Internet 学会是 1992 年成立的非营利组织，监视相关政策与协议的建立，确定 Internet 如何使用和交互。

每个连接 Internet 的计算机都是网络的一部分，即使你的家用计算机也是。例如，可能用 Modem 拨一个市话，连接 Internet 服务提供商（ISP）。上班时，你可能在局域网（LAN）中，但通常还是用公司连接的 ISP 连接 Internet。连接 ISP 时，就成为其网络的一部分。然后 ISP 可以连接大型网络，成为其网络的一部分。Internet 就是网络的网络。

大多数大型通信公司都有自己的专用主干，连接不同地区。在每个地区，公司有一个存在点（POP）。本地用户从 POP 访问公司的网络，通常使用市话线或专用线。这里的奇怪之处是没有总体控制的网络，而是几个高速网络通过网络访问点（NAP）相互连接。

当你连接到因特网时，你的计算机就成为了网络的一部分。

14.1 网络例子

下面举一个例子。假设公司 A 是个大型 ISP，在每个大城市，都有一个 POP。每个城市的 POP 是个机架，有许多 Modem，让 ISP 的客户拨入。公司 A 从电话公司租用光缆线，将 POP 连接起来。

假设公司 B 是公司 ISP，在各大城市建有大楼，公司把 Internet 服务器放在这些大楼中。公司 B 很大，大楼之间有自己的光缆，因此它们都是互联的。

在这个布局中，公司 A 的所有客户可以相互通信，公司 B 的所有客户可以相互通信，但公司 A 的客户与公司 B 的客户无法相互通信。因此，公司 A 和 B 同意连接不同城市的 NAP，两个公司间的通信流在 NAP 处的网络间流动。

在实际的 Internet 中，几十个大型 Internet 提供者在各个城市的 NAP 处互联，其中有各个网络间的几万亿字节数据流。Internet 是大公司网络的集合，协定在 NAP 处进行所有相互通信。这样，Internet 上的每台计算机都能相互连接。

14.2 沟通分割

所有这些网络利用 NAP、主干和路由器相互通信。在此过程中令人惊异的是，消息可以在几分之一秒内从一台计算机经过多个不同网络传递到地球另一边的计算机上。

路由器确定将一台计算机发往另一计算机的信息发到哪里。路由器是专用计算机，发送每个 Internet 用户的消息，使其沿几千种路径到达目的地。路由器有两个独立而相关的任务：

- 保证信息不发到不需要的地方。这样可以避免大量数据使"无辜旁人"的连接产生拥塞。
- 保证信息到达所要目的地。

执行这两个任务时，路由器对处理两个独立计算机网络极为有用。它连接两个网络，从一个网络向另一网络传递信息。它还在网络间提供保护，防止一个网络的通信流不必要地溢出到另一网络。不管连接多少个网络，路由器的基本操作与功能都是相同的。由于 Internet 是由几万个小网络构成的大网络，因此路由器的使用是必不可少的。

14.3 主干网络

美国国家科学基金会（NSF）于 1987 年建立了第一个高速主干网，称为 NSFNET，是 T1 线路，把 170 个小网连接起来，以 1.544Mbps（每秒百万位）速度操作。IBM、MCI 与 Merit 公司和 NSF 一起建立这个主干网，于次年开发了 T3（45Mbps）主干网。

主干网通常是光纤主干线路，主干线路把多个光缆结合起来，提高容量。光缆指定 OC，表示光纤载波，如 OC-3、OC-12 或 OC-48。OC-3 线路的传输速度是 155Mbps，而 OC-48 线路的传输速度是 2488Mbps（2.488Gbps）。与此相对比，典型 56K modem 只能传输 56 000bps。由此可见，现代主干网多么快捷。

如今，许多公司有自己的高容量主干网，通过全世界的各个 NAP 相互连接。这样，Internet 上的每个人都能和地球人其他人通信，不管他们在哪里，利用哪个公司的服务。整个 Internet 都遵从巨大的和不断增大的公司间自由互联协定。

14.4 Internet 协议：IP 地址与域名系统

Internet 上的每台机器有唯一的标识号，称为 IP 地址。IP 表示 Internet 协议，是计算机在 Internet 上通信时使用的语言。协议是从使用者与服务通信时要用的预定语言。使用者可以是人，但通常是 Web 浏览器之类的计算机程序。

典型的 IP 地址如下：

216.27.61.137

为了便于记住 IP 地址，通常表示成上述十进制格式，称为点号十进制数。但是，计算机是以二进制形式通信的。这个 IP 地址的二进制形式如下：

11011000.00011011.00111101.10001001

IP 地址中的 4 个数称为 8 位位组，因为其二进制形式各有 8 位。如果把所有位相加，则得到 32，因此 IP 地址是个 32 位数。由于 8 位中每一位各有两种状态（1 或 0），因此总的组合数为 2^8=256。因此，每个 8 位位组可以包含 0～255 的值，4 个 8 位位组组合起来，可以得到 2^{32} 或 4 294 967 296 个不同值。

在这 43 亿种组合中，有些值只限于作为特定 IP 地址。例如，IP 地址 0.0.0.0 留作默认网络，而 255.255.255.255 留作广播地址。

8 位位组不仅分开数字，而且可以生成 IP 地址类，根据对方规模分配给某个公司、政府或其他实体。8 位位组分成两个部分：网络与主机。网络部分总是包含第一个 8 位位组，用于标识计算机所属的网络。主机（也称为节点）标识网络上的实际计算机。主机部分总是包含最后一个 8 位位组。IP 类有 5 个，还有一些特殊地址。

在 Internet 发展的初期，只是少量计算机通过 Modem 和电话线连接起来，只能在连接时提供要连接的计算机的 IP 地址。例如，典型 IP 地址。例如，典型 IP 地址可能是 216.27.22.162，如果又有几台主机，这样做是可以的，但随着联机系统不断增加，渐渐会出问题。

要解决这个问题，第一个方案是由网络信息中心维护简单文本文件，将名称映射到 IP 地址。这个文本文件很快会变得很大，很难管理。1983 年，威斯康星大学建立了域名系统（DNS），能自动将名称映射到 IP 地址。

14.5 统一资源定位器

使用 Web 或发送电子邮件消息时，用的是域名。例如，统一资源定位器（URL）。http://www.Yiren.com 包含域名 Yiren.com，下列电子邮件地址也是：example@Yiren.com。每次使用域名时，要用 Internet 的 DNS 服务器将可读域名变成机器可读的 IP 地址。

顶级域名也称为一级域名，包括.COM、.ORG、.NET、.EDU 与.GOV。每个顶级域中有大量二级域。例如，在一级域 COM 中，包括：

- Yiren。
- Yahoo。
- Microsoft。

一级域 COM 的每个名称要唯一。最左边的单词（如 WWW）是主机名，指定域中特定主机的名称（具有特定 IP 地址）。给定域中可能有几百万个主机名，只要它们在域中全部唯一。

DNS 服务器从程序和其他名称服务器中接受请求，将域名变成 IP 地址。收到请求时，DNS 服务器可以做 4 件事：

- 答复 IP 地址请求，因为它已经知道所请求域的 IP 地址。
- 联系另一个 DNS 服务器，试找到所请求名称的 IP 地址，可以进行多次。
- 可以让它不知道请求域的 IP 地址，但能提供一个知识更丰富的 DNS 服务器的 IP 地址。
- 可以返回一个错误消息，因为请求的域名无效或不存在。

14.6 客户机、服务器与端口

Internet 服务器实现了 Internet，Internet 上的所有机器都是服务器或客户机。向其他机器提供服务的机器称为服务器，而连接这些服务的机器称为客户机。Web 服务器、电子邮件服务器、FTP 服务器等满足了全球所有 Internet 用户的需求。

连接 www.Yiren.com 页面进行浏览时，你是客户机端的用户，访问 Yiren 公司的 Web 服务器。服务器机器找到请求的页面并发送给你。客户机访问服务器时是有特定意图的，因此客户机将请求定向到服务器机器上运行的特定软件服务器。例如，如果机器上运行 Web 浏览器，则与服务器机器上的 Web 服务器通信，而不是与电子邮件服务器通信。

服务器使用静态 IP 地址，不经常改变。另一方面，通过 Modem 拨号的家用计算机则通常在每次拨号时由 ISP 分配一个 IP 地址。这个 IP 地址在本次会话中唯一，但下次拨号时可能不同。这样，ISP 只要对支持的每个 Modem 提供一个 IP 地址，而不必对每个客户提供一个 IP 地址。

所有服务器都使用编号端口提供服务，服务器上提供的每个服务器有一个端口。例如，服务器机器运行 Web 服务器和 FTP（文件传输协议）服务器时，Web 服务器通常在端口 80 提供，而 FTP 服务器通常在端口 21 提供。客户机连接特定 IP 地址和特定端口号的服务。

客户机连接特定端口的服务后，用特定协议访问这个服务。协议通常是文本，只是描述客户机与服务器如何进行会话。Internet 上的每个 Web 服务器都遵循超文本传输协议（HTTP）。

网络、路由器、NAP、ISP、DNS 和强大的服务器使 Internet 得以实现。所有这些信息在几个毫秒的时间内就可以在全球发送，真是让人惊奇。这些组件对现代生活非常重要，如果没有它们，就没有 Internet。而如果没有 Internet，我们的生活就会黯然失色。

注释

1. The Internet Society, a non-profit group established in 1992, oversees …

译文：Internet 学会是 1992 年成立的非营利组织，监视相关政策与协议的建立，确定 Internet 如何使用和交互。

2. Company B is such a large company that it runs its own fiber optic lines between its buildings so that they are all interconnected.

译文：公司 B 很大，大楼之间有自己的光缆，因此它们都是互联的。

3. Therefore, Company A and Company B both agree to connect to NAPs in various cities, and traffic between the two companies flows between the networks at the NAPs.

译文：因此，公司 A 和 B 同意连接不同城市的 NAP，两个公司间的通信流在 NAP 处的网络间流动。

4. What is incredible about this process is that a message can leave one computer and travel halfway across the world through several different networks and arrive at another computer in a fraction of a second!

译文：在此过程中令人惊异的是，消息可以在几分之一秒内从一台计算机经过多个不同网络传递到地球另一边的计算机上。

5. Routers are specialized computers that send your messages and those of every other Internet user speeding to their destinations along thousands of pathways.

译文：路由器是专用计算机，发送每个 Internet 用户的消息，使其沿几千种路径到达目的地。

6. Called NSFNET, it was a T1 line that connected 170 smaller networks together and operated at 1.544 Mbps (million bits per second).

译文：称为 NSFNET，是 T1 线路，把 170 个小网连接起来，以 1.544Mbps（每秒百万位）速度操作。

7. An OC-3 line is capable of transmitting 155Mbps while an OC-48 can transmit 2488Mbps (2.488Gbps).

译文：OC-3 线路的传输速度是 155Mbps，而 OC-48 线路的传输速度是 2488Mbps（2.488Gbps）。

8. The entire Internet is a gigantic, sprawling agreement between companies to intercommunicate freely.

译文：整个 Internet 都遵从巨大的和不断增大的公司间自由互联协定。

9. When the Internet was in its infancy, it consisted of a small number of computers hooked together with modems and telephone lines.

译文：在 Internet 发展的初期，只是少量计算机通过 Modem 和电话线连接起来。

10. Every time you use a domain name, you use the Internet's DNS servers to translate the human-readable domain name into the machine-readable IP address.

译文：每次使用域名时，要用 Internet 的 DNS 服务器将可读域名变成机器可读的 IP

地址。

11. A given domain can, potentially, contain millions of host names as long as they are all unique within that domain.

译文：给定域中可能有几百万个主机名，只要它们在域中全部唯一。

12. Clients that come to a server machine do so with a specific intent, so clients direct their requests to a specific software server running on the server machine.

译文：客户机访问服务器时是有特定意图的，因此客户机将请求定向到服务器机器上运行的特定软件服务器。

13. A home machine that is dialing up through a modem, on the other hand, typically has an IP address assigned by the ISP every time you dial in.

译文：另一方面，通过 Modem 拨号的家用计算机则通常在每次拨号时由 ISP 分配一个 IP 地址。

14. If a server machine is running a Web server and a file transfer protocol (FTP) server, the Web server would typically be available on port 80…

译文：服务器机器运行 Web 服务器和 FTP（文件传输协议）服务器时，Web 服务器通常在端口 80 提供。

第 15 章 网络搜索引擎工作原理

对于因特网和它的重要组成万维网来说，好处是那里有着数以百万计的网页，包含各种各样的信息供人查阅，然而不幸的是，这些成千上万的网页中，大部分的标题源于其作者的突发奇想，几乎所有的网页在其服务器上都有个怪名字。那么，要如何找到所需要的网页呢？大多数人都会去求助于网络搜索引擎。

网络搜索引擎是一个特殊的站点，它是为了帮助人们查询所需的在其他站点上的信息而设计的。不同的搜索引擎工作的方式是有不同的，但它们都包含以下 3 部分内容：

- 它们根据搜寻的关键字搜索整个网络，或网络的一部分。
- 它们把所找到的关键字，以及它所出现的位置，作为索引保存下来。
- 它们允许用户在该索引表中查询各种关键字或关键字的组合。

早期的搜索引擎能够搜索几十万个网页和文档，每天大约为一两千个请求提供服务。现在，一个顶级的搜索引擎能够搜索上亿的网页，并且每天响应数千万的搜索请求。在这篇文章里，我们将告诉你这些主要的工作是如何实现的，以及搜索引擎是如何把信息汇总起来以帮助你在网站上找到所需的信息。

15.1 搜索网络

当大多数人谈到搜索引擎时，他们所指的是万维网的搜索引擎。在万维网成为因特网最为重要的组成部分之前，已经有搜索引擎在帮助人们在网上搜寻所需要的信息了。一些像 gopher、archie 的程序记录了存储在网络服务器上的文件的索引，并极大地减少了查询程序和文档所需的时间。在 20 世纪 80 年代晚期，能在网络上得到有价值的东西意味着要知道如何使用 Gopher、Archie、Veronica 和其他一些搜索引擎。

如今，大多数网络用户都局限于网页的搜索，所以我们这篇文章集中讲述网页搜索引擎。

1．一个简单的开始

在告诉你所寻找的文档的位置之前，搜索引擎首先要找到它。要在上百万的网页中找到所需的信息，搜索引擎需要一个叫做"蜘蛛"的特别的软件"机器人"来帮助它建立在网站上搜索到的关键字的列表。蜘蛛建立列表的过程成为"网络爬行"。为了建立并维护一个有用的关键字列表，搜索引擎的蜘蛛们需要查询大量的网页。

蜘蛛是如何开始它的网络爬行的呢？通常的起点是那些最常用的服务器和最流行的网页。蜘蛛会从一个流行的站点开始，将它网页上的关键字建立列表并且跟踪在这个网页上的每个链接。通过这个方式，蜘蛛系统迅速地开始扩展，在网络上最为常用的网页上形成广泛的分布。

Google 网最初是一个学术搜索引擎。在一篇描述该系统是如何建立的论文中，Sergey Brin 和 Lawrence 举了一个例子来说明他们的蜘蛛工作的速度有多快。在最初建立的系统中，他们用了多个蜘蛛，通常是 3 个同时使用，每个蜘蛛能够保证一次在一个打开的网页上记录下 300 个链接。最好性能下，在用 4 个蜘蛛的时候，他们的系统能在每秒搜索超过 100 个网页，生成大约 600KB 的数据。

为了保证其快速运行，系统必须为蜘蛛提供必要的信息。早期的 Google 系统有一个服务器专门来为蜘蛛提供地址信息。为了将时间延迟较少到最低，Google 使用自己的域名服务器而不是网络上所提供的服务器。

当 Google 蜘蛛在一个网页上搜索时，它做两个标记：

- 在这个网页上的关键字。
- 这个词是在哪里出现的。

如图 15.1 所示，爬虫采集网页的内容，并创建关键的搜索词，以便在线的用户可以找到他们要寻找的网页。

出现在标题、副标题、Meta 标签和其他一些相关的重要位置的词被特殊标记，提供给用户作为辅助搜索。Google 的蜘蛛把网页上每个重要的词建立列表，忽略那些文章里 a、an、the 之类的冠词。其他的蜘蛛则采取不同的方法。

这些不同的方法通常是要使蜘蛛运行得快一些，或者搜寻到的信息更为有效，或两者皆是。比如，有的蜘蛛将出现在标题、副标题、链接上的词以及这个网页上最常用的 100 个词和文章前 20 行的每个词都保存下来。据说 Lycos 就是采用这种蜘蛛来搜索网络的。

其他的系统，比如 AltaVista，用的是另一种方法。它把网页上每个字，包括冠词和其他"无关紧要的"词都建立列表。之所以采用这种完全记录的方法，是为了配合那些想要把隐藏于网页上的 meta 标签也找出来的系统。

2．Meta 标签

Meta 标签允许网页的设计者在将要建立索引的网页上指定关键字和内容。这是很有用的，特别是在该词在那页有双重甚至三重意思时，Meta 标签能指引搜索引擎几种意思中哪个是正确的。然而过度依赖 Meta 标签也存在一些风险，因为一些粗心的或者不道德的网页设计者会在 Meta 标签中加入现在流行的主题，而实际上内容却和那些主题毫不相关。为了防止这种情况，蜘蛛将 Meta 标签和网页的内容相关联，丢弃那些实际上网页上的词不相关的 Meta 标签。

上述这些都是基于网页设计者想要把自己的页面包含在搜索结果的列表上的假设的。很多情况下，网页设计者并不想让他的网页出现在主要搜索引擎的搜索页面上，或者不想蜘蛛访问他的网页。比如一个新建立的游戏，每当打开一个页面或者链接时，都要将其激活。如果一个网络蜘蛛访问了这样的一个页面，并开始跟踪所有的链接，这个游戏会把当前的行为错误地认为是一个速度很快的玩家而超速运转。为了避免这种情况，发展出了 robot 排除协议（robot exclusion protocol）。这个协议在网页开始的 Meta 标签段中执行，告诉蜘蛛不要访问这个网页——既不要对该网页上的字建立索引，也不要跟踪该网页上的链接。

15.2 建立列表

一旦蜘蛛完成了在网络上搜寻信息的工作（我们应该注意到实际上这个工作是不可能真正完成的——网络上的内容是在不断变化的，这意味着蜘蛛也将不停地搜索），搜索引擎必须以某种方式存储信息以保证它的有效性。为了让用户能得到所搜寻到的数据，需要解决两个关键问题：

- 存储在数据上的信息。
- 信息被列表的方法。

最简单的情况下，搜索引擎只要将所搜寻到的关键字以及它所在的地址保存下来。实际上，这样的搜索引擎是用处不大的。因为无从得知这个关键字在这个网页上是重要的还是微不足道的内容，这个关键字出现了几次，这个网页上有没有包含这个关键字的其他网页的链接。换句话说，也就无法将所得到的结果按最符合用户需要的顺序来排列。

为了让结果更有用，大多搜索引擎都不止存储关键字和地址。一个引擎可能会存储关键字在网页中的出现次数。引擎将根据每个条目在文中的位置分配权重，看它是出现在文档中、副标题中、连接中、meta 标签中还是网页的标题中。每个商业搜索引擎都有不同的规则来指定出现在列表中的关键字的重要性。这也是在不同的搜索引擎上搜索同一个关键字得到的列表网页的排列顺序不同的原因之一。

不管搜索引擎将额外的信息块联接得如何紧密，数据都要被编码来节省存储空间。比如 Google 就是用 2 个字节（每字节 8 位）来存储额外信息——如大小写、字体大小、位置和其他一些用来帮助排序的信息。每个信息占用 2~3 位，2 位为一组。这样，大量的信息就可以被很紧凑地存储。当信息被压缩好后，就可以建立索引了。

索引只有一个目的：让信息被尽可能快地被找到。用来建立索引的方法有很多，其中最有效的方法之一是建立哈希表（hash table）。在建立哈希表的时候，用一个公式来将每个词和某一个数值建立映射关系。这个公式是要用预定的数值分界来均匀地分派全局。这种数值的分界不同于字母表上的单词的分界，它也是哈希表效力的关键所在。

在英语中，以某些字母作为开头的单词有很多，相对的，以另一些字母开头的单词很少。比如，你会发现在字典中，以 M 开头的单词部分比以 X 开头的单词部分厚很多。这种不平衡意味着查找一个有比较多以它开头的单词的字母会花费比较多的时间。哈希表将这种不平衡性均匀分布，从而减少了找到条目的平均时间。它还把索引和真实条目分离开来。哈希表保存着被处理过的单词以及指向真实条目的指针。这种有效的索引和存储使它能够快速地得到结果，即使当用户进行一个很复杂的搜索的时候。

15.3 建立搜索

在索引上搜索包括用户向搜索引擎提交搜索信息。这个信息也许很简单，但最少是一个单词。建立一个复杂的搜索需要用到布尔操作，这样允许你提炼和扩展搜索的条件。

布尔操作经常以下列形式出现：

- AND——用 AND 来连接的条件一定要出现在页面或文档上。一些搜索引擎用 "＋" 号来替代单词 AND。
- OR——在页面或文档至少出现用 OR 进行连接中的一种条件。
- NOT——NOT 后所出现的条件一定不能出现在页面或文档中。一些搜索引擎用 "–" 号来替代 NOT。
- FOLLOWED BY——其中一个条件必须服从其他条件。
- NEAR——其中的一个条件必须包含其他条件特定数量的单词。
- Quotation Marks——在 quotation 标记之间的单词被看作短语，而且这个短语必须出现在页面或文档中。

15.4 未来的搜索

布尔操作所定义的搜索是字面上的搜索引擎查询符合用户输入的单词或短语的信息。当输入的词有歧义时就会出现问题。拿 "bed" 这个单词来说，它可以是指睡觉的地方，也可以是指用来种花的地方，也可以是指车库，也可以是指鱼产卵的河床。如果你只关心这其中的一个意思，也许就不想看到其他意思的信息。你可以在搜索时输入不会产生歧义的关键字来避免，但是如果搜索引擎本身能解决这个问题就更好了。

搜索引擎研究的领域之一就是"基于概念"的搜索。这些研究中有一些用统计学来分析包含你所查询的单词或短语的网页，以此来找到其他一些你可能感兴趣的网页。显然，对于"基于概念"的搜索引擎来说，关于每个页面所存储的信息量更大了，对每个搜索所进行的处理也更多了。尽管如此，还是有许多团队致力于改进这类搜索引擎的结果和性能。其他一些团队则致力于另一个称为"自然语言查询"的研究领域。

"自然语言查询"的想法是让你像对一个坐在你身边的人询问你想要知道的问题一样输入问题，而不需要了解布尔操作和复杂的结构。自然语言查询网站 AskJeeves.com 只能查询一些简单的条件，如果要做一个能处理复杂的查询的自然语言查询的搜索引擎，挑战还是很大的。

注释

1. The bad news about the Internet is that there are hundreds of millions of pages

available, most of them titled according to the whim of their author, almost all of them sitting on servers with cryptic names.

原译：关于因特网的坏消息是：这数以百万计的网页中，大多数网页的标题都是由网页设计者们突发奇想编出来的，他们往往表意隐秘。

译文：然而不幸的是，这些成千上万的网页中，大部分的标题源于其作者的突发奇想，几乎所有的网页在其服务器上都有个怪名字。

分析：原译在意思上转义得太过头了，虽然意思上也没什么错，但我们已经见不到其原来的表达方式了。

2. Programs with names like "Gopher" and "Archie" kept indexes of files stored on servers connected to the Internet, and dramatically reduced the amount of time required to find programs and documents.

原译：一些像 Gopher、Archie 的程序记录了那些由存储在网络上的服务器中的文件建立的索引，并极大地减少了查询程序和文档所需的时间。

译文：一些像 Gopher、Archie 的程序记录了存储在网络服务器上的文件的索引，并极大地减少了查询程序和文档所需的时间。

分析：文中有多个 of，各个层次要分清楚：程序记录的是文件的索引，文件又存储在服务器上，服务器则连在网络上（这里就称为网络服务器以使表达简洁）。

3. These different approaches usually attempt to make the spider operate faster, allow users to search more efficiently, or both. For example, some spiders will keep track of the words in the title, sub-headings and links, along with the 100 most frequently used words on the page and each word in the first 20 lines of text. Lycos is said to use this approach to spidering the Web.

译文：这些不同的方法通常是要使蜘蛛运行得快一些，或者搜寻到的信息更为有效，或两者皆是。比如，有的蜘蛛将出现在标题、副标题、链接上的词以及这个网页上最常用的 100 个词和文章前 20 行的每个词都保存下来。据说 Lycos 就是采用这种蜘蛛来搜索网络的。

分析：此句翻译时应注意的是，当一系列短语列举出来时如何表达得顺畅。译文就做了不错的示范。

4. This protocol, implemented in the meta-tag section at the beginning of a Web page, tells a spider to leave the page alone-to neither index the words on the page nor try to follow its links.

译文：这个协议在网页开始的 meta 标签段中被执行，告诉蜘蛛不要访问这个网页——既不要将该网页上的字建立索引，也不要跟踪该网页上的链接。

分析：句子比较长，关键是要抓主干，implemented in the meta-tag section at the beginning of a Web page 是后置定语，tells 才是谓语部分。

5. The engine might assign a weight to each entry, with increasing values assigned to words as they appear near the top of the document, in sub-headings, in links, in the meta tags or in the title of the page.

原译：引擎将为每个条目分配一个重要性的权重，看它是出现在文档中、副标题中、连接中、meta 标签中或者网页的标题中。

译文：引擎将根据每个条目在文中的位置分配权重，看它是出现在文档中、副标题中、连接中、meta 标签中还是网页的标题中。

分析：原译表达不是很清楚，译文采用了增译的办法。

第 16 章 加　　密

互联网正以不可思议的速度发展着，这使企业和消费者都欢欣鼓舞，因为互联网向我们保证它将改善我们生活和工作的方式。然而，人们最关心的是互联网的安全问题，特别是在网络上发送一些敏感信息的时候。

计算机和网络提供了各种信息保护的方法，一个简单但直接实用的方法是把敏感信息存放在类似软盘的移动存储介质中。但最常用的方法是密码保护。给信息编码加密后，只有拥有密钥的人或计算机才能解码。

本文主要介绍有关编码原理和身份鉴定。公共密匙和对称密钥以及哈希算法也会提及。

16.1　关于密钥

计算机加密是以密码学为基础的，而密码学已经具有很长的历史。在数字时代来临前，政府是最大的密码学的应用者，尤其是在军事领域。加密信息的出现可以追溯到罗马帝国时代。现在大部分加密技术依靠计算机完成，因为人为的加密码很容易被计算机破译。

大部分计算机加密系统属于以下两类。

1．对称密钥

使用对称密钥加密的计算机都有一个秘密钥匙（编码器），它能将信息在发送到另一台计算机上之前编码。对称密钥系统要求收发信息双方的计算机都装有该密钥。对称密钥系统本质上是一个密码：两台计算机中的每一台都必须知道。这个密码提供了信息解码的钥匙。

举个例子：你给一个朋友写了一封加密的信件，其中每个字母都用其在英文字母表中后 2 位的字母代替，因此字母 A 用 C 代替，字母 B 用 D 代替。你已经将解码的规则——后移两个字母告诉了那位朋友。你的朋友收到了这个信息并能解读信件。不懂规则的人只能看到一段无意义的符号。

2．公共密钥

在公共密钥加密中，需要公钥和私钥结合。私钥只有发出信息的计算机知道，而公钥提供给要与其安全通信的计算机。要解译加密信息除了要有私钥之外，还需要发出信息的计算机提供的公钥。一个十分通用的公共密钥加密方法叫优化加密法（Pretty Good Privacy，PGP），它几乎可以加密所有信息。更多有关 PGP 的知识可以在 PGP 的网站上了解。

建立一个安全的服务网站需要使用大量的公钥，极为烦琐，需要一种更为通用的方式，于是数字证书应运而生。数字证书是由一个独立、具有极高信誉的数字认证机构颁发的。网络服务器只要拥有数字认证机构颁发的数字证书，那么它就是可信任的，同一把公钥就可在可信任的计算机间通用。

现在常用的公钥加密系统叫加密套接字协议层（Secure Sockets Layer，SSL），它由美国 Netscape 公司开发，SSL 是一种网络安全协议，用于浏览器和网络服务器传送机密信息。SSL 已经成为全球性的网络安全协议 Transport Layer Security（TLS）的一部分。

如图 16.3（图略）所示，挂锁符号可以知道你正在使用加密。

16.2　构造散列数据

公共密钥编码是以散列函数值为基础的，输入计算机的原数据经过散列算法计算输出的就成了散列函数值。本质上，散列函数值就是原数据的概要。如果不知道构造散列函数值的数据，几乎不可能取得原输入数据。举个简单的例子：

输入数值	散列运算法则	散列函数值
10667	输入#×143	1525381

16.3　可信度

在早期，一台计算机要安全发送信息到另一台计算机上，需要将所有数据编码加密，而且只有接收的那台计算机可以破译信息。现在出现了另一种方式——身份验证，它用于鉴别信息来源是否可靠。基本上，如果信息被验证为可信的，接收者就可以知道是谁发出信息，而且那些信息自从产生以来就没经过任何改变。现在这两种方式双管齐下，一起维护网络安全。

下面介绍几种计算机验证身份或信息的方式：

- 口令法——输入用户名和密码是最常用的验证身份法，计算机将登录者输入的用户名与密码和用户注册资料核对，如果不符合，用户就会无法登录。
- 通行卡——通行卡可以是用磁条的简单的卡，也可用是信用卡或一张精密复杂的嵌入式计算机芯片。
- 数字签名——数字签名可以确保电子文档（电子邮件、电子表格、电子文本）的可信度。数字签名标准 DSS 的构造方式是基于使用数字签名算法的公共密钥加密方法（DSA）。DSS 是美国政府的数字签名格式。DSA 算法由一个只有签名人知道的私钥和一个公钥组成。公钥有 4 个部分，这些部分将在本文中论述。数字签名后如果文档中任何东西被改变，将会相应改变数字签名，这将使数字签名无效。

现在有更多先进的身份鉴定方式开始出现在家庭和办公系统中。最新的是生物鉴定法，它通过辨认生理信息来区分个体。

为了确保数据在传送和加密过程中不出错，有两种常用方式：

- 校验和法（Checksum）——可能是最老的数据校验方法。校验和还提供一种验证形式，因为如果数据有任何损坏，都会被它识别出来。校验和以两种方式中的一种来检定信息。以一个校验包长度为 1 个字长为例。1 个字长有 8 位字节，每个字节有 2 种状态，因此会有 256（2^8）种排列组合。如果第一个组合等于 0，那么 1 个字长所存储位的最大值为 255。

- 循环冗余码校验 CRC——CRC 和校验和（Checksum）类似，但 CRC 使用多项式除法来确定循环冗余码校验码的值，其长度一般为 16 或 32 比特。CRC 的优点是它非常精确，一旦单个字节信息出错，CRC 值将会出现不匹配情况，校验和 CRC 都可以避免信息传送时随机错误，但不能很好地保护信息不被恶意攻击。对称和公钥加密技术更为安全。

注释

1. The existence of coded messages has been verified as far back as the Roman Empire. But most forms of cryptography in use these days rely on computers, simply because a human-based code is too easy for a computer to crack.

原译：加密信息的出现被证实早在罗马帝国时代。近代计算机出现后，人为的加密码很容易被计算机破译，所以计算机加密被大规模使用。

改译：加密信息的出现可以追溯到罗马帝国时代。现在大部分加密技术依靠计算机完成，因为人为的加密码很容易被计算机破译。

分析：原译的第一句犯了死译的毛病，第二句又意译得过头了，显得啰唆，并且与第一句连接不紧密。改译后，更明了，两句之间的逻辑连接更好。

2. Think of it like this: You create a coded message to send to a friend in which each letter is substituted with the letter that is two down from it in the alphabet. So "A" becomes "C," and "B" becomes "D". You have already told a trusted friend that the code is "Shift by 2". Your friend gets the message and decodes it. Anyone else who sees the message will see only nonsense.

原译：这样想：你给一个朋友写了一封加密的信件，其中的字母 A 表示字母 C，字母 B 表示 D，你告诉你的朋友读信规则，即读信时每个字母都用其在英文字母表中位置的后 2 位的字母代替（读到字母 A 用 C 代替，字母 B 用 D 代替），遵守此规则才能看懂信，不懂规则的人只能看到一段废话。

改译：举个例子：你给一个朋友写了一封加密的信件，其中每个字母都用其在英文字母表中后 2 位的字母代替，因此字母 A 用 C 代替，字母 B 用 D 代替。你已经将解码的规则——后移两个字母告诉了那位朋友。你的朋友收到了这个信息并能解读信件。不懂规则的人只能看到一段无意义的符号。

分析：原译将原文的逻辑顺序调整，反而更让人费解，且有重复的毛病，一些词语的翻译也不准确。可见，科技英语翻译时不能随意脱离原文的逻辑顺序并全用自己的话表述，而是应该尽量忠实于原文。

3. To decode an encrypted message, a computer must use the public key, provided by the originating computer, and its own private key.

原译：为了解译加密信息，计算机必须用公钥以及自己的私钥，公钥由发信的计算机提供。

改译：要解译加密信息除了要有私钥之外，还需要发出信息的计算机提供的公钥。

分析：改译时对句子的结构进行了调整，更符合中文习惯。

第 17 章　近看 DCE

美国花旗银行正在致力于研究一种集成了操作系统与非网络服务的称作分布计算环境（DCE）的技术，以获得系统间的透明连接。利用 DCE，花旗银行的开发人员可以将精力集中于设计应用程序本身，而不是集中精力设计应用程序的启动程序。

DCE 是开放软件基金会（包括 OSF）于 1990 年 5 月宣布的，它是一种可能使很多用户中止向系统和网络供应商购买使用许可的技术。大多数重要的公司都已承诺采用 DCE 作为未来系统与网络软件的基础。

目前，DCE 由一套服务组成，它们分为两类：

第一类是基础的分布服务，向程序员提供开发分布计算应用程序必要的工具。它们中有线程、远程过程调用（RPC）、目录、时间与安全服务。

另一类是数据共享服务，向程序员提供开发建立与网络有效信息利用之上的能力。它们包括一个分布文件系统和由 MS-DOS 文件与打印机支持服务组成的个人计算机的集成。

将来，OSF 计划增加假脱机、事务处理与分布式面向目标环境等功能。

17.1　公共线程

DCE 的线程服务提供了可移植功能，让程序员编制能同时完成多个动作的应用程序。例如，一个线程执行 RPC 的同时，另一个线程可以处理用户输入。与之相反，应用程序通常与单线程控制打交道。线程服务也被 DCE 的其他成分使用，如 RPC、安全、命名、时间和分布文件系统等。

线程服务也包括了在单个进程中生成和控制多线程执行的操作，以及在一个应用程序内使全局数据进行同步存取。

这个线程能力，例如在 RPC 的上下文关系中特别重要。从本质上讲，RPC 是同步操作。客户机发出对远程功能的调用，然后等待，直到这个请求被完成。然而，利用多线程，一个线程可以发出请求，而另一个线程开始处理来自不同请求的数据。线程技术能大大地改善一个分布应用程序的性能。

线程服务对程序员技能的要求比其他技术，如显式异步操作或共享存储器要低些。异步接口虽然在有些环境中已存在了一些时候，但它需要的再培训可能会造成很大的成本开支，一项新技术所需的重新培训应该越少越好。

17.2 远程调用

远地过程调用（RPC）是实现分布处理，经实践证明可取的方式之一。其功能是使一个应用程序中的过程在网络中任何一处的计算机上运行。

RPC 处理分布应用最基础的工作，如调用的语义、客户机与服务器的结合或者通信故障。理论上，程序员不必成为通信专家就能编写分布的网络应用程序。程序员使用接口规范语言，详述远程操作。然后对这些子程序进行编译产生既可供客户机，又可供服务器运行的程序代码。

RPC 这类方法的好处是使程序员工作得以简化。RPC 尽可能地保持接近本地过程模式，同时以简洁的方法提供应用程序的分布式功能。换言之，它给开发人员少减少概念上的变化，从而减少重新培训的时间。这对公司内部的开发队伍来说尤为重要。

不管采用何种传输协议，RPC 在应用程序中都能提供一致的表现形式并使系统连接管理不显露出来。这意味着开发人员不必重写应用程序以支持不同的传输服务。RPC 接口同时支持各种传输，并能引入新的传输与协议而不影响应用程序的编写。

17.3 目录服务

在分布网络中确定目标位置是 DCE 的分布命名服务（DNS）的任务，目标包括用户、服务器、数据和应用程序等。此服务能使程序员和最终用户按名字识别诸如服务器、文件、磁盘或打印队列等资源，而无须知道它们在网络中的位置。

OSF 为解决单元间的和全球范围内的通信规定了 DNS 的两层体系结构。这里所谓的单元是为 DCE 中系统而定的组织机构的基本单元。它们可以是社会的、政治的或是组织机构的区域范围，由必须频繁相互通信的计算机组成。如工作组、公司部门间的相互通信。通常一个单元中的计算机在地理上是靠得很近的，每个单元可以由两台到数千台计算机组成。

DCE 的 DNS 有 4 个部分：单元目录服务（CDS）、整体目录代理（GDA）、整体目录服务（GDS）和 X/Open 目录服务（XDS）。

DCE 的 CDS 处理来自单元中客户机的目录查询。例如，它看一下文件名的第一部分，确定一下文件名的第一部分，确定一下所需数据是否在此单元中。如果在，它便提供数据；如果不在，它把此请求传给 GDA，由 GDA 在 GDS 中查找，并通过 CDS 将它返回给客户机。然后，请求的客户机可以向 DCS 发出一个带有文件地址数据的直接调用。CDS 一般都留驻在网络上各单元的多个服务器上。

17.4 分布安全服务

DCE 的分布安全服务（DSS）提供了多级安全措施。DCE 使用了麻省理工学院在 Athena 计划中开发的 Kertberos 鉴别系统。

Kerberos 采用了私钥加密技术，提供三级保护。最低一级只需要在连接开始时就确定用户身份真实性，并假定后来的信息都会在参加认证的基本实体间传达；下一级对每个网络信息都要进行认证；在最高级，安全信息都是专用信息，每个都要经过加密以及认证。

最终用户受这种基于网络服务的影响应为最小。换言之，你不必记住数十个口令或代码。这种安全带来的大量好处来自此项管理用户访问或认证的网络服务。

DSS 也有一项验证服务。DCE 支持基于符合"可移植操作系统接口"的访问控制表（ACL）和连接 RPC 的鉴别接口的合法性检查。OSF 也允许 DCE 的应用程序从 Kerberos 移植到公共密钥验证方案上。

17.5 分布文件系统

DCE 的分布文件系统（DFS）指在获得了许可的情况下，对网络任何节点上的任何文件提供透明的访问。

DFS 是基于 Transare 公司的第四版 Andrew 文件系统（AFS），OSF 摒弃了 Sun 公司的网络文件系统（NFS）而选择 AFS 的 DFS 软件留驻在每个网络节点上。DFS 将每个节点上的文件系统与 DCE 的目录服务相结合，确保存储在 DFS 中的所有文件有统一的命名规则。

OSF 选择 AFS 是因为它让用户在网络的任何地方都可以以相同的通道名查阅文件，而不管用户使用何种计算机。

它使用配有 ACL 的 DCE 安全系统，而此安全系统控制对每个人文件夹的访问。RPC 的数据流功能允许 DFS 在很大范围内瞬间移动大量的数据，而不是一些小包的一点点移动。由于广域网中潜在固有的延迟性，上述功能非常重要。

为最大限度地提高文件访问的性能，DFS 在工作站本地磁盘内高速缓存频繁的访问文件。当用户访问文件服务器上的数据时，此数据副本就在本地高速缓冲存储器存储起来。当用户用完此数据后，本地高速缓存管理程序便将此数据写回到服务器。为避免不同计算机上多个用户存取和修改同一数据而产生问题，DFS 使用一种令牌管理方案来协调文件的修改。

当客户机本地高速缓存数据时，服务器就把文件令牌分配给客户机。如果客户机希望修改文件时，它必须向服务器请求允许获得其修改的写令牌。此设置保证拥有只读令牌的客户机在文件修改后，将被告知其文件不再有效。

17.6 分布时间服务

分布网络系统需要一致的时间服务，以使网络上各计算机的操作同步进行。

在 DCE 中，为实现同步而采用了一个时间服务器，给其他系统提供时间服务。任何一个非时间服务器系统都被称作职员机。分布时间服务（DTS）采用 3 种类型的服务器来协调网络时间表。局部服务器与同一局域网上的其他局部服务器同步工作。在扩展的局域网或广域网上可使用整体服务器。传信机是一台指定的局部服务器，它有规则地与整体服务器进行协调。

服务器通过 DTS 规则周期性地与局域网上每台局部服务器进行同步。

17.7 DCE 的扩展与使用

这并不是说 DCE 已经十全十美了，它还存在着许多改进的可能，其中有些将由 OSF 承担，有些则将由供应商与用户来解决。

由于过去几年来对 DCE 的讨论，很容易让人忘掉这种软件才刚刚开始对市场产生影响。工具箱软件正在出现，早期的用户，如花旗银行，才刚开始其试验性开发。

随着用户开始探索此技术，供应商推出支持 DCE 的新产品，看来工业界正在真正透明的互用性道路上迈出了第一步。

注释

1. Citibank is relying on an integrated set of operating system and net-independent services called the Distributed Computing Environment (DCE) to achieve transparent connections between the systems.

原译：美国花旗银行正在致力于一种叫分布计算环境（DCE）的独立于网络服务的整套操作系统，以获得系统间的透明连接。

译文：美国花旗银行正在致力于研究一种集成了操作系统与非网络服务的称作分布计算环境（DCE）的技术，以获得系统间的透明连接。

分析：分析句子结构可以发现 DCE 是一种技术，集成了 operating system 和 net-independent services 两项内容。这样，在意思上就没错了。

2. Asynchronous interfaces, although they have existed in some environments for some time, can be a major cost drain, the less retraining a new technology requires, the better.

原译：异步接口虽然在有些环境中已存在了一些时候，但实现起来可能还很复杂。在商业世界中，对程序员的再教育可能是一项大的成本开支，一项新技术所需的重新培训越少就越好。

译文：异步接口虽然在有些环境中已存在了一些时候，但它需要的再培训可能会造成很大的成本开支，一项新技术所需的重新培训应该越少越好。

分析：此句原文省略了太多内容，原译试图将其意思表达完整，加入了过多自己的话。增译其实也要惜字如金。

3. Kerberos uses private key encryption to provide three levels of protection. The lowest requires only that user authenticity be established at the initiation of a connection, assuming that subsequent network messages flow from the authenticated principal. The next level requires the authentication of each network message. On the level beyond these, safe messages are private messages, where each is encrypted as well as authenticated.

原译：Kerberos 采用了私钥加密技术，提供三级保护。最低一级只需要在连接开始时就确定用户身份真实性，并假定后来的信息都会在参加认证的基本实体间传达。下一级对每个网络信息都要进行确认，在这一级，安全信息都是专用信息，每个都要经过加密以及鉴别。

译文：Kerberos 采用了私钥加密技术，提供三级保护。最低一级只需要在连接开始时就确定用户身份真实性，并假定后来的信息都会在参加认证的基本实体间传达；下一级对每个网络信息都要进行认证；在最高级，安全信息都是专用信息，每个都要经过加密以及认证。

分析：改正了一些错误。此外，原译中 authentication 三次同样的意思却没有统一口径，这也是要注意的一点。

4. The cell is a fundamental organizational unit for systems in the DCE. They can map to social, political or organizational boundaries and consist of computers that must frequently communicate with one anther-such as in work groups, departments of divisions of companies. Generally, computers in a cell are geographically close and each cell ranges in size from two to thousands of computers.

译文：这里所谓的单元是为 DCE 中系统而定的组织机构的基本单元。它们可以是社会的、政治的或是组织机构的区域范围，由必须频繁相互通信的计算机组成，如工作组、公司部门间的相互通信。

第 18 章 什么是 Wi-Fi 以及它是如何运作的

Wi-Fi 日益成为全球互联网连接的首选模式。你只需要在计算机上配置无线适配器，就能够连接互联网。热点是提供 Wi-Fi 上网的发射点，你可使用如无线网管理工具等高级软件检测和接入热点，连接互联网。当要建立无线连接时，需要确认无线路由器已连上网和其他设备都正确安装。

无线技术近年来已广泛传播。无论在家、图书馆、学校、机场、酒店或餐馆，你都可以随时随地连接互联网。

无线网络，称之为 Wi-Fi，Wi-Fi 是无线高保真的缩写。由于 Wi-Fi 覆盖 IEEE 802.11(1)技术，亦被视为 802.11 无线网络的代名词。Wi-Fi 的最大优势是几乎兼容所有操作系统、游戏设备和高级打印机。

18.1 Wi-Fi 是如何工作的

Wi-Fi 和手机一样，通过无线电波连接互联网来传输信息。在配置无线适配器的计算机上把数据转换传输成无线电信号。相同的电信号将通过天线传输到路由器进行解码，解码后的数据会通过有线以太网传送到互联网。无线网络是双向运作的，从互联网接收的数据也将通过路由器编码成无线电信号，最终传送给计算机上的无线适配器。

18.2 用途

当你要连接到无线局域网的时候，计算机必须配备无线网络接口控制器。计算机和接口控制器的组合称为"站"。如果所有站点共享一个射频通信通道，那么这个通道上传输的信息会在一定射程内被所有站点接收。信息传输不能保证完全传送，所以被视为尽力而为的运输装置。用载波传输的数据以字节每单位,成为"以太网帧"。

1. 网络连接

具 Wi-Fi 功能的设备可以从范围内的无线网络连接到网络。一个或多个(互联)接入点(称为热点)可以组成一个面积由几间房间到数千平方米范围的上网空间,覆盖的面积大小可能取决于接入点的重叠范围。例如,在英国伦敦,Wi-Fi 技术已被用于无线网状网络。

组织和企业,如机场、饭店、餐厅等经常为来访者提供免费热点,以吸引客户。商家会依客户需求提供服务,有时也为在某些领域推销企业而提供免费的 Wi-Fi 站点。

路由器,结合了调制解调器和 Wi-Fi 接入点,通常在家或其他场所设置。路由器提供互联网连接以及为所有设备提供无线或有线的网络互联。

现在,许多智能手机也可充当小型无线路由器,供周围的设备连接互联网。但运营商经常禁用该特性,或者收取额外的费用来使用它,特别对有无限数据计划的客户。一些笔记本有无线调制解调器卡也可作为移动互联网无线接入点。通常没有网络连接的地方,如厨房和花园,也可通过 Wi-Fi 连接网络。

2. 城市 Wi-Fi 覆盖

21 世纪初期,世界各地的许多城市都宣布计划构建全市 Wi-Fi 网络。事实上,城市 Wi-Fi 已有很多成功的例子。例如在 2004 年,迈索尔成为了印度第一个提供全市 Wi-Fi 的城市。一家名叫 Wi-Fi Net 的公司在迈索尔建立热点,覆盖整个城市和一些附近的村庄。在 2005 年,圣克芬德、佛罗里达和加州森尼维尔成为在美国的第一批提供全市免费 Wi-Fi 的城市。明尼阿波利斯为供应商带来每年 120 万美元的利润。2010 年 5 月,伦敦市长鲍里斯·约翰逊承诺在 2012 年伦敦将实施全市覆盖 Wi-Fi,几个自治市镇包括威斯敏斯特和伊斯灵顿已经有了广泛的 Wi-Fi 覆盖。

3. 校园的 Wi-Fi 覆盖

现在,大多数校园已设置无线上网,特别是发达国家的传统大学。卡内基·梅隆大学于 1993 年在其匹兹堡校区建立了世界上第一个校园无线网络,成为无线安德鲁,比起源于 1999 年的 Wi-Fi 品牌还要早。1997 年 2 月,卡内基·梅隆大学已实现 Wi-Fi 全覆盖。在国际认证的硬性基础设施下,许多大学现已合作为学生和员工提供无线网络连接。

4. 计算机和计算机直接通信

Wi-Fi 无线通信也可以不需要通过接入点,直接从一台计算机传到另一台。这就是所谓 Ad-Hoc 模式的 Wi-Fi 传输。这种无线 Ad-Hoc 网络模式受到掌上游戏机(如任天堂 DS 游戏机)、PSP、数码相机和其他消费性电子设备的欢迎。有些设备还可以使用 Ad-Hoc 网络连接,成为热点或"虚拟路由器"。同样,Wi-Fi 联盟推动一个新的安全方法规范,称为 Wi-Fi Direct,直接进行文件传输和媒体共享。Wi-Fi 联盟还推出了名为 TDLS(Tunneled Direct Link Setup,通道直接链路建立)的无线标准,这项标准不需通过接入点,允许两款设备通过 Wi-Fi 网络进行点对点直连。

18.3 频率

无线网络采用 2.4GHz 和 5GHz 频段以适应用户发送的数据量。802.11 标准主要依赖用户的不同需求，说明如下：

- 802.11a 定义了一个在 5GHz ISM 频段上的数据传输速率。正交频分复用，缩写为 OFDM。其思想是将未输送到路由器的高速数据信号转换成并行的低速子数据流，以提供接受率。你可以发送最多 54Mbps 的数据。
- 802.11b 在相对较慢的 2.4GHz 频段上进行数据传输。你可以发送最多 11Mbps 的数据。
- 802.11g 可以 2.4GHz 频段的数据传输速度，发送最多 54Mbps 的数据。这是因为 802.11g 同样采用正交频分复用（OFDM）技术。
- 更先进的 802.11n 可以在 5GHz 频率下传送最多 600Mbps 的数据传输。

18.4 优势和挑战

1. 商业优势

Wi-Fi 部署区网（LAN）降低网络部署的成本。许多空间不能架设电缆，如在户外地区和历史建筑中，可运用无线区网来改善。现在大多数笔记本电脑制造商已经内置无线网络设备。Wi-Fi 的价位持续下跌，使之渐渐普及，已成为企业普遍的基础设施。不同的无线接入点品牌和客户端网络接口在基本水平中实现互操作服务。根据 Wi-Fi 联盟指定，"Wi-Fi 认证"的产品是向后兼容的。不同于移动电话，任何 Wi-Fi 标准设备将在世界上任何地方正确运行。

2. 限制

Wi-Fi 在全球各地的频率分配和操作限制并不相同。美国和欧洲有 13 个频道，比标准在 2.4 GHz 频带的 11 个频道多出 2 个，而日本要多 3 个，共 14 个频道。

一个 Wi-Fi 信号在 2.4 GHz 频段实际上占用五个频道，两个频道编号之差大于 5 的频道，如 2 和 7，不会发生频道重叠，因此在美国只有 3 个非重叠频道：1、6、11。在欧洲和日本有 3 个或 4 个非重叠频道：1、6、3 和 1、5、9、13。

3. 传递的距离

Wi-Fi 网络范围有限。一个使用 802.11b 或 802.11g 的典型无线路由器和天线，在无任何障碍物下可覆盖范围可达到 100 米（330 英尺）。IEEE 802.11n 可到达超过这个范围两倍的距离，范围随频率的波段调整。Wi-Fi 传递的距离同时受频段影响。Wi-Fi 在 2.4 GHz 的频率区段范围比 5 GHz 的频率区段稍微好些。覆盖范围也可通过提升天线质量得以扩展。如通过使用定向天线，室外覆盖范围可提高数公里或以上。在一般情况下一个 Wi-Fi 设备的最高功率传输是受限于地方法规。如等效各向同性辐射功率（EIRP）在欧盟的规定是 20dBm（100 兆瓦）。

为达到无线区网的应用要求，和其他设备相比 Wi-Fi 显得相当耗电。其他技术如蓝牙（可支持无线 PAN 应用）提供了一个小于 100 米的传播范围，因此耗电量较低。其他低耗电技术，如 ZigBee 有相当长的范围，但传输速率却很低。Wi-Fi 的高耗电特性使得电池寿命的问题渐受重视。

研究人员已经开发出"无须新线"的技术，试图弥补 Wi-Fi 室内范围不足、安装新的电线如 CAT-6 不具成本效益的问题。例如，ITU-TG.hn 标准的高速局域网使用同轴电缆，电话线和电源线等现有的家庭线路来达成。虽然 G.hn 不具备 Wi-Fi 可移动和供户外使用等优势，它的设计应用，如 IPTV 分配仍在室内范围发挥功能。

典型 Wi-Fi 的频率由于电波传播的复杂性，特别是树和建筑物影响信号的反射，只能大约测出 Wi-Fi 有关地区的发射器的信号强度。但这不包括远距离 Wi-Fi，因为远距离 Wi-Fi 是使用塔台或高建筑顶上的天线所架设的。

基本上，Wi-Fi 的实际应用范围受到限制，例如，仓库中或零售空间的盘点机、结账条码阅读器、收发台。无线路由器的最高覆盖率只可达 80 平方米，苹果公司的 AirPort 技术可达 100～140 平方米。

18.5 网络安全

相比传统的有线网络以太网，无线网络简化了互联网连接。其特性形成了无线网络安全的主要成因。有线网络的连接必须通过实物的接入，或突破外部的防火墙。而无线网络，你只要在 Wi-Fi 的覆盖范围内即可连接。大多数商业网络不允许外部访问，试图保护敏感的数据和系统。如果网络使用不足或没有加密，启用无线连接会降低安全性。

黑客获得访问无线网络路由器可以立即发起域名系统（DNS）攻击，使网络其他使用者措手不及。

1. 数据安全风险

最常见的无线加密标准——有限等效保密（WEP），即时正确配置，已被证明是很容易被破解的。Wi-Fi Protected Access（WPA 和 WPA2）的保护系统在 2003 年推出，旨在解决这一问题。Wi-Fi 的接入点通常默认为加密模式，新手用户受益于这种快速使用的零配置设备，但这个默认不保证任何无线安全，提供开放的无线局域网访问。用户需要配置设备来保障安全，通常是通过一个软件图形用户界面（GUI）。未加密的无线网络连接设备可以检测和记录包括个人信息等数据。这种网络只能通过获得其他手段的保护，如虚拟私人网络（VPN）、安全传输层协议上的超文本传输协议安全（HTTPS）。

无线保护加密（WPA2）使用增强认证码，被认为是安全的。WPA-OPT 或 WPA3 是 WPA2 的改进版，存储的芯片只产生一次性密钥。芯片连接所有设备，它们通过强加密散列的数据发送或接收来定期更新。本质上，这将是不可能使用任何计算机系统，包括量子计算机系统进行破解的。如果正确实施，散列数据是随机的，没有固定的模式可被检测。其主要缺点是，它需要多吉字节（GB）的存储芯片，使消费者不能承担昂贵的费用。

2. 保护方法

隐藏访问点名称，是阻止未经授权用户访问的一个常见措施。用户可通过关闭无线网络标识广播（SSID）来实现。虽然对普通用户来说，该方法有效。但它其实是无效的安全方法，因为 SSID 可以通过客户名称查询找到。另一个方法是只允许有媒体访问控制

（MAC）地址的计算机加入网络，但坚定的窃听者可以通过欺骗一个授权地址加入网络。

有线等效隐私（WEP）加密的目的是防止随意窥探，但已不再认为是安全的。一些工具如 Air Snort 或 Aircrack-ng 已快速地恢复 WEP 的加密密钥。由于 WEP 的弱点，Wi-Fi 联盟推出了提供临时密钥完整性协议（TKIP）的 Wi-Fi 网络安全存取（WPA）。WPA 专门设计，通过固件升级处理旧设备。虽然 WPA 比 WEP 更安全，但也有弱点。

WPA2 在 2004 年引入，使用高级加密标准，提供更安全的保护。WPA2 已被大多数 Wi-Fi 设备支持，与 WPA 完全兼容。

2007 年，一个缺陷添加到 Wi-Fi 的特性，称为 Wi-Fi 保护设置。该设置允许忽略 WPA 和 WPA2 的安全保护，或在许多情况下有效地中断。截至 2011 年年末，唯一的补救方法就是关闭 Wi-Fi 保护设置，但这并不总是可能的。

注释

1. Like mobile phones, a Wi-Fi network makes use of radio waves to transmit information across a network. The computer should include a wireless adapter that will translate data sent into a radio signal.

译文：Wi-Fi 和手机一样，通过无线电波连接互联网来传输信息。在配置无线适配器的计算机上把数据转换传输成无线电信号。

2. Wi-Fi also allows communications directly from one computer to another without an access point intermediary. This is called ad hoc Wi-Fi transmission. This wireless ad hoc network mode has proven popular with multiplayer handheld game consoles, such as the Nintendo DS, PlayStation Portable, digital cameras, and other consumer electronics devices.

译文：Wi-Fi 无线通信也可以不需通过接入点，直接从一台计算机传到另一台。这就是所谓 Ad-hoc 模式的 Wi-Fi 传输。这种无线 Ad-hoc 网络模式受到掌上游戏机（如任天堂 DS 游戏机）、PSP、数码相机和其他消费性电子设备的欢迎。

3. A wireless network will transmit at a frequency level of 2.4 GHz or 5GHz to adapt to the amount of data that is being sent by the user.

译文：无线网络采用 2.4GHz 和 5GHz 频段以适应用户发送的数据量。

4. An attacker who has gained access to a Wi-Fi network router can initiate a DNS spoofing attack against any other user of the network by forging a response before the queried DNS server has a chance to reply.

译文：黑客获得访问无线网络路由器可以立即发起域名系统（DNS）攻击，使网络其他使用者措手不及。

第 19 章　Shockwave 三维技术

近年来，你一定听说过 Internet 上操纵三维图形的新技术。许多 Web 站点用这类软件已经有一段时间，但由于缺乏通用的三维查看程序，因此市场仍然接近空白。Macromedia 公司和 Intel、NxView 等公司一起通过最新版 Shockwave 播放器和

Shockwave 编写程序 Director 把这个技术带给更多 Web 用户。

如果你在 Web 上花较多时间，则可能已经遇到 Shockwave，这是动画和交互展示的图形格式。Shockwave 文件是由 Director 程序生成的，最初是为光盘使用而设计的。这个格式在 Web 所有者中非常普及，使其可以生成精彩的 Web 内容，可以在 Internet 上相当迅速地传输。

在旧版 Shockwave 和 Director 中，Web 艺术家只能生成二维动画。二维动画有两种形式：

- 帧动画相当于传统卡通，由一系列二维静止图形顺序显示产生动感。你的视角由电影创作者设置。
- 矢量动画使用二维实体（圆、矩形、直线），进行相对移动。由于矢量动画基于简单几何方程，因此使艺术家可以生成文件很小而相当复杂的电影。

最新版 Director 加进了 Intel 公司的 Internet 三维技术，是由 Intel 体系结构实验室开发的。这个程序使 Web 艺术家可以生成交互式三维动画，将其发表到 Web 上。最新版 Shockwave 播放器使大多数 Internet 用户可以浏览这些交互动画，即使使用拨号连接。

利用 Shockwave 三维技术，用户可以实际下载和操纵三维模型，可以成为导演移动镜头。可以用两种方式看这个问题：

- 可以下载实体和在镜头前面旋转实体，从不同角度观看。
- 可以下载环境，通过其移动镜头。这与玩第一人称视频、游戏相似，通过移动控制这个世界中的镜头，程序将你带进虚拟三维世界中。通过让镜头左右移动或前后移动。

根据你的操作，计算机可以从新的角度画出场景的新帧。

这个操作相当复杂：三维软件要从用户那里接收输入，解释输入和确定如何重画图形，生成所要的运动感。玩游戏时，计算机或游戏控制台很容易处理这个操作，但通过 Internet 发送这个信息时，问题会更复杂些。此外，标准 Web 浏览器没有自动装备成处理这些模型，即不是每个人都能访问三维内容。Macromedia 公司最新的 Shockwave 播放器可以解决这些问题，使大多数 Web 用户可以方便地访问三维文件。

19.1 使用 Shockwave 技术

在 Shockwave 中增加三维功能使人们可以访问各种新 Web 格式。其中最明显的应用之一是基于 Web 的三维游戏。第一人称冒险类游戏和其他具有完全真实三维世界的游戏在 PC 和游戏控制台市场上已经主导了将近 10 年。新的 Shockwave 功能使这种游戏可以在 Web 上玩。

基于 Web 的三维游戏引起了巨大的关注，但只是这个新技术的一个市场。三维功能也许更适合促进电子商务。Web 商家可以让客户看到产品的三维图形，从而使客户更清楚地了解其产品目录中的产品。利用三维模型，联机商店更像进店购买，客户可以旋转货物，从不同角度检查。

客户还可以根据特定需求修改三维模型。一个最有用的应用是买衣服。如果联机购物时输入自己的尺寸，则三维软件可以生成这个人的身材模型，可以将特定衣服穿到这个三维模型身上。这是实际试衣室的虚拟版。

这类用户交互对教育站点也大有帮助。发动机的三维模型可以旋转和互动，可以比二维模型更清晰地演示其工作原理，就像亲手拿着发动机进行检查一样。

如图 19.3（图略）所示，这个彩弹枪的三维模型使彩弹和 BB 枪的工作原理一目了然。

如图 19.4（图略）所示，要了解彩弹枪的工作原理，可以用一个三维模型，可以看到这些机械如何配合和发射。

在所有这些应用中，三维的最大好处是更大的用户参与性。可以自己决定要看什么，而不只是浏览预先设置的电影。这种差别就像看电视与玩视频游戏的差别。

19.2 提供三维内容

最后，Shockwave 的新播放器是生成与查看交互式三维 Web 内容的新格式。把这类内容发表到 Web 上不是什么新思想，但技术公司与 Web 站点并没有成功地把三维内容推向更多浏览者。主要原因有两个：

- 低带宽连接上传输三维"运动"很费时间；
- 我们通常要在每次查看另一站点的三维内容时下载新插件。

新的 Shockwave 播放器解决了这些障碍，最终使三维内容成为 Web 上的重要组成部分。大多数 Web 用户已经安装 Shockwave 播放器，只要下载最新更新就可以增加三维功能。Macromedia 公司与许多 Web 公司建立了伙伴关系，使人们使用它的技术。过去，Macromedia 公司在 Shockwave 和 Flash 格式上取得了巨大成功，因为它们与各大浏览器都能够很好地配合，很容易安装和更新。Intel、NxView 和其他公司都与 Macromedia 公司结成伙伴，因为这个公司能很好地传播它的播放器技术。

新格式的设计适合各种带宽连接，即使连接速度只有 28.8kBps。为此它使用了两种方法：

- 在 Web 上浏览二维动画时，Web 站点向你的计算机发送每个连续帧。这样，动画中的一切要通过 Internet 一一传输。使用 Shockwave 三维技术时，Web 站点一次性地向你发送完整图像。然后，要移动图形时，站点只发送进行移动所要的信息，告诉你的计算机如何调整外层线框，计算机即会进行其余工作，填充多边形和纹理。
- 过去 5 年生产的大多数个人计算机，其处理器都能处理高级视频游戏中的复杂三维世界，因此非常适合这个工作。通过利用主要位于客户端（你的 PC）上的功能，可以大大减少从服务器（存放 Web 站点的计算机）传输的信息，只有装入初始图形时才会有效并大量地下载，然后站点只要传输调整的内容，不需要太大的带宽。

但最初的下载怎么办？Shockwave 的新播放器用适应性三维几何技术解决了这个问题。适应性三维几何是一组复杂算法，对特定 Internet 连接自动调整三维模型比例。在低速连接中，Web 站点传输简化纹理和减少多边形个数的图形，而在高速连接中，则传输更复杂的图像。

- 如图 19.5（图略）所示，简化三维模型的多边形更少，只有 862 个多边形。
- 如图 19.6（图略）所示，要生成更多细节的模型，就要增加多边形，这个手有 3444 个多边形。

利用这些元素，就可以访问三维内容，不管使用何种 Internet 连接。但怎样自己利用 Shockwave 三维技术呢？下节介绍如何生成 Shockwave 三维动画，Web 主人如何把三维内容放进站点中。

19.3 开发三维内容

我们有机会和 Miriam Geller 交流，它是 Macromedia 公司 Director 与 Shockwave 播放器的产品总经理。要生成像上例中自动化传输一样的三维实体，就要使用 3 种不同工具：

用标准三维建模软件包生成三维实体。例如，可以用 Director。利用这些工具，可以生成线框图形，指定盖住线框的多边形。可以用新的 W3D 文件格式导出三维建模软件包内容。

（1）将 W3D 文件装入 Macromedia 应用程序 Director Shockwave Studio 中。这个应用程序可以帮你准备要发布到 Web 上的三维实体。例如，可以：

- 采用不同技术限制用户机器上三维实体所算的带宽量或处理功能，如多分辨率网格或细分表面。
- 增加用户交互特性。例如，可以让三维实体不同部分响应用户请求而移动。
- 在实体中增加雾、雨之类效果。

（2）我们从 Director Shockwave Studio 导出正常的 DCR 文件，将其放在 Web 服务器中。

（3）然后用户用浏览器和 Shockwave 播放器（V8.5 以上）下载和浏览 DCR 文件。

这个过程并不简单，但如果已经熟悉 3D Studio Max 之类三维建模程序，则不难学会。

注释

1. In the past year or so, you may have heard about a new technology that lets you manipulate 3-D images over the Internet. Many Web sites have been using this sort of software for a while, but it has mostly remained a niche market due to a lack of universal 3-D viewer programs.

译文：近年来，你一定听说过 Internet 上操纵三维图形的新技术。许多 Web 站点用这类软件已经有一段时间，但由于缺乏通用的三维查看程序，因此市场仍然接近空白。

2. Macromedia, in conjunction with Intel, NxView and others, hopes to bring this technology to many more Web users with the newest versions of the Shockwave player and the Shockwave authoring program Director.

译文：Macromedia 公司和 Intel、NxView 等公司一起通过最新版 Shockwave 播放器

和 Shockwave 编写程序 Director 把这个技术带给更多 Web 用户。

3. The format is very popular with webmasters because it allows them to create elaborate Web content that can be transmitted fairly quickly over the Internet.

译文：这个格式在 Web 所有者中非常普及，使其可以生成精彩的 Web 内容，可以在 Internet 上相当迅速地传输。

4. The newest version of the Shockwave player allows most Internet users, even ones with dial-up connections, to view these intricate animations.

译文：最新版 Shockwave 播放器使大多数 Internet 用户可以浏览这些交互动画，即使使用拨号连接。

第 20 章 3D 体感摄影机：Kinect

Kinect（开发中的代号为 Project Natal）是一款由微软开发供 Xbox360、Xbox One 以及 Windows 计算机使用的运动感知输入设备。基于摄像头式的附加外设，它使用户能够在不需要游戏手柄的情况下，通过基于手势和语音命令的自然交互界面来直接对他们的控制台/计算机进行操控和交互。第一代 Kinect 于 2010 年 11 月首次推出，其目的是在原有用户的基础上拓宽 Xbox360 的受众范围。用于 Windows 的版本发布于 2012 年 2 月 1 日。Kinect 与其他运动控制的家用游戏机如为 Wii 和 Wii U 设计的 Wii Remote Plus，为 PlayStation 3 设计的 PlayStation Move 和 PlayStation Eye，以及为 PlayStation 4 设计的 PlayStation Camera 产生了竞争。

2011 年 6 月 16 日，微软发布了 Windows 7 版本的 Kinect 的软件开发工具包。该 SDK（software development kit）是为了让开发人员可以在 C++/CLI、C# 或 Visual Basic .NET 上编写 Kinect 的应用程序。

20.1 科技

Kinect 是基于微软游戏工作室的子公司 Rare 内部开发的软件技术，同时还基于一系列由以色列开发的 PrimeSense 摄像技术。其开发了一个可以理解特定手势的系统，该系统通过红外线投影仪和摄像头和一个特殊的微芯片来跟踪在三维空间的物体和个人的运动，这使得不用双手来操控电子设备变成了可能。这种被称为光编码的三维扫描系统采用了大量的基于图像的三维重构。

Kinect 感应器是一个外型类似网络摄影机的装置。Kinect 感应器是连接到一个带有水平杠的电动支点的小型基底，并被设计在视频显示器的上方或下方纵向定位。该器件具有"RGB 摄像头，深度传感器，多阵列麦克风运行专有软件"，它提供全方位的身体的 3D 动作捕捉以及面部识别和语音识别功能。刚刚发售时，语音识别功能只在日本、英国、加拿大和美国的产品中提供。后来在 2011 年春季欧洲大陆可以使用该功能。现在语音识别支持澳大利亚、加拿大、法国、德国、爱尔兰、意大利、日本、墨西哥、新西兰、英国和美国。Kinect 感应器的风阵列式麦克风允许 Xbox 360 进行声源定位和环境噪声的消除，并支持 Xbox Live 的无佩戴聊天。

深度传感器是由一个红外激光投影仪和单色 CMOS 传感器组合而成，其可以在任何环境光条件下捕获 3D 视频数据。深度传感器的感应范围是可调的，并且 Kinect 软件能够根据游戏和玩家的物理环境来自动校准传感器，并且不受家具或其他障碍物的存在的干扰。

软件技术实现了高级手势识别，面部识别和语音识别，被微软工作人员形容为 Kinect 的初级创新。根据提供给零售商的信息，Kinect 是能够同时跟踪多达六人，其中包括对两个玩家每个人 20 个关节的特征提取的运动分析。然而，PrimeSense 指出，设备可以"看到"（但不作为玩家处理）的人的数目仅受限于有多少人出现在摄像机的限定的感受区域。

逆向工程使得 Kinect 可以在帧率 9～30Hz 根据具体分辨率进行多种的视频输出。默认的 RGB 视频流使用的是 8 比特的 VGA 分辨率（640×480 像素）并用 Bayer 滤波器。但是硬件可以将分辨率提升到 1280×1024（在低帧的情况下）并且支持如 UYVY 等彩色格式。单色深度传感视频流是在 VGA 分辨率（640×480 像素）与 11 比特的深度，它提供 2048 级别的灵敏度。Kinect 还可以将来自红外照相机的以较低的帧速率 640×480 或用 1280×1024 的速率流出（在此之前它已被转化为深度图）。与 Xbox 软件使用时，Kinect 感应器有 1.2～3.5 米（3.9～11.5 英尺）距离的实际范围的限制。标准所需的面积是约 6 m²，虽然该传感器能保持在大约 0.7～6 米（2.3～19.7 英尺）的扩张范围的跟踪。该传感器的视角最大范围包括水平 57°和垂直 43°，而电动支点能够倾斜传感器，高达向上或向下各 27°。因此在 0.8 米（2.6 英尺）的最小观看距离 Kinect 传感器的探测范围是水平 87 厘米（34 英寸）垂直 63 厘米（25 英寸），因此，刚刚超过分辨率 1.3 毫米每个像素（0.051 英寸）。麦克风阵列具有四个麦克风并以 16kHz 的采样率各信道处理 16 比特的音频操作。

Kinect 感应器的电动倾斜功能需要的电力超出了 Xbox 360 的 USB 端口可以提供的上限，因此设备利用一个专用连接器的 USB 接口供给额外的电力。重新设计的 Xbox 360 S 型号包括一个特殊的 AUX 口用于容纳连接器，而老款型号则需要特殊的电源线（包括传感器），其分割连接到单独的 USB 和电源的连接；电力从电源通过 AC 适配器的方式提供。

20.2 历史

2009 年 6 月 1 日，Kinect 的首次以代号 Project Natal 于 E3 发布。如以往微软的传统使用城市作为代号，Project Natal 是以巴西的城市纳塔尔命名来表彰促成了该计划的出生于巴西的微软主管亚历克斯 Kipman。如此命名也是因为单词 Natal 与新生有关，反映了微软对于该项目的观点是"次世代家庭娱乐的诞生"。

在 2009 年的 E3 展览会上，Kinect 给出了 3 个样本：Ricochet、Paint Party 和 Milo & Kate。而另一个基于 Burnout Paradise 的样本在微软的展示上出现。在 2009 年 E3 展示的骨骼映射技术可以同时追踪四个人，并且可以以 30Hz 的频率提取一个人 48 个骨骼点的人体特征。

据传，此次推出的 Project Natal 会同新的 Xbox 360 控制台的版本一起推出。微软在公开场合否认了这些报道并反复强调，Natal 计划将与所有的 Xbox 360 游戏机完全兼容。微软表示，Kinect 对其来说是一次如同 Xbox Live 一样意义重大的开始，并会发布一个类似于 Xbox 的新游戏主机。Kinect 甚至在芝加哥俱乐部的演讲上被微软首席执行官史蒂夫·鲍尔默称为"新的 Xbox"。当被问及这项引进是否会延迟下一代次世代主机平台推出的时间（平台之间的历史大约 5 年），微软企业副总裁 Shane Kim 重申，该公司认为，Xbox 360 的生命周期将持续到 2015 年（10 年）。

在 Kinect 的发展过程中，项目组成员实验性地适应着众多游戏的 Kinect 基础控制方案，以帮助评估可用性。在这些游戏当中，"美丽块"和"太空侵略者极端"在 2009 年 9 月的东京游戏展进行了展示。创意总监工藤角田表示，对预先存在的游戏添加基于 Kinect 的控制，会涉及重要代码的改变，因此不太可能通过软件更新来增加 Kinect 的功能。

尽管最初的时候是计划在探测器元件上安装一个可以处理系统内骨架映射的微型处理器。但在 2010 年 1 月却被揭露出其不会使用装有专用的处理器。作为代替，处理过程会由 Xbox 360 的 Xenon 处理器的一个核来完成。据 Alex Kipman 称，Kinect 系统大约占据了 Xbox 360 百分之十到十五的计算量。但是在 11 月的时候 Kipman 宣称"新的系统会从之前占据的 10%～15%降为个位数"。

2010 年 3 月 25 日，微软为一项"世界首演的 360 体验计划 Project Natal 于 E3 上发出了标有日期的传单。事件发生在 2010 年 6 月 13 日（周日）的晚上在盖伦中心并伴有太阳剧团的表演。据称该系统将正式被称为 Kinect，是单词"运动"和"连接"混合，它描述了创新的关键点。微软还宣布，Kinect 在北美的推出日期将是 2010 年 11 月 4 日。尽管以前的声明驳回了新的 Xbox 360 推出新的控制系统的猜测，微软在 E3 2010 上宣布推出重新设计的 Xbox 360，配有一个专门连接 Kinect 的端口。此外，于 2010 年 7 月 20 日，微软宣布了一个与重新设计的 Xbox 360 捆绑的 Kinect，可在 Kinect 发售时获得。

2011 年 6 月 16 日，微软宣布了非商业用途的 SDK 的官方发布。2011 年 7 月 21 日，微软宣布，有史以来的第一款白色的 Kinect 感应器将作为"Xbox360 限量版的 Kinect 星球大战捆绑包"的一部分，其中还定制包括 Kinect 大冒险的星球大战为主题的控制台和控制器，以及 Kinect Adventures 和 Star Was Kinect 的副本。而之前的所有 Kinect 传感器都是亮黑色的。

2011 年 10 月 31 日，微软宣布的商业版本的 Kinect for Windows 程序和用于公司的 SDK 的发布。微软产品经理大卫·丹尼斯说，"我们现在合作的有数百个组织，我们将帮助他们确定什么是可以通过这项技术达到的。"

2012 年 2 月 1 日，微软发布了 Kinect 的 Windows SDK 的商业版本，并告诉记者，来自 25 个国家 300 多家公司都在开发 Kinect 的应用程序。

20.3 发售

微软在 Kinect 发售的广告预算达到了五亿美元，这比在 Xbox 上投入的预算更多。意在扩大受众的营销活动"你就是遥控器"包含了在 Kellogg 的麦片盒和 Pepsi 的饮料瓶上的广告以及在不同的节目（如 Dancing with the star Gleel）中插入的广告和在不同杂志

（如 People and InStyle）上的宣传。

在 10 月 19 日，微软在 The Oprah Winfrey Show 上通过赠予观众 Xbox 360 和 Kinect 的方式来推广 Kinect。两周后同 Xbox 捆绑的 Kinect 被送给了 Late Night with Jimmy Fallon 的观众。10 月 23 日，微软在 Beverly Hills 举办了一场预售派对。该派对是由 Ashley Tisdale 举办，并有贝克汉姆及其三个儿子一起参加。客人们尝试了 Dance Central 和 Kinect Adventure 并在 Tisdale 的指挥下和 Nick Cannon 进行了一次 Kinect 的语音聊天。在 11 月 1 日到 28 日，汉堡王每十五分钟便会送出一个 Kinect 摇杆。

11 月 3 日一个大事件被安排于时代广场。歌手 Ne-Yo 同上百名舞者参与了 Kinect 的午夜发布。在这一活动中，微软送出了 T 恤和 Kinect 游戏。

2010 年 11 月 4 日 Kinect 在北美发售，在 2010 年 11 月 10 日于欧洲发售，于 2010 年 11 月 18 日在澳大利亚，新西兰和新加坡发售，于 11 月 20 日在日本发售，传感器外设的购买选项包括捆绑套装 Kinect Adventure 和 Kinect 控制台以及 4 GB 或 250 GB Xbox 360 控制台和 Kinect Adventure 捆绑发售。

20.4 评价

1. 对于 Xbox 360 上 Kinect 的评价

在 Kinect 发布后，其囊括了来自评论家和批评家各种积极的意见。IGN 给出了 7.5 分的评价（满分 10 分）表示 Kinect 可以给休闲玩家带来极大的乐趣，并且创造性的无手柄操作的概念有着不可否认的吸引力。但同时也补充道"149.99 美元的价格对于 Xbox360 的外设追踪设备会偏高，尤其是 Xbox 360 本身才 199.99 美元"。Gamer Informer 给 Kinect 的评分是 8（满分 10 分），除了称赞科技之外，也表示 Kinect 需要一段时间来适应操作以及对空间的要求都有可能影响游戏体验。Computer and Video Games 将 Kinect 称为科技之宝并表扬了其语音控制，但同时也批评了连线发售和 Kinect Hub。

CNET 指出了 Kinect 是怎么样通过对玩家全身动作的感知来让玩家活动起来，但同时也批评了他的学习曲线以及老款的 Xbox 360 需要额外的电源支持和空间需求。Engadget 同样也将需要大的场地列为负面的评价。同时还有 Kinect 发售阵列以及收拾控制界面的响应缓慢。评论表扬了系统有强大科技还有如瑜伽、舞蹈等有发展潜力的游戏。Kotaku 首先赞美了该设备的革命性创新。但同时也指出了游戏有时无法识别姿势或者响应迟缓。得出 Kinect 不是必须拥有的，而更像是一定要最后拥有的。TechRader 表扬了其声音控制也指出了设备在空间需求和响应速度的问题。Gizmodo 也看到了 Kinect 的潜力并表达了对主流媒体会如何评价该科技表示好奇。Ars Techinica 的观点关注了 Kinect 的核心特征——没有游戏手柄，会阻碍那些有固定玩家或者自动控制的游戏的发展。

主流媒体也评价了 Kinect。USA Today 将其与少数派报告终的未来控制器相提并论并表述了玩游戏感觉很棒的观点，同时给出了 3.5 颗星的评价（满分 4 颗星）。Davrd Progue 预测玩家在第一次玩 Kinect 的时候会感受到一种"疯狂的魔法般的冲击性的感觉"。尽管没有 Wii 的调用精度高，但 Progue 认为"Kinect 创造了一个全新的拥有社交性质，跨年龄段的活动"。The Globe and Mail 将 Kinect 称为"动作控制的新标准"，轻微的动作延迟和 Kinect 注册问题对于大多数游戏来说都不是主要问题。并且评论称 Kinect

是一个好的创新作品，在满分 4.0 的情况下评获 3.5 分。

2．对于 Xbox One 上 Kinect 的评价

尽管相比于老版的 Kinect 有所提升，但是下一代仍受限于混合的响应。Kinect 被称赞其广角摄像和快速的响应时间以及高质量的摄像头，但对于有口音的英语的识别问题 Kinect 仍饱受诟病。而且微软将 Xbox One 捆绑销售的做法也引起了一些争议。尽管最初公布的时候，传感器是在 Xbox One 中但最后关于这项决定却被修改了。也有一些关于隐私的一些担忧。在 2014 年 5 月，微软宣布将于 6 月 9 日在原有的基础上开始提供没有 Kinect 控制台的 Xbox One；一个单独的 Kinect 感应器将在日后公布。

20.5 销售

截至 2013 年 2 月，Kinect 已经邮寄出 2400 万台。因在市场上的前 60 天售出了 800 万台，Kinect 已经创下了作为"销售最快的消费电子设备"的吉尼斯世界纪录。据 Wedbush 的分析师 Michael Pachter，Kinect 的捆绑版占了在 2010 年 12 月所有的 Xbox 360 游戏机的销售的一半左右，截至到 2011 年 2 月，已经超过了三分之二。超过 75 万台 Kinect 仪器在 2011 年 01 月黑五的一周内售出。

20.6 荣誉

- 关于 Kinect 的内人体运动捕获的机器学习作品荣获 2011 麦克罗伯特奖工程创新。
- Kinect 在 T3 荣获了 2011 年的"年度小工具"奖。它还荣获"年度最佳游戏小工具"的大奖。
- "微软 Kinect for Windows 软件开发工具包"在纽约市大众机械突破奖颁奖典礼"十大最具创新科技 2011 年产品"被排在第二位。

注释

1. Based around a webcam-style add-on peripheral, it enables users to control and interact with their console/computer without the need for a game controller, through a natural user interface using gestures and spoken commands.

译文：基于摄像头式的附加外设，它使用户能够在不需要游戏手柄的情况下，通过基于手势和语音命令的自然交互界面来直接对他们的控制台/计算机进行操控和交互。

分析：本句中有些复合型形容词 如 webcam-style、add-on。看到这种词的时候应该拆词并联系上下文。像本段中便翻译为"基于摄像头式的，附加的"。同时段落中会出现如 with 引导的插入成分，翻译时应该灵活处理。

2. Kinect builds on software technology developed internally by Rare, a subsidiary of Microsoft Game Studios owned by Microsoft, and on range camera technology by Israeli developer PrimeSense, which developed a system that can interpret specific gestures, making completely hands-free control of electronic devices possible by using an infrared projector and

camera and a special microchip to track the movement of objects and individuals in three dimensions.

译文：Kinect 是基于微软游戏工作室的子公司 Rare 内部开发的软件技术，同时还基于一系列由以色列开发的 PrimeSense 摄像技术。其开发了一个可以理解特定手势的系统，该系统通过红外线投影仪和摄像头和一个特殊的微芯片来跟踪在三维空间的物体和个人的运动，这使得不用双手来操控电子设备变成了可能。

分析：本段插入语和定语从句较多造成了一定阅读难度，如 which 引导的定语从句和 making 做伴随的状语又加上了 by 引导的短语。同时有公司名称可能会造成误读如 Rare 和 PrimeSence 虽然都有对应的英文，但不能直译否则会引起歧义。

3. Reverse engineering has determined that the Kinect's various sensors output video at a frame rate of ~9 Hz to 30 Hz depending on resolution. The default RGB video stream uses 8-bit VGA resolution (640 × 480 pixels) with a Bayer color filter, but the hardware is capable of resolutions up to 1280 × 1024 (at a lower frame rate) and other colour formats such as UYVY.

译文：逆向工程使得 Kinect 可以在帧率 9～30Hz 根据具体分辨率进行多种的视频输出。默认的 RGB 视频流使用的是 8 比特的 VGA 分辨率（640×480 像素）并用 Bayer 滤波器。但是硬件可以将分辨率提升到 1280×1024（在低帧的情况下）并且支持如 UYVY 等彩色格式。

分析：本段最大的难点是专业信息词汇的理解，例如 RGB, UYVY 这种图片颜色格式。以及不同的颜色对应不同的解析度。翻译前应明确各种专业词汇，才能保证翻译的质量。

4. Because the Kinect sensor's motorized tilt mechanism requires more power than the Xbox 360's USB ports can supply, the device makes use of a proprietary connector combining USB communication with additional power. Redesigned Xbox 360 S models include a special AUX port for accommodating the connector, while older models require a special power supply cable (included with the sensor) that splits the connection into separate USB and power connections; power is supplied from the mains by way of an AC adapter.

译文：Kinect 感应器的电动倾斜功能需要的电力超出了 Xbox 360 的 USB 端口可以提供的上限，因此设备利用一个专用连接器的 USB 接口供给额外的电力。重新设计的 Xbox 360 S 型号包括一个特殊的 AUX 口用于容纳连接器，而老款型号则需要特殊的电源线（包括传感器），其分割连接到单独的 USB 和电源的连接；电力从电源通过 AC 适配器的方式提供。

分析：本文中的逻辑连词比较多，例如 because while 等。在翻译过程中对于不同的连词应该有不同的译法。例如本段调整了 because 的语序保证了其阅读上的通顺，而 while 的转折则调整为"而"来表述。

5. Although the sensor unit was originally planned to contain a microprocessor that would perform operations such as the system's skeletal mapping, it was revealed in January 2010 that the sensor would no longer feature a dedicated processor. Instead, processing would be

handled by one of the processor cores of Xbox 360's Xenon CPU.

译文：尽管最初的时候是计划在探测器元件上安装一个可以处理系统内骨架映射的微型处理器。但在 2010 年 1 月却被揭露出其不会使用装有专用的处理器。作为代替，处理过程会由 Xbox 360 的 Xenon 处理器的一个核来完成。

分析：不同的词在不同的语境中译法也不同。例如文中的 instead 不是译为"而是"或者"反过来"最终译为了"作为代替"。一词多义在译文中也经常发生，需要反复检验，以保证语义完整。

习 题 答 案

Chapter 1

Multiple or single choices
1. ABC   2. AC   3. ACD   4. AC   5. AD

True or False
1. T   2. T   3. F   4. F   5. T

Chapter 2

Multiple or single choices
1. ACD   2. BC   3. BC   4. BCD   5. AC

True or False
1. F   2. F   3. T   4. T   5. T

Chapter 3

Multiple or single choices
1. ABCD   2. ABC

True or False
1. T   2. F   3. T   4. T   5. T

Chapter 4

Multiple or single choices
1. A   2. ACD   3. C   4. ACD   5. BCD

True or False
1. T   2. F   3. T   4. T   5. T

Chapter 5

Multiple or single choices
1. AB   2. C   3. C   4. B   5. D

True or False
1. T   2. F   3. T   4. T   5. F

Chapter 6

Multiple or single choices
1. AB   2. ABD   3. ACD   4. ABCD

True or False
1. T   2. T   3. T   4. T   5. T

Chapter 7

Multiple or single choices
1. C   2. ABD   3. ACD   4. ABCD   5. ABCD

True or False
1. F   2. T   3. T   4. T   5. T

Chapter 8

Multiple or single choices
1. ABCD   2. B   3. A   4. AD   5. AC

True or False
1. T   2. F   3. T   4. F   5. T

Chapter 9

Multiple or single choices
1. BD   2. AC   3. D

True or False
1. T   2. F   3. T

Chapter 10

Multiple or single choices
1. AC   2. AB   3. ABD   4. ACD

True or False
1. T   2. T   3. F   4. F   5. F

Chapter 11

Multiple or single choices
1. ABD   2. ABCD   3. AC   4. ACD   5. ABCD

True or False
1. T   2. T   3. T   4. T   5. F

Chapter 12

Multiple or single choices
1. ACD 2. ABD 3. ACD 4. CD 5. AC

True or False
1. T 2. T 3. F 4. F 5. F

Chapter 13

Multiple or single choices
1. ACD 2. ABC 3. BD 4. ACD 5. ABD

True or False
1. T 2. F 3. T 4. T 5. T

Chapter 14

Multiple or single choices
1. BC 2. BC 3. ABCD 4. ABCD

True or False
1. T 2. F 3. T 4. F 5. F

Chapter 15

Multiple or single choices
1. ABD 2. C 3. AC 4. ABC 5. BC

True or False
1. T 2. T 3. T 4. T 5. F

Chapter 16

Multiple or single choices
1. ABD 2. D 3. ABC 4. ABCD

True or False
1. T 2. T 3. T

Chapter 17

Multiple or single choices
1. ABCD 2. ACD 3. ABD

True or False
1. T 2. F 3. T 4. T 5. T

Chapter 18

Multiple or single choices
1. ABD   2. AB   3. ABCD   4. ABCD

True or False
1. F   2. F   3. T   4. T   5. F

Chapter 19

Multiple or single choices
1. ABD   2. ABCD   3. ABCD

True or False
1. T   2. T   3. F   4. T

Chapter 20

Multiple or single choices
1. ABD   2. ACD   3. ABCD

True or False
1. T   2. F   3. T   4. F